To Gene with appreciation

Twila Bratcher

LIVING TEREBRAS OF THE WORLD

A Monograph of the Recent Terebridae of the World

LIVING TEREBRAS OF THE WORLD

A Monograph of the Recent Terebridae of the World

By Twila Bratcher
and
Walter Cernohorsky

AMERICAN MALACOLOGISTS, INC.
MELBOURNE, FLORIDA
BURLINGTON, MASSACHUSETTS

LIVING TEREBRAS OF THE WORLD

by

Twila Bratcher
Hollywood, CA 90046

and

Walter Cernohorsky
Auckland Institute and Museum
Auckland 1, New Zealand

Edited by
R. Tucker Abbott

Library of Congress catalog card number 86-72677

Published by AMERICAN MALACOLOGISTS, INC.
Editorial Office: P. O. Box 2255
Melbourne FL 32902-2255
Order and Shipping Dept.: P. O. Box 1192
Burlington MA 01803

Produced by MADISON PUBLISHING ASSOCIATES, New York, NY

FOREWORD

At last a popular coverage of this large yet greatly neglected family of the colorful marine terebras has been created for the benefit of the scientific community as well as for the legions of serious shell collectors.

Two eminent experts in the field of living Terebridae have combined their knowledge and efforts in order to bring together and organize over two hundred years of scientific mollusk literature. Both authors have spent several years in the field observing and collecting these tropical mollusks, and both of them have closely examined many thousands of specimens of Terebridae in large private collections and in the leading natural history museums of the world. Twila Bratcher has taken great pains to obtain and present in this book photographs of all of the holotypes of *Terebra* known to be still preserved in international museums.

The authors are well-known for their sensible and conservative approach to systematics and speciation, and have refrained from manufacturing a plethora of new species for the many color, sculptural and ecologic variations that characterize some of these beautiful gastropods. Their original work and careful synthesis will long be a guide for future research and a welcome and useful tool for worldwide collectors.

R. Tucker Abbott, Ph.D.
American Malacologists, Inc.
Melbourne, Florida

CONTENTS

INTRODUCTION

The family Terebridae, popularly known as terebras, auger shells and marlin spikes, are a dominant molluscan element of lagoons, sand bars and deep-waters of the tropics and subtropics. Early European expeditions to the Indian and Pacific Oceans and the Caribbean brought back the first species, which were later incorporated into the conchological cabinets of the rich and titled, and formed the basis of the descriptions and illustrations of early works on conchology.

Despite being a lesser known family of the toxoglossan order of mollusks, the Terebridae are nevertheless collector's favorites, together with the cowries, cones, volutes, murex and olive shells. Popular modern monographic treatments of the latter five families have been available to collectors for many years, but no recent satisfactory identification guide has been forthcoming for the Terebridae. Information on the group can be found in obscure short monographs or original descriptions by 18th and 19th century authors, or in Tryon's outdated monograph on the family (1885), or in restricted faunal monographs by modern authors (e.g. Cernohorsky, 1967; Bratcher & Burch in Keen, 1971; Kay, 1979, etc.). The lack of up-to-date reference books, combined with a certain amount of nomenclatorial chaos, is no doubt responsible for the scores of unnamed or misidentified specimens of Terebridae deposited in public museums and private collections.

In preparation of this work, all original descriptions since 1758 to the present day have been gathered and evaluated. It was necessary to examine all extant type-specimens in worldwide museums, compile an extensive bibliography, photograph specimens and types, and form an index of species, together with synonymies and references. In order to appreciate the influences of habitat, ecology, and the range of variation within a species, we have personally collected Terebridae in all major faunal provinces by means of intertidal exploration, snorkling, scuba diving, hand dredging and deep-water dredging.

In studying all available type specimens of Terebridae we have made extended personal visits to the major museums and institutions. Museums we were unable to visit in South Africa and India loaned us types for examination and photography, and museums in Vienna, Oxford, Argentina, and East Berlin provided us with photographs of type specimens. Our search for Chemnitz's terebrid types in the University Zoological Museum in Copenhagen, and Lorois' types in the National Museum of Paris, was largely unsuccessful.

It is quite natural that some taxonomic changes in terebrid nomenclature will be necessary at some time in the future, particularly as more material becomes available and our understanding of several species increases. We believe that our present assessment of the living species will serve those interested in this intriguing family.

Acknowledgments

This study would have been impossible without access to type-collections and kind permission to examine and photograph types and specimens from mollusk collections in museums, universities, and scientific institutions. We are indebted to the following people: E. Binder and C. Vaucher, Museum d'Histoire Naturelle, Geneva; K. J. Boss, Museum of Comparative Zoology, Harvard University, Massachusetts; P. Bouchet, Museum National d'Histoire Naturelle, Paris; F. Climo and B. Marshall, National Museum of New Zealand, Wellington; H. Coomans, Zoologisch Museum, Universiteit van Amsterdam; A. D'Attilio, San Diego Museum of Natural History; G. Davis and R. Robertson, Academy of Natural Sciences of Philadelphia; W. E. Emerson and the late W. Old, Jr., American Museum of Natural History, New York; E. Gittenberger, Rijksmuseum van Natuurlijke Historie, Leiden; J. van Goethem, Institut Royal des Sciences Naturelle de Belgique, Brussels; T. Gosliner, South African Museum, Cape Town and now at the California Academy of Sciences, San Francisco; F. G. Hochberg, Santa Barbara Museum of Natural History; R. Kilburn, Natal Museum, Pietermaritzburg; R. Kilias Museum für Naturkunde, Humboldt University, E. Berlin; J. Knudsen, Universitetets Zoologiske Museum, Copenhagen; S. Kosuge, Institute of Malacology, Tokyo; J. McLean and G. Sphon, Los Angeles County Museum of Natural History; G. Oliver, National Museum of Wales, Cardiff, W. F. Ponder and staff, Australian Museum, Sidney; the late T. Pulley and C. Boone, Houston Museum of Natural History; N. V. Subba Rao, Zoological Survey of India, Calcutta; the late J. Rosewater and H. A. Rehder, National Museum of Natural History, Washington; B. Roth, California Academy of Sciences, San Francisco; J. Taylor and K. Way, British Museum (Natural History), London.

For photographs of types we would like to thank J. Gallardo, Museo Argentina de Ciencias Naturales, Buenos Aires; R. Kilias, Humboldt University, E. Berlin; O. Paget and E. Wawra, Naturhistorisches Museum, Vienna; and J. Hull, University Museum, Oxford.

Many individuals, too many to list, have helped with this book by donating or loaning specimens for study and discussing terebrid problems, but a few must be mentioned: B. Bailey; P. T. Caddey; C. Cardin; P. Clover; E. R. Cross; V. Dan; J. Ernest; F. Hernandez; P. Homes; M. Morrow; O. K. McCausland; S. Pras: B. Parkinson; R. Purdy; H. Rehder; E. C. Rios; D. Robinson; R. Salisbury; O. Schoenberg; C. Skoglund; and J. Trondle.

We particularly appreciate the work done by P. Bouchet on West African Terebridae, and we are grateful for his permission to use many of his protoconch drawings. We are also grateful for the sketches drawn by A. D'Attilio.

We are indebted to David K. Mulliner for the color photographs. Other photographs taken by the following have added immeasurably to our own: B. Draper, A. Foubert, S. Kaicher, Scott Johnson (color pl. A) and H. Mienis.

To the late R. D. Burch, whose study-collection of Terebridae, collection of literature on the Terebridae, and shared knowledge of the family formed the nucleus of this book, we give our special acknowledgment.

Museum Abbreviations

The abbreviations used for a number of institutional collections are:

AIM —Auckland Institute and Museum, Auckland.
AMS —Australian Museum, Sydney.
ANSP —Academy of Natural Sciences of Philadelphia.
BM(NH) —British Museum (Natural History), London.
CAS —California Academy of Sciences, San Francisco.
IMT —Institute of Malacology, Tokyo.
IRSN —Institut Royal des Sciences Naturelles, Brussels.
LACM —Los Angeles County Museum of Natural History, Los Angeles.
MCNA —Museo de Ciencias Naturales de Argentina, Buenos Aires.
MCZ —Museum of Comparative Zoology, Harvard University, Cambridge, Mass.
MNHU —Museum für Naturkunde Humboldt University, East Berlin.
MHNG —Museum d'Histoire Naturelle, Geneva.
MNHNP—Museum National d'Histoire Naturelle, Paris.
NM —Natal Museum, Pietermaritzburg, South Africa.
NHMV —Naturhistorisches Museum, Vienna.
NMV —National Museum of Victoria, Melbourne, Australia.
NMWC —National Museum of Wales, Cardiff, Wales.
RMNHL—Rijksmuseum Naturlijke Historie, Leiden.
SAM —South Australian Museum, Adelaide.
SBMNH —Santa Barbara Natural History Museum, Santa Barbara.
USNM —National Museum of Natural History, Washington.
UZMC —Universitetes Zoologisk Museum, Copenhagen.
ZMUA —Zoologisch Museum, Universiteit van Amsterdam, Amsterdam.
ZSI —Zoological Survey of India, Calcutta.

Color Plate A. Living specimen of *Hastula inconstans* (Hinds, 1844) from the Hawaiian Islands. Shell 30 mm in length. Siphon is extended at right; small, translucent brown operculum is seen on the upper central part of the extended foot. Photo courtesy of Scott Johnson.

NATURAL HISTORY OF THE TEREBRAS

Shell Morphology

Terebridae have slender, pointed, multi-whorled shells (see text fig. 1). In size, mature specimens range from about 6 mm to 274 mm (11 inches).

Coiling is dextral. The sculpture of some species is limited to spiral rows of grooves, cords, nodes or punctations. Some species have only axial sculpture. Most have a combination of the two, with some being cancellate, having axial and spiral sculpture of equal strength. Some species are completely smooth, and many have variable sculpture. Most species have a subsutural band marked by a groove or a series of punctations just below the suture.

Terebridae have a small aperture, which may be slender and elongate, quadrate with the outer lip flaring, or almost triangular as in the subgenus *Impages* (see text figs. 2-4). The outer lip is not thickened. There is a sturdy columella which is almost straight in some species and extremely twisted in others. It may or may not bear a plication. The plications seen on that part of the columella visible in the aperture is not necessarily indicative of the plications on the interior columella (see text fig. 5). When the outer lip is broken, often well-developed plications are revealed which may not be visible when the outer lip is intact. Some species have a parietal callus, while others do not. All Terebridae have a siphonal fasciole and a short anterior canal.

The protoconch may have only 1½ whorls or it may have as many as 4½ or 5 (see text figs. 6-28). The multi-whorled protoconch denotes a long planktonic existence which may account for the wide dispersal of some species. *H. cinerea*, like several other molluscan species, is found on both sides of the Atlantic, although it is rare in West Africa. It also is found on both sides of Panama, along with *T. petiveriana* and *T. dislocata*. It is not suggested that the veligers of these three species came through the Panama Canal but rather that they existed on both sides before the closing of the Isthmus in recent geologic times.

Terebridae are members of the superfamily Conacea along with the families of Conidae and Turridae, but confusing them with members of these other two families is quite unlikely as the shell shape differs appreciably. However, the shape of some of the Pyramidellidae, Turritellidae, and Cerithiidae is similar to that of the Terebridae, and they are sometimes mistakenly added to *Terebra* collections, although they are in no way related. The shells of all three may be distinguished from the Terebridae by their lack of a siphonal fasciole. In addition, the Pyramidel-

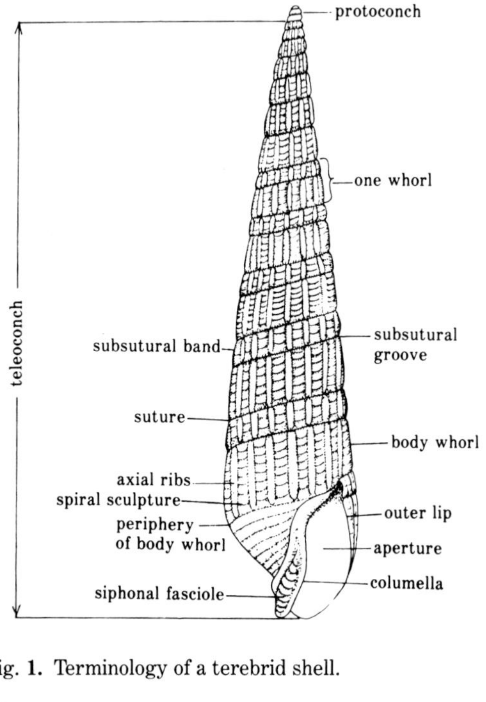

Fig. **1**. Terminology of a terebrid shell.

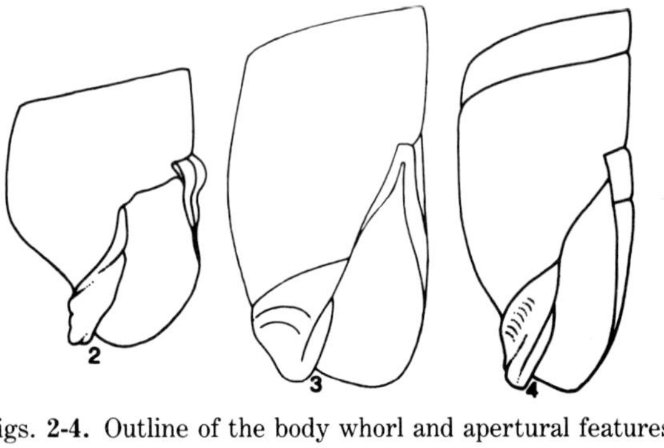

Figs. **2-4**. Outline of the body whorl and apertural features of Terebridae. **2**, *Terebra armillata* Hinds. **3**, *Hastula (Impages) hectica* (Linnaeus). **4**, *Duplicaria raphanula* (Lamarck).

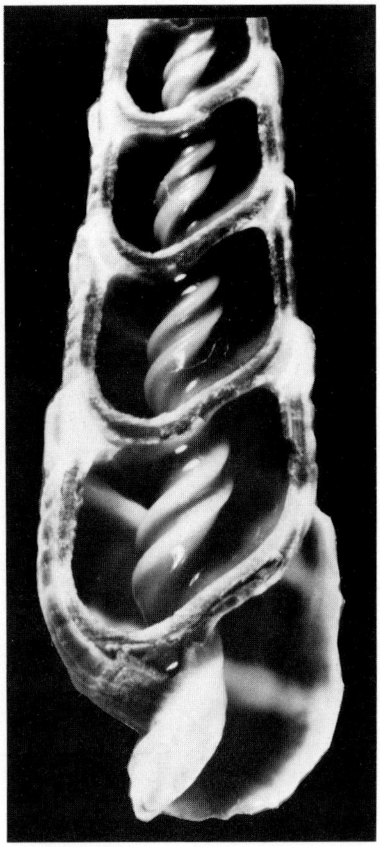

Fig. 5. Sectioned shell of *Terebra variegata* Gray showing the inside of the whorls and the columella plaits, or folds. X **2.**

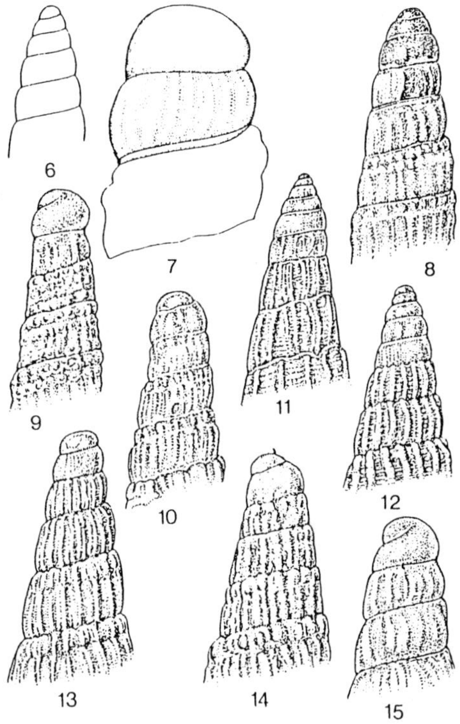

FIGS. **6-15** Protoconchs of Terebridae, **6,** *Terebra brandi* Bratcher & Burch. **7,** *T. fenestrata* Hinds. **8,** *T. corrugata* Lamarck. **9,** *T. guineensis* Bouchet. **10.** *T. fatua* Hinds. **11,** *T. grayi* E.A. Smith. **12,** *T. reticulare* (Pecchioli in Sacco). **13,** *T. senegalensis* Lamarck. **14,** *T. twilae* Bouchet. **15,** *T. histrio* Deshayes. (Figs. **8-15** after Bouchet, 1983).

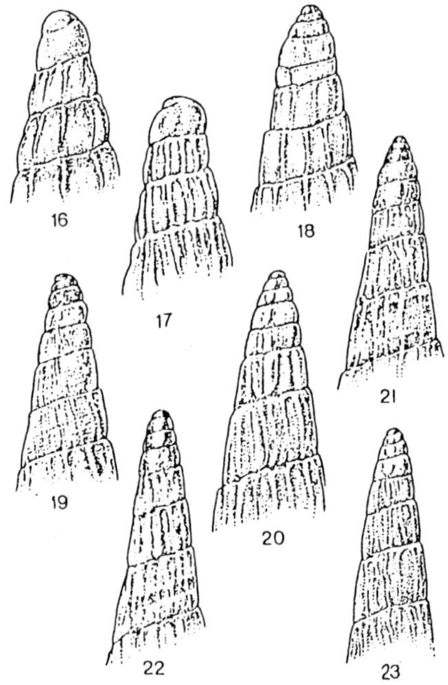

FIGS. **16-23.** Protoconchs of Terebridae. **16,** *Terebra brevicula* Deshayes. **17,** *T. fernandesi* Bouchet. **18,** *Hastula knockeri* (E. A. Smith). **19,** *H. lepida* (Hinds). **20,** *H. leloeuffi* Bouchet. **21,** *H. cuspidata* (Hinds). **22,** *H. (Impages) cinerea* (Born). **23,** *H. (I.) aciculina* (Lamarck).

lidae have heterostrophic embryonic whorls and often have lirate or denticulate apertures, which separate them from the Terebridae. The Turritellidae have a round operculum with small bristles, while that of the terebrids is oblong or elliptical in shape and annular. The Cerithiidae often have varices, but there are none in the Terebridae.

Colors vary within the Terebridae, sometimes within a species. Some are patterned with spiral or axial stripes, spots, or other markings, and those patterns usually vary little within a species. Other species may show a great variation in color. Marcus and Marcus (1960) described the formation of the color in *Hastula cinerea* found in Brazil, "The larval shell is brown due to its thick conchiolinous coat. It is filled up by secondary calcification as are the 3 or 4 following whitish whorls. The remaining whorls of living snails are bluish gray, sometimes olive-gray. These colours are brought about by colourless layers of calcium carbonate alternating with dark brown conchiolinous ones."

In 1895 Dall made the following observations; "This genus is one of the most difficult to handle from the inexhaustible tendency to variation the species exhibit, and which renders it frequently almost impossible to come to any satisfactory conclusion as to the relative rank and permanency of the mutations exhibited."

Functional Morphology, Feeding and Reproduction

Few papers have been published on the general morphology of the soft parts. Risbec (1953) studied and illustrated the anatomy of *T. succincta* (Gmelin) which he listed as its synonym, *T. cancellata* Quoy & Gaimard. He found that in the radular sac the teeth are arranged like those of *Conus* (see text figs. 29-36). Each tooth is very tapered, and the strongest part is near the posterior extremity. The tooth is obliquely truncated at the base and is attached to the radular ribbon by a curved band, larger than its shaft. The sample studied showed 35 well-formed teeth arranged in two series, those reserved in the radular sac lying lengthwise of the partition of the sac. The well-formed teeth occupy only the anterior half of the sac while the posterior half show only transluscent granules. The venom gland faces the radular sac. At the base of the shaft he found teeth oriented in two directions and could not understand how they could be turned and used.

While some members of the Terebridae possess both radular teeth and a poison gland typical of the Toxoglossa, others lack these features.

Marcus & Marcus (1960) discovered that the labial tubes can be extroverted. In living and preserved animals the tube was turned inside out. The animals possibly seize their food, such as polychaetes, with it. The hollow teeth are 0.54 × 0.08 mm at the base. Sometimes loose teeth with backward-directed tips were seen in the oesophagus. Probably such teeth had been engulfed together with the prey.

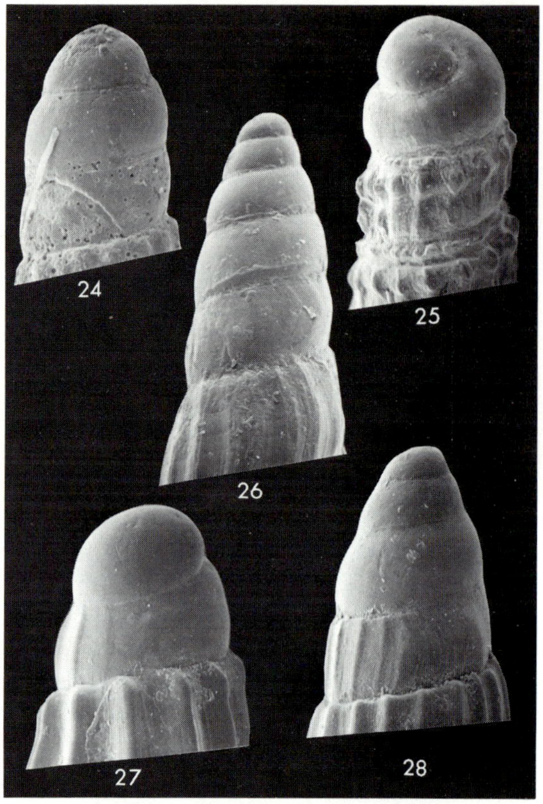

FIGS. **24-28.** Protoconchs of Terebridae (SEM photographs). **24,** *Terebra subulata* (Linnaeus). Guam I, Marianas Is; × 35. **25,** *T. mammilata* Watson. Zamboanga, Philippines × 15. **26,** *T. brandi* Bratcher & Burch. Manzanillo, Mexico; × 35. **27,** *Duplicaria duplicata* (Linnaeus). Carita Beach, W. Java, Indonesia; × 50. **28,** *Hastula strigilata* (Linnaeus). Kauai, Hawaiian Is; × 35.

Miller (1971) divided the Terebridae into the following feeding types with similar life histories

---

Color Plate B

1: *Terebra senegalensis* Lamarck; Senegal; 69.2 mm. (sp. **197**).
2: *Terebra felina* (Dillwyn); Australia; 68.5 mm. (sp. **17**).
3: *Terebra chlorata* Lamarck; Australia; 76.7 mm. (sp. **16**).
4: *Terebra variegata* Gray; West Mexico; 66.8 mm. (sp. **152**).
5: *Terebra robusta* Hinds; W. Mexico; 66.8 mm. (sp. **137**).
6: *Terebra commaculata* (Gmelin); Andaman Is.; 88.8 mm. (sp. **40**).
7: *Terebra pretiosa* Reeve; Japan; 125.0 mm. (sp. **43**).
8: *Terebra taurina* (Lightfoot); Florida; 137.2 mm. (sp. **142**).
9: *Terebra achates* Weaver; Hawaii; 91.9 mm. (sp. **3**).
10: *Terebra subulata* (Linnaeus); Guam; 93.2 mm. (sp. **2**).
11: *Terebra areolata* (Link); Japan; 87.0 mm. (sp. **1**).
12: *Terebra formosa* Deshayes; W. Mexico; 70.6 mm. (sp. **138**).
13: *Terebra ornata* Gray; W. Mexico; 61.0 mm. (sp. **139**).
14: *Terebra corrugata* Lamarck; Senegal; 71.9 mm. (sp. **144**).
15: *Terebra vicdani* Kosuge; Philippines; 64.6 mm. (sp. **41**).
16: *Terebra tessellata* Gray; E. Africa; 62.8 mm. (sp. **5**).
17: *Terebra consobrina* Deshayes; Red Sea; 54.2 mm. (sp. **4**).
18: *Terebra eximia* Deshayes; India; 55.2 mm. (sp. **7**).

Color Plate B. *Terebra* (various oceans).

and animal morphology. He described them as follows:

Type I: species, of which there are two subtypes, have a long labial tube and a short buccal tube. Buccal organs are limited to a pair of salivary glands. They have no poison gland and hence are incapable of rapid prey immobilization and capture.

Type **IA:** terebrids that are specialized to feed on slow moving prey, with all known species selecting hemichordate worms. Feeding is elicited by contact of the propodium of the foot with the prey. The labial tube everts, enters the hemichordate burrow, and grasps the worm. As the labial tube retracts and pulls the worm into the labial cavity, the buccal tube then grasps the worm and pulls it in further. This sequence is continued until the prey is completely ingested. Feeding occurs on or just under the surface of the sand. The species he studied which fall in this category are *Terebra areolata* (Link), *T. crenulata* (Linnaeus), *T. dimidiata* (Linnaeus), *Duplicaria gouldi* (Deshayes), *T. dislocata* (Say), *T. glauca* Hinds, *T. variegata* Gray, and *T. armillata* Hinds.

Type **IB:** terebrids that have an extremely long labial tube and a broad fleshy foot specialized for deep-burrowing in sand. All type **IB** species feed on capitellid polychaetes living in loosely compacted sand. When these species contact a capitellid in its burrow, the foot digs down and pulls the shell into the sand at a 90 degree angle. At the same time, the long slender labial tube everts, enters the capitellid burrow, and extends into the burrow until it grasps the prey. Feeding is then similar to that of Type **IA.** The species Miller placed in this category are *T. chlorata* Lamarck, *T. felina* (Dillwyn), *T. maculata* (Linnaeus), *T. strigata* Sowerby, and *T. ornata* Gray.

Species with the **Type II** proboscis are the only terebrids exhibiting typical toxoglossan

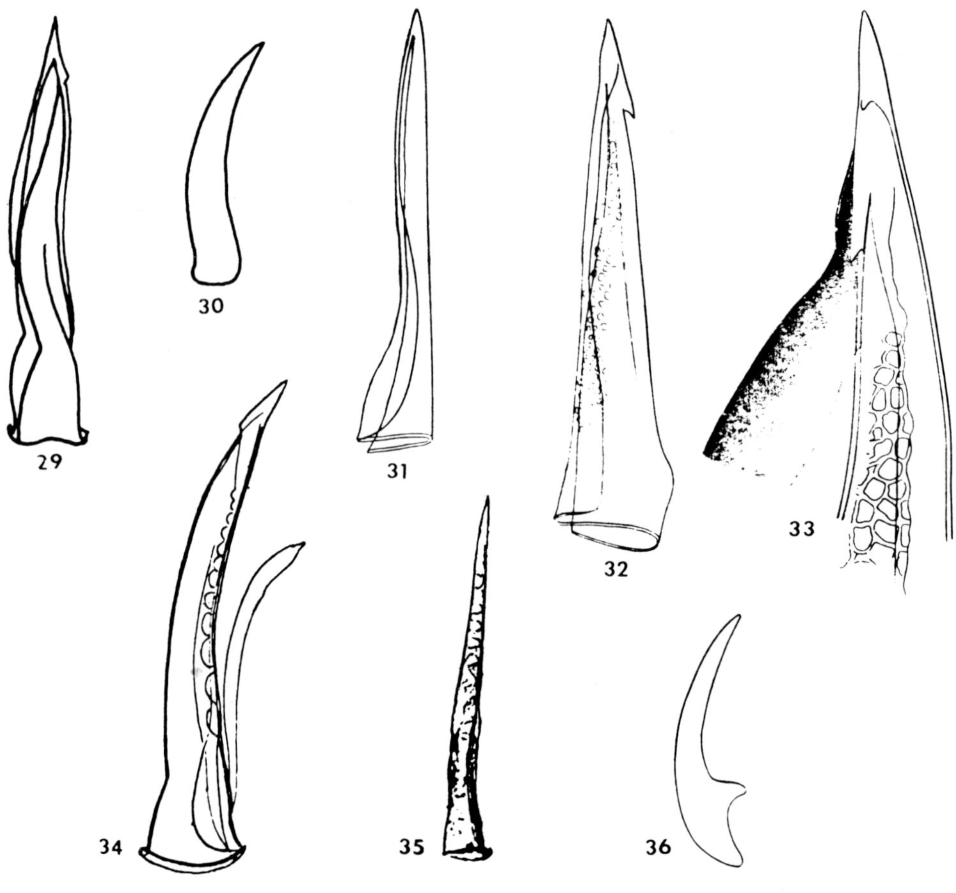

FIGS. **29-36.** Radular teeth of Terebridae. **29,** *Terebra babylonia* Lamarck. **30,** *T. tristis* Deshayes (both after Rudman, 1969). **31,** *Hastula cuspidata* (Hinds). **32,** *Hastula (Impages) hectica* (Linnaeus). **33,** Magnified view of radular tooth of *H. (I.) hectica* (Linnaeus) [all after Troschel, 1865]. **34,** *H. (I.) cinerea* (Born) [after Marcus & Marcus, 1960]. **35,** *H. (I.) salleana* (Deshayes) [after Morrison, 1968]. **36,** *Duplicaria duplicata* (Linnaeus) [after Troschel, 1865].

characteristics. They have a long retractile buccal tube, and contain within the cephalic hemocoel a poison bulb, poison gland, and two rows of harpoonlike radular teeth. There are also two kinds within this type.

Type **IIA:** these terebrids occur in great abundance on surf-washed beaches throughout the tropics. Some species live in the surf zone, occupying the same microhabitat as *Donax*, and others are found just beyond the breakers. Prey, evidently worms, are initially detected by chemo-reception, and this elicits labial tube eversion. When the propodium of the foot comes in contact with the prey, the labial tube everts completely and begins scanning movements over the sand. The buccal tube holding the radular tooth everts at the same time, and when contact is made with the prey, the animal lunges, the tooth is injected, and poison flows into the wound. The labial tube then engulfs the prey, and the snail then burrows into the sand to complete feeding. Prey-capture is usually completed between the passage of two successive ocean waves. The species of this type are *Impages hectica* (Linnaeus), *Hastula strigilata*, (Linnaeus), *I. inconstans* (Hinds), *H. penicillata* Hinds, *H. solida* (Deshayes), *H. matheroniana* Deshayes, *H. albula* (Menke), *H. lanceata* (Linnaeus), *T. nitida* Hinds, *I. cinerea* (Born), *I. salleana* (Deshayes), *I. maryleeae* R. D. Burch, and *T. plumbea* Quoy & Gaimard.

Type **IIB** species differ significantly from Type IIA. All have a small foot, live in deep calm areas, are very secretive, and use the small poison apparatus to immobolize their prey, thus preventing it from retracting into its burrow. The animals are slow-moving and do not burrow deeply into the sand during feeding. Most feed on small tube-dwelling polychaetes. The species included are *T. subulata* (Linnaeus), *T. guttata* (Röding), *T. argus* Hinds, *T. funiculata* Hinds, *T. textilis* Hinds, *T. cingulifera* Lamarck, *T. babylonia* Lamarck, *T. robusta* Hinds.

Type III species lack the radular apparatus as is the case of type IA terebrids, and many have lost the salivary glands and buccal tube as well. They differ from all other terebrids in that they have an accessory feeding organ contained in the labial cavity, indicating that the animal uses the feeding organ to grasp the tentacles of cirratulid polychaetes and pull them into the labial cavity. These species are *T. conspersa* Hinds, *T. flavofasciata* Pilsbry, *T. paucistriata* E. A. Smith, *T. columellaris* Hinds, *T. undulata* Gray, *T. affinis* Gray, and *T. lucana* Dall.

In studying three species of *Terebra*, *T. affinis*, *T. maculata*, and *T. dimidiata*, having neither radula nor poison gland, Rudman (1969) found the pseudoproboscis very long and speculated that there is little doubt that in these cases the prey is caught, not by the typical toxoglossan method of harpooning with a poison tooth, but by the long pseudoproboscis being everted and sucking in the prey.

All Terebridae are carnivorous. Many are prey-specific, enabling several species to live in the same area without direct competition. Although some species have harpoon-shaped radular teeth (figs. 29-36) and a poison gland like the Conidae, there have been no reports of death or injury to humans from *Terebra* stings.

The eyes sometimes look like tiny black points on or at the base of small stalks or weak tentacles. In some species they are placed on the anterior dorsal part of the lobe of the head, but some species lack eyes altogether. It has been suggested that these species locate their prey by chemosensory means.

The foot and siphon are the principal soft parts which extend from the shell. The squarish foot is cleft anteriorly and the sole is densely ciliated. In some species of Terebridae it is more than half the length of the shell. As in other burrowing prosobranchs, such as the Naticidae, the anterior border of the foot is an important sense organ. The operculum, which is attached to the back of the foot, is chitinous, usually yellow or brownish, and may be thin and almost transluscent or heavy and moderately thick. It is annular and has a terminal nucleus. The operculum is not large enough to seal the aperture in most species. Even when the animal is completely withdrawn, the tip of the siphon and a small portion of the foot remains visible in some species.

The mantle, contained within the body whorl, is wide, and the mantle cavity is very deep, water being drawn into it through the anteriorly located inhalant siphon and passed over the gills. Positioning of the proboscis, oesophagus, salivary gland, reproductive and other internal organs may vary from species to species.

The sexes are separate, and fertilization take place by copulation. The penis is scimitar-shaped or hatchet-shaped and its length is from 15% to 30% of the length of the shell. Some species mate beneath the sand, others on the surface. The jellylike eggs are placed on grains of sand, small stones or basalt chips. The eggs develop into planktonic veligers in the majority of Terebridae although many species have direct development. Of the 204 species examined, 123 have planktonic development, and 81 have direct development. The protoconchs of these gastropods give us much information as to how they develop and consequently the ease of dispersal. The paucispiral protoconch (fig. 25) indicates a species with direct development while the multispiral protoconch of four and one-half whorls (fig. 26) indicates that the species has a long planktonic life and therefore may be carried long distances by the currents before settling to the bottom and adding the whorls of the teleoconch. *H. (I.) cinerea*, with a protoconch of four and a half whorls, is a good example, being found on both sides of the Atlantic.

Except in the subjects of feeding and reproduction, little is known about the behavior of these animals.

Habitats

Terebridae are sand-dwellers living from the intertidal zone to a depth of 260 meters. Some species are limited to clean coral sand while others prefer a mixture of sand and mud. A few species are found in mud with little or no sand. They are all nocturnal. Some species live a few mm. beneath the sand by day, often crawling on the surface of the sand at night where they can be found by scuba divers or snorklers with underwater lights. Others live on surf-swept beaches, appearing briefly, usually at the turn of the tide. As the wave draws back, they glide against the receding water with the narrow posterior of the spire pointing with the current flow. By the time the next wave arrives, they have disappeared into the sand or have dug deep enough to remain in their place.

Terebridae thrive in warm tropical waters, with the greatest number being found in the central Indo-Pacific area. A few species are found in the cool waters off the coasts of California, New Zealand, Australia, the Peruvian province, and Argentina. *T. gemmulata* Kiener is distributed in frigid water from Rio de Janeiro around the tip of South America to Chile. The type locality for the two junior synonyms of *T. gemmulata* are Patagonia for *T. patagonica* Orbigny and Chile for *T. chilensis* Deshayes.

Fossil Terebridae

Their geologic record goes back to the Eocene, and fossil Terebridae are found in all faunal provinces. The fossil beds in Europe, particularly in Italy, are rich in fossil terebrids although the family is not included among the living fauna found there today. Sinistral (left-handed coiling) fossil species, have been recorded from Belgium (*T. inversa* Nyst) and from England (*T. inversa oakleyana* Finlay). The first record of the genus *Hastula* in the new world is a Miocene fossil from California, *H. gnomen* Keen, 1943, although the genus is unknown in California today. Cossman (1896) reports *Hastula s. s.* (*H. plicatula* Lamarck) from Eocene deposits of France and *Hastula (Impages)* (*H. fuscata* Brocchi)

Color Plate C

1: *Terebra cerithina* Lamarck; Hawaii; 54.5 mm. (sp. **19**).
2: *Hastula hectica* (Linnaeus); Gulf of Oman; 54.0 mm. (sp. **226**).
3: *Hastula hectica* (Linnaeus); Hawaii; 45.7 mm. (sp. **226**).
4: *Terebra cumingii* Deshayes; Philippines; 76.8 mm. (sp. **44**).
5: *Terebra triseriata* Gray; Formosa; 82.1 mm. (sp. **45**).
6: *Terebra jenningsi* R. D. Burch; Borneo; 56.5 mm. (sp. **49**).
7: *Terebra argus* Hinds; Marshall Is.; 70.8 mm. (sp. **13**).
8: *Terebra succinea* Hinds; Qld., Australia; 73.3 mm. (sp. **12**).
9: *Duplicaria trochlea* Hinds; Marquesas Is.; 79.2 mm. (sp. **253**).
10: *Terebra strigata* Hinds (without dark stripes); W. Mexico; 75.2 mm. (sp. **141**).
11: *Terebra crenulata* (Linnaeus) (noded); Marquesas; 91.7 mm. (sp. **11**).
12: *Terebra crenulata* (Linnaeus) (smooth); Borneo; 110.4 mm. (sp. **11**).
13: *Terebra dimidiata* (Linnaeus); Papua New Guinea; 89.8 mm. (sp. **10**).
14: *Terebra dimidiata* (Linnaeus); Western Samoa; 100.7 mm. (sp. **10**).
15: *Terebra dimidiata* (Linnaeus) (pathological specimen); Philippines; 90.8 mm. (sp. **10**).
16: *Terebra strigata* Sowerby (smooth); Panama; 90.0 mm. (sp. **141**).
17: *Terebra strigata* Sowerby (ribbed); W. Mexico; 97.7 mm. (sp. **141**).

Color Plate C. Indo-Pacific and Eastern Pacific *Terebra*.

from the Miocene of Europe. We have included fossil genera and species only when the same genera and species are among the recent material as well. As there have been more names proposed for fossil Terebridae than for recent species, it would take additional years of research before they could be included in this book.

Predation

There are two types of predators of Terebridae, primarily crabs and drilling gastropods. The fact that the soft parts of Terebridae can be withdrawn one to two whorls back from the outer lip often accounts for unsuccessful attempts by crabs or fish which would have to break back the strong shell from the outer lip. The scars induced by such attempts are evidenced in many specimens. Spire removal without injury to the lip is a common method of breakage used by hard-bottom crabs on rocky shore gastropods. Terebridae have been found with the apex removed, but the truncated apical end of the shell is closed off in all cases. Terebridae generally have the apical portion filled in; therefore, if a predator is to break the spire in order to extract the soft parts, it must do it at some distance from the apex. Incomplete drill holes are common in large species and suggest that the terebrid shell is well-adapted to withstand attacks by Naticidae and other drilling gastropods and octopods, (Vermeij, Zipser, & Dudley, 1980).

Collecting and Cleaning

Being nocturnal, most of the Terebridae are found buried under a few inches of sand during daylight hours. Consequently, night collecting for these mollusks is most productive. They will be found crawling on the surface of the sand or so close to the surface that the shell or outline of it and their trails will be visible under the beam of an underwater light. This is effective for intertidal collecting, snorkling, or scuba diving.

The next best time for hunting these shells by snorkling is in the early hours of daylight before the trails left by the nocturnal animals have been washed away by the tide. This necessitates looking for a hump at one end of the trail.

While snorkling or scuba diving for sand-dwellers when no trails are visible, one may locate shells often by fanning the sand. Locating them by this method is more difficult but is successfully employed by many collectors.

A good method for collecting all sand-dwelling mollusks is by the use of a hand dredge. The mesh covering the frame of the dredge should not be smaller than 1/4 inch in diameter (as it would be clogged by the sand) nor so much larger that it will allow small specimens to slip through the mesh. Copper mesh is the most suitable, as it is far more resistant to salt water than galvanized iron mesh, and is not as brittle or as heavy as stainless steel. Ordinary chicken wire mesh can be used but will need to be replaced regularly. Care should be taken not to set the jaws on the dredge-plate at too steep an angle, as the dredge will bite into the sand rather suddenly and tip over. The dredge should be dragged through the sand until it is half or entirely full. Then, holding the dredge by the handle, the excess sand is washed from it by shaking it up and down in the water. The handle is supported on the sea floor if one is intertidal collecting, and the dredge will rest on the swivel arm at right angles to the handle and allow easy access to sorting the shells. The dredge makes collecting available to non-swimmers as it can be used in water up to waist or chest deep. For scuba diving, it is best to dispense with the removable handle.

A good collecting bag is a necessity. There are many types to choose from in dive shops. A screw-cap medicine vial filled with sea water in the collecting bag will protect the more delicate shells and help keep outer lips from chipping and protoconchs from being broken by contact with larger shells. More than one shell can be carried in the collecting bag. A man's sock pinned or sewn inside the bag so it will fall short of the bottom is also a help for protecting shells from contact with larger ones. For the non-swimmer a bag made of sturdy material about 9 to 12 inches in length by 6 to 7 inches in width is ideal with two belt loops in the back and a flap in front.

Removing the animal from the many-whorled terebrid shells is almost an impossibility unless one has a microwave oven available while the shells are fresh. They can be placed in a plastic bag which has a slit in it, and microwaved for a few seconds. The entire animal will be ejected by the steam created in the spire, but only when the terebrid is fresh. Other methods used for cleaning *Cypraea*, *Conus* and other mollusks with broader, fewer-whorled shells will not be successful with terebrids. It is best to place them in alcohol (a mixture of 70% grain alcohol and 30% water being suitable) for a few days, depending on the size of the shell, and then dry them. The alcohol helps draw the moisture from the animal tissues. They may be kept in alcohol until one is ready to return from a collecting trip. Then the damp shell may be wrapped in plastic or cotton for protection and placed in a screw-top plastic jar or several layers of plastic bags for transportation. Very fine plastic tubing may be cut in short lengths and placed over the apex of terebrids to protect the protoconch and early whorls from breakage. Plastic jars insure odor-free luggage. With larger terebrids, it will be some time before the tissue dries, smaller ones a few days. Animals may be stored in a 70% alcohol solution after returning from the collecting trip and used later for anatomical and radular studies, although it usually involves breaking the shell to extract the animal. There is much to be learned about the ecology, habits, and food preferences of various terebrid species. Observations by collectors in the field may offer valuable contributions.

STUDY METHODS and EXPLANATION OF TEXT

Except for original descriptions of new species of Terebridae, the citations given in the synonymy under each species are only those in which the species were illustrated, because without figures it is impossible to know which species the author had in mind. For the same reason, only illustrated articles or those with original descriptions or that deal with anatomy, radula, or habits are cited in the bibliography.

In our synonymy citations of valid species and junior synonyms, we give the date of publication; the original genus; subgenus, if any; species; subspecies, if any; author of the species; any subsequent illustrated publication, including author of the publication; volume number; issue or part number in parentheses; page number; plate number, and figure number.

The species descriptions include in order the size, color, outline of whorls, protoconch, subsutural band, axial sculpture, spiral sculpture, aperture, columella, and any unusual features. In our description of the species, if a parietal callus is not mentioned it may be assumed that the species is without the callus. The siphonal fasciole is mentioned only when it is unusual. If the protoconch or operculum is not mentioned it is unknown to us. The operculum is unknown in many species.

For each valid species and junior synonym we cite the type locality, in quotations, as stated in the original description. Any remarks outside the quotation marks are our own. The type locality given for several of the earlier species was erroneous because of collecting data becoming mixed or neglected during the long voyages on sailing vessels.

The distribution included here is that known to us personally from labels in museum and private collections and from our own personal experience. We realize there are many range extensions unknown to us. Where the depth range is omitted under distribution, it is unknown to us.

For each valid species and junior synonym cited we give the institution in which the type is deposited, the size of the type specimen and whether it is a holotype or lectotype. The type of each species, where extant, both valid and synonymized, is illustrated. Cernohorsky (1969) listed 216 holotypes or designated lectotypes of *Terebra* located in the British Museum (Natural History) all of which are illustrated in this book.

Terebra, *Hastula*, and *Hastula (Impages)* are split into faunal provinces to assist collectors and curators in identifying their specimens. Three species of Indo-Pacific terebrids are found in the Revillagigedo Islands of Mexico along with a mixture of other Indo-Pacific and Eastern Pacific molluscan fauna. There are no *Duplicaria* in the Eastern Pacific or Atlantic.

We have omitted remarks on the abundance of most species because species rare in one area frequently are abundant in another, and those plentiful at one time may be scarce at another season or year.

We have chosen to use black and white illustrations and enlargements in order to show the sculpture in greatest detail as sculpture is one of the best criteria for identifying this family of mollusks.

The species sequence is arranged so that species with similar shell characteristics from the same faunal province are together and may be compared.

SYSTEMATICS OF THE TEREBRIDAE

Outline of Classification
Phylum **Mollusca**
Class **Gastropoda**
Subclass **Prosobranchia**
Order **Neogastropoda**
Suborder **Toxoglossa**
Superfamily **Conacea**

Family TEREBRIDAE Mörch, 1852

1852 Terebridae Mörch, Cat. Conchyl. Yoldi 1:74.
1853 Acusidae Gray, Ann. Mag. Nat. Hist. (ser. 2), 11 (62): 129× (based on *Acus* Gray, 1847 [non Lacepède, 1803]).
1853 Terebrinae H. & A. Adams, Gen. Rec. Moll. 1:224.
1969 Pervicaciidae Rudman, Veliger 12 (1): 63.

The gastropod family Terebridae belongs to the subclass Prosobranchia, order Neogastropoda, along with the superfamilies Muricacea, Buccinacea, Volutacea and Mitracea. The Terebridae belong to the superfamily Conacea which is included in the suborder Toxoglossa (poison tongue), the most highly specialized of the prosobranchs. The superfamily Conacea also includes the Conidae and Turridae. Most species in this superfamily have radular teeth and a poison gland to immobilize their prey. Some species of the Terebridae have harpoon-shaped radular teeth like those of *Conus* and a poison gland. Other species have a radula but no poison gland while still others lack both.

Although Terebridae are one of the most confusing groups among the larger gastropods, relatively little work has been done on the taxonomy

---

Color Plate D

1: *Hastula bacillus* (Lamarck); Philippines; 18.3 mm. (sp. **232**).
2: *Hastula bacillus* (Lamarck); Philippines; 24.5 mm. (sp. **232**).
3: *Hastula aciculina* (Lamarck); Ghana; 21.9 mm. (sp. **236**).
4: *Hastula cuspidata* (Hinds); W. Africa; 19.7 mm. (sp. **225**).
5: *Hastula inconstans* (Hinds); Hawaii; 28.6 mm. (sp. **228**).
6: *Hastula celidonota* Melvill & Sykes; Reunion Is.; 13.7 mm. (sp. **218**).
7: *Hastula anomala* (Gray); Papua New Guinea; 33.1 mm. (sp. **229**).
8: *Hastula rufopunctata* (E. A. Smith; W. Australia; 19.9 mm. (sp. **210**).
9: *Hastula rufopunctata* (E. A. Smith) W. Australia; 21.3 mm. (sp. **210**).
10: *Hastula rufopunctata* (E. A. Smith); W. Australia; 21.3 mm. (sp. **210**).
11: *Hastula rufopunctata* (E. A. Smith); W. Australia; 21.2 mm. (sp. **210**).
12: *Hastula rufopunctata* (E. A. Smith); W. Australia; 27.7 mm. (sp. **210**).
13: *Terebra miranda* E. A. Smith; Thailand; 23.6 mm. (sp. **63**).
14: *Terebra waikikiensis* Pilsbry; Hawaii; 23.6 mm. (sp. **54**).
15: *Terebra dillwynii* Deshayes; Senegal; 25.0 mm. (sp. **198**).
16: *Terebra undulata* Gray; Solomon Is.; 38.0 mm. (sp. **69**).
17: *Terebra parkinsoni* Cernohorsky & Bratcher; Solomon Is.; 34.3 mm. (sp. **70**).
18: *Terebra nebulosa* Sowerby; Hawaii; 30.2 mm. (sp. **68**).
19: *Terebra columellaris* Hinds; Papua New Guinea; 29.7 mm. (sp. **66**).
20: *Terebra kilburni* Burch; Papua New Guinea; 24.9 mm. (sp. **67**).
21: *Terebra paucistriata* E. A. Smith; Fiji Is.; 26.1 mm. (sp. **71**).
22: *Terebra picta* Hinds; Philippines; 23.0 mm. (sp. **81**).
23: *Terebra cinctella* Deshayes; India; 15.2 mm. (sp. **96**).
24: *Terebra capensis* (E. A. Smith); S. Africa; 14.3 mm. (sp. **121**).

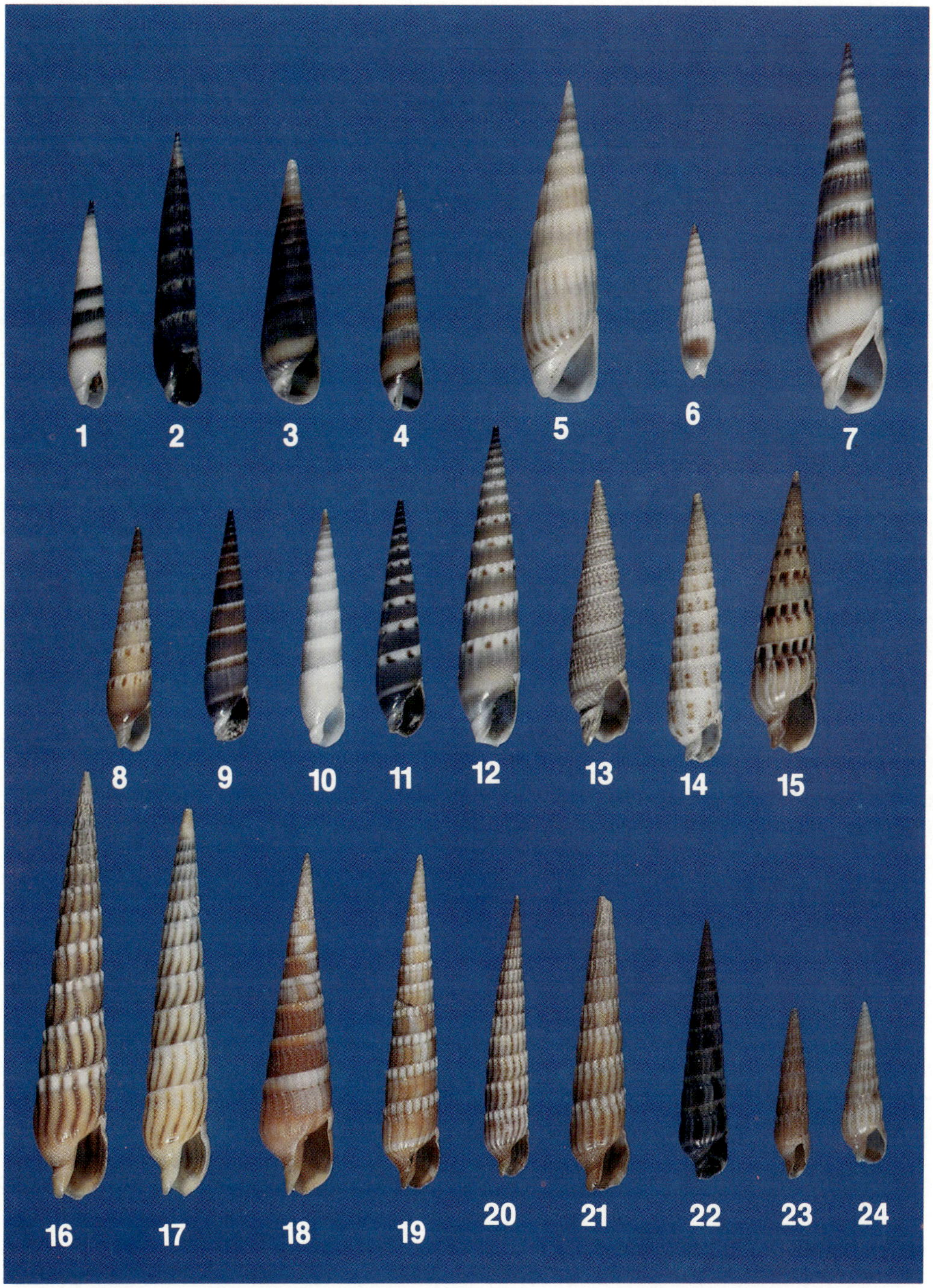

Color Plate D. Indo-Pacific and African *Terebra.*

of the family in the last hundred years. There were four comprehensive illustrated reviews or monographs in the preceding century. In 1844 Hinds, in Sowerby's monograph of the genus *Terebra (Thesaurus Conchyliorum,* vol.1), listed 109 species he considered valid. Deshayes' monograph in 1859 *(Proceedings of the Zoological Society of London)* included 221 species of Terebridae plus a number of doubtful species. A year later Reeve's monograph of the genus *Terebra (Conchologia Iconica,* vol. 12) contained 155 valid names. In 1885 Tryon, (*Manual of Conchology,* vol. 7) placing many names in synonymy, reduced the figure to 120 species. That was the last monograph of the Terebridae. With the advent of scuba diving, the comparative ease in reaching formerly isolated areas, and collecting by dredging, trawling, and netting, we now have on record 265 species of living Terebridae.

Linnaeus (1758), with the beginning of binomial nomenclature, placed all *Terebra* in the genus *Buccinum,* family Buccinidae. When Bruguière proposed the genus *Terebra* for this group in 1789, it was accepted by most authors, although *Buccinum* was still used by some. In 1798 Röding placed all terebrids in the genus *Epitonium,* but this placement was not accepted by subsequent authors. Since Mörch proposed the family Terebridae in 1852, there have been 40 genera and subgenera proposed for living Terebridae, two of which subsequently have been moved to the Columbellidae upon discovery that they have a columbellid radula.

So great is the variability within this family that shell characters blend into each other by gradual transition. It is impossible to devise groupings based on shell characters alone where there is little or no doubt as to which group a species belongs. Under many of the genera and subgenera proposed, some of which are based on a single shell character, a specimen can fall in one genus when young and another when mature, or the two ends of variability within a single species may fall in two different genera. Bartsch separated his *Punctoterebra* from *Hastula* on the basis of having a punctate subsutural groove, but some species, such as *H. penicillata* may be found both with and without the punctate subsutural groove.

Bartsch published a key to the family Terebridae in 1923, but he omitted five of the genera which had been proposed up to that time and included several errors that make his arrangement unusable. He changed the spelling of Dall's *Acuminia* to *Acuminea* and *Oxymeris* to *Oxomeris,* and also perpetuated Dall's error of citing Lamarck, instead of Linnaeus, as the author of *Terebra maculata* and *duplicata.*

Hoping to find a generic division that would encompass both shell and animal morphology, we considered the works of B. A. Miller (1966-1979) in which he separated species according to their feeding methods. We attempted to correlate his groupings with those features of shell morphology which do not change within a species. The anatomical features cut across shell characters and could not be correlated with a single shell feature.

There are differences among many species of Terebridae in the soft parts, such as the proboscis, digestive system, reproductive system, and placement of the eyes. Much more anatomical work needs to be done before a meaningful generic division can be made. Unfortunately the soft parts of many species are not available for study. Unlike some families of mollusks, the shell of *Terebra* must be broken and destroyed to obtain the animal for dissection or radular study, and collectors are reluctant to have a rare shell destroyed even for scientific purposes. Until more work is done or new methods are devised, we propose to arrange the living species of Terebridae in the following conservatively defined groups as aids for identification only:

Order **Neogastropoda**

Genus **Terebra** Bruguière, 1789

Type species: *Terebra subulata* (Linnaeus, 1758).

Shells generally large, narrowly elongate, with the last whorl relatively small, the columellar not elongated, aperture squarish ovate. Sculpture smooth or with strong spiral and/or axial riblets, often beaded.

As it is impossible to devise additional groupings on shell characters alone that will not overlap, we are including here all terebrids with shell characteristics which do not have the requirements for our concept of the genera *Hastula, Duplicaria,* and *Terenolla.*

Genus **Hastula** H. & A. Adams, 1853
Type species: *Hastula strigilata* (Linnaeus, 1758).

Hastula, *sensu stricto*

The shell characteristics of the species we place in this genus are the following: smooth, shiny, almost slippery-feeling shells; axial sculpture of fine crenulations below the suture, or fine ribs which do not swell into nodes; no spiral sculpture, except for a subsutural groove or subsutural punctations. Last whorl usually drawn out, the aperture somewhat elongate.

Subgenus **Impages** E. A. Smith, 1873
Type species: *Terebra caerulescens* Lamarck, 1822.
Now *Hastula hectica* (Linnaeus, 1758).

The shell characters of the species we place in this subgenus fit quite well in the genus *Hastula*, but the shell may be less shiny and slippery than that of typical *Hastula*. In addition, it has an almost triangular aperture which flares broadly at the anterior and narrows to a point at the posterior, with the point being somewhat drawn out. (See fig. 3, a sketch of *Impages* aperture). It has an almost straight columella.

Until it was discovered that *Hastula (Impages) stylata* (Hinds) has neither radula nor poison sac, we believed that *Impages* was a full genus based on both shell and animal morphology, and that all species with these shell characters possessed a typical toxoglossate radula with harpoon-shaped teeth and a poison gland. We had planned to use the genera, *Terebra, Hastula, Duplicaria,* and *Terenolla,* merely as a tool for grouping shells, to give greater ease in locating the proper identification, but also to designate *Impages* as a true biological genus. However, with the discovery that at least one of the species in that group lacks radulae and poison gland, we must consider *Impages* to be an artificial designation and use it also for expediency only at the subgeneric level.

Genus **Duplicaria** Dall, 1908
Type species: *Duplicaria duplicata* (Linnaeus, 1758).

The characteristics of the species we place in this genus are the following: shell axially ribbed; subsutural groove cutting through ribs; suture well-marked; no spiral sculpture or very weak, microscopic spiral lines.

We are using the generic division based entirely on shell characters for the convenience of identification. Some of the species listed under *Duplicaria* have no radula, others have a radula which differs from that of *Hastula*.

Without being able to correlate shell characters with soft anatomical characters, there may be species that could fall in more than one of our divisions.

Genus **Terenolla** Iredale, 1929
Type species: *Terenolla pygmaea* (Hinds, 1844)

The shells of this genus are small and straight-sided, with the outer lip and columella rimmed with brown. We know of no anatomical work done on this species, and it is questionable whether this genus belongs in the Terebridae, but we are tentatively retaining it here. It possibly may belong to the Columbellidae.

HIGHER CATEGORY NAMES PREVIOUSLY PROPOSED IN THE TEREBRIDAE

Family **Terebridae** Mörch, 1852

1852 Terebrina Mörch, Cat. Conchyl. Yoldi 1:74
1853 Acusidae Gray, Ann. Mag. Nat. Hist. (ser. 2), 11 (62): 129 (based on *Acus* Gray, 1847) [non Lacepede, 1803]
1853 Terebrinae H. & A. Adams, Gen. Rec. Moll. 1:224.
1969 Pervicaciidae Rudman, Veliger 12 (1):63.

The following genera and subgenera have been proposed for recent Terebridae:

Abretia H. and A. Adams, 1853, Gen. Rec. Moll. 1: 235. [non Rafinesque, 1815]. Type species, *Terebra cerithina* Lamarck, 1822, recent, Indo-Pacific. "Whorls longitudinally ribbed; columella simple, straight, produced anteriorly; outer lip not sinuated at the fore part. Examples: *cerithina* Lam., *fictilis* Hinds, *lepida* Hinds, *mera* Hinds, *nassoides* Hinds, *pygmaea* Hinds, *tenera* Hinds."

Abretiella Bartsch, 1923, Nautilus 32 (2): 61. [replacement name for *Abretia* H. & A. Adams, 1853, non Rafinesque 1815]. Type species, *Terebra cerithina* Lamarck, 1822, recent, Indo-Pacific.

Acuminia Dall, 1908, Nautilus 21 (1): 124. Type species, *Terebra lanceata* Linnaeus, 1767 = *Hastula lanceata* (Linnaeus), recent, Indo-Pacific. "Sulcus wholly absent. Adult slender, smooth."

Acus H. & A. Adams, 1853 Genera Rec. Moll. 1:224. Type species, *Terebra albida* Gray, 1834, recent, Indo-Pacific. "Eyes on the tips of the tentacles; shell subulate, whorls numerous, simple; aperture elongate; emarginate anteriorly, not produced into a canal; columella simple, recurved, not tortuous; outer lip simple, acute, without a sinus at the fore part." Examples: *albida* Gray, *caerulescens* Lam. *casta* Hinds, *chlorata* Lam., *cingulata* Lam., *crenulata* Linn., *dimidiata* Linn., *duplicata* Linn., *dussumieri* Kien., *eburnea* Hinds, *fatua* Hinds, *hastata* Gmel., *laurina* Hinds, *maculata* Linn., *muscaria* Lam., *nitida* Hinds, *plumbea* Quoy, *raphanula* Lam., *senegalensis* Lam., *strigata* Sow., *stylata* Hinds, *tigrina* Gmel., *varicosa* Hinds.

Brevimyurella Oyama, 1961, Venus 21 (2): 187. Type species, *Terebra lischkeana* Dunker, 1877 = *T. japonica* E. A. Smith, 1873, recent, Japan.

Cinguloterebra Oyama, 1961, Venus 21 (2): 183. Type species, *Terebra hedleyana* Pilsbry, 1905 = *T. adamsi* E. A. Smith, 1873, recent, Western Pacific.

Clathroterebra Oyama, 1961, Venus 21 (2): 188. Type species, *Terebra fortunei* Deshayes, 1957, recent, Western Pacific.

Decorihastula Oyama, 1961, Venus 21 (2): 185. Type species, *Terebra affinis* Gray, 1834, recent, Indo-Pacific.

Dimidacus Iredale, 1929, Australian Zoologist 5: 341. Type species, *Terebra cingulifera* Lamarck, 1822, recent, Indo-Pacific. Replacement name for *Terebrina* Bartsch, [non Rafinesque, 1815], a new subgenus.

Diplomeriza Dall, 1919, Nautilus 33 (1): 32. Type species, *Terebra duplicata* Lamarck, 1822 = *D. duplicata* (Linnaeus, 1758), recent, Indo-Pacific. This was an unnecessary replacement name for *Duplicaria* Dall, 1908, which was erroneously believed to be a homonym of *Diplicaria* Rafinesque, 1833.

Duplicaria Dall, 1908, Nautilus 21 (11): 124. Type species, *Terebra duplicata* Lamarck, 1822 = *D. duplicata* (Linnaeus, 1758), recent, Indo-Pacific. "Sculpture persistent, suture channeled. Shell axially ribbed, sulcate."

Egentelaria Rehder, 1980, Smithsonian Contributions to Zoology no. 289: 93. Type species, *Terebra stylata* Hinds, 1844, recent, Pacific. "Shell moderately long, subulate, with straight sides and flattened whorls; whorls with more or less distant axial ribs that may become evanescent towards the base of whorls; with a pale band below the suture and the periphery of the body whorl."

Euryta H. and A. Adams, 1853, Gen. Rec. Moll. 1: 225. [non *Euryta* Gistel, 1848]. Type species, *Terebra aciculata* Lamarck, 1822, recent, Atlantic. Examples: *aciculata* Lam., *cosentini* Phil., *fulgurata* Van., *granulosa* Lam." This genus is *Mazatlania* Dall, 1900, which belongs to the Columbellidae as it has a columbellid radula.

Euterebra Cotton & Godfrey, 1932, S. Austr. Nat. 13 (2): 65. Type species, *Terebra inconspicua* Pritchard & Gatliff, 1902 = *T. tristis* Deshayes, 1859, recent, Western Pacific. "Small, regularly tapering; axial ribs numerous, faint; spiral striae, microscopic; protoconch blunt, oblique, of 1½ smooth whorls; adult whorls numerous, telescopically arranged, flatly convex, slightly shouldered; suture well defined."

Gradaterebra Cotton & Godfrey, 1932, S. Austr. Nat. 13 (2): 66. Type species, *Terebra scalariformis* Cotton & Godfrey, 1932, recent, Australia. "Tapering, narrow; axial ribs numerous, flexuous; protoconch blunt, two turns, smooth, round; last whorl cylindrically oblong, contracted at the base; columella slightly convex; anterior notch a little everted. Closely related to the genus *Euterebra* but distinguished by the development of the axials and the convex whorls."

Granuliterebra Oyama, 1961, Venus 21 (2): 188. Type species, *Terebra bathyrhaphe* E. A. Smith, 1875, recent, Indo-West Pacific.

Color Plate E

1: *Terebra fortunei* Deshayes; Japan; 69.8 mm (sp. **111**).
2: *Terebra mactanensis* Bratcher & Cernohorsky; Philippines; 63.7 mm. (sp. **112**).
3: *Terebra mactanensis* Bratcher & Cernohorsky; Philippines; 63.4 mm. (sp. **112**).
4: *Duplicaria australis* (E. A. Smith); W. Australia; 65.2 mm. (sp. **246**).
5: *Duplicaria raphanula* (Lamarck); Papua New Guinea; 50.8 mm. (sp. **256**).
6: *Terebra quoygaimardi* Cernohorsky & Bratcher; Philippines; 48.2 mm. (sp. **30**).
7: *Terebra tricolor* Sowerby; Okinawa; 42.0 mm. (sp. **31**).
8: *Terebra montgomeryi* Burch; Australia; 44.7 mm. (sp. **22**).
9: *Hastula solida* (Deshayes); Philippines; 27.3 mm. (sp. **216**).
10: *Terebra troendlei* Bratcher; paratype; Marqueses Is.; 30.4 mm. (sp. **18**).
11: *Terebra succincta* (Gmelin); Solomon Is.; 42.4 mm. (sp. **98**).
12: *Terebra pseudopertusa* Bratcher & Cernohorsky; holotype; Reunion Is.; 36.1 mm (sp. **265**).
13: *Terebra pertusa* (Born); Hawaii; 34.2 mm. (sp. **77**).
14: *Terebra fatua* Hinds; Cape Verde Is.; 38.4 mm. (sp. **199**).
15: *Terebra fenestrata* Hinds; Philippines; 47.3 mm. (sp. **52**).
16: *Terebra adamsii* E. A. Smith; Japan; 42.4 mm. (sp. **38**).
17: *Terebra lucana* Dall; W. Mexico; 48.9 mm. (sp. **149**).
18: *Terebra tuberculosa* Hinds; Ecuador; 47.0 mm. (sp. **146**).
19: *Terebra petiveriana* Deshayes; Panama; 33.9 mm. (sp. **148**).
20: *Terebra pustulosa* (E. A. Smith); Japan; 39.5 mm. (sp. **62**).
21: *Terebra bifrons* Hinds; Japan; 48.5 mm. (sp. **61**).
22: *Terebra armillata* Hinds; W. Mexico; 31.2 mm. (sp. **156**).
23: *Duplicaria badia* (Deshayes); Formosa; 39.5 mm. (sp. **248**).
24: *Duplicaria spectabilis* (Hinds); Borneo; 35.6 mm. (sp. **242**).
25: *Terebra babylonia* Lamarck; Philippines; 35.9 mm. (sp. **32**).
26: *Terebra larvaeformis* Hinds: W. Mexico; 41.8 mm. (sp. **183**).

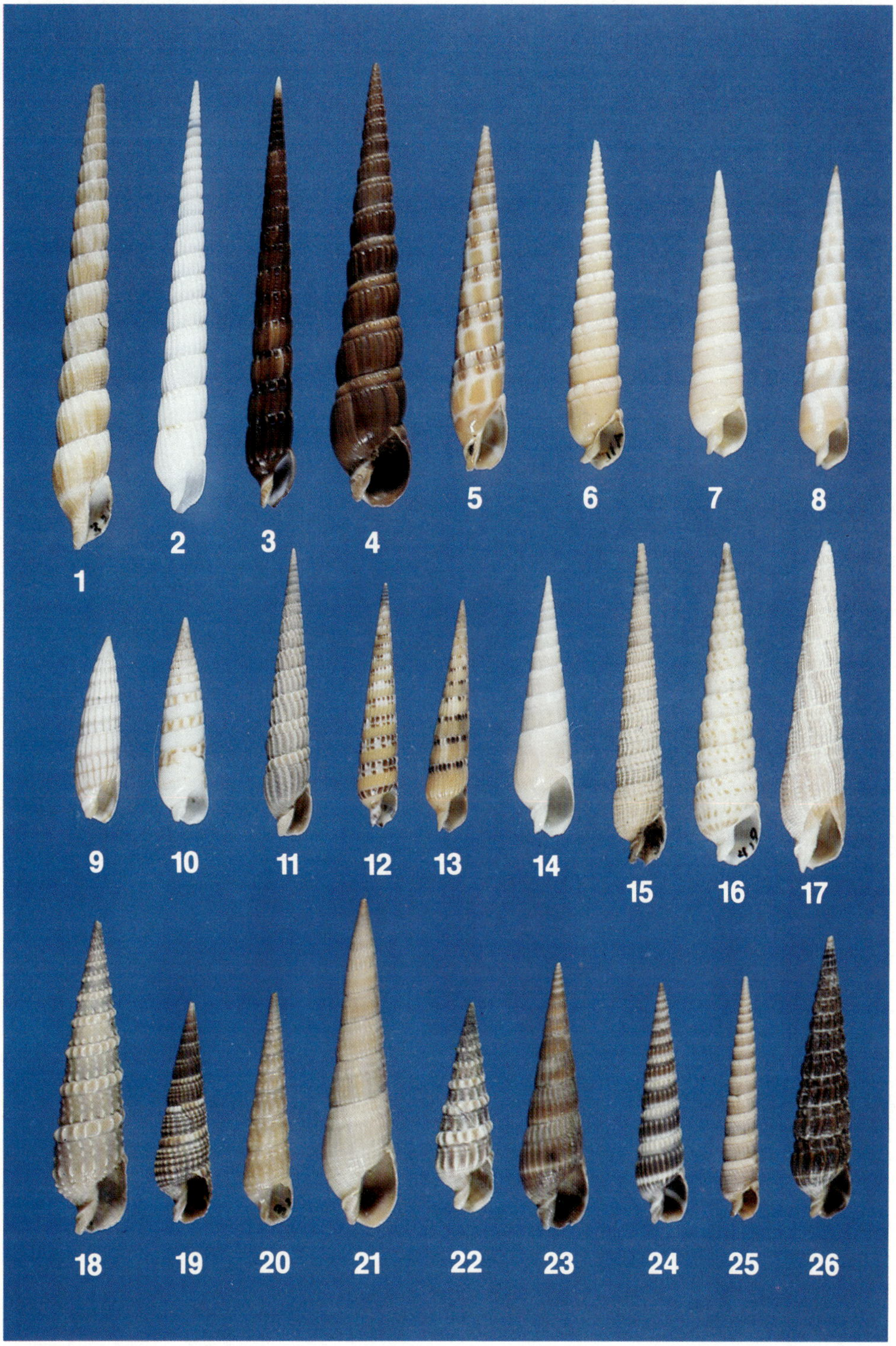

Color Plate E. *Terebra* and *Duplicaria*.

Hastula H. and A. Adams, 1853, Gen. Rec. Moll., 1: 225. Type species by subsequent designation by Cossmann, 1896, *Terebra strigilata* Linnaeus, 1758, recent, Indo-Pacific. "Whorls smooth; columella simple, straight, produced anteriorly; outer lip simple, not sinuated at the fore part." Examples: *albula* Mke., *anomala* Gray, *cinerea* Born, *cuspidata* Hinds, *inconstans* Hinds, *lanceata* Lam., *laevigata* Gray, *luctuosa* Hinds, *penicillata* Hinds, *rustica* Hinds, *spectabilis* Hinds, *strigilata* Lin., *venosa* Hinds.

Hastulina Oyama, 1961, Venus 21 (2): 183. Type species, *Terebra casta* Hinds, 1844 = *Hastula albula* (Menke, 1843), recent, Indo-Pacific.

Hastulopsis Oyama, 1961, Venus 21 (2): 181. Type species, *Terebra melanacme* E. A. Smith, 1875, recent, Japan.

Impages E. A. Smith, 1873, Ann. Mag. Nat. Hist., ser. 4, vol. 11: 263. Type species, *Terebra caerulescens* Lamarck, 1822 = *H. hectica* (Linnaeus, 1758), recent, Indo-Pacific. "A callous band encircling the whorls above, but contiguous to the suture." Described as a subgenus of *Terebra*.

Mazatlania Dall, 1900. New name for *Euryta* H. & A. Adams [non Gistel, 1848]. Type species, *Terebra aciculata* Lamarck, 1822, Recent, Atlantic. This genus was removed from the Terebridae when it was found to have a columbellid radula.

Microtrypetes Pilsbry & Lowe, 1932, Proc. Acad. Nat. Sci. Philadelphia, vol. 84: 43. Type species, *Terebra iola* Pilsbry & Lowe, 1932, recent, Eastern Pacific. "Axial ribs continuous; no subsutural band; one weak fold on the columella." Subgenus of *Terebra*.

Myurella Hinds, 1844, Thes. Conch. p. 171. Type species, *Terebra affinis* Gray, 1834, recent, Indo-Pacific. "Numerous whorls; noded subsutural band; columella produced in mature shells."

Myurellisca Bartsch, 1923, Nautilus 37 (2): 63. Type species, *Terebra (Myurellisca) duplicatoides* Bartsch, 1923 = *Duplicaria duplicata* (Linnaeus, 1758), recent, Indo-West Pacific. "Axial ribs strongly developed on all whorls; spiral sculpture consisting of subsutural groove only." Subgenus of *Terebra*.

Myurellina Bartsch, 1923, Nautilus 37 (2): 63. Type species, *Terebra ornata* Gray, 1834, recent, Eastern Pacific. "Columella with two folds. Spiral sculpture of subsutural groove only. Axial ribs not strongly developed on all whorls. Axial ribs evanescent on the later turns." Subgenus of *Myurella*.

Nototerebra Cotton, 1947, Rec. S. Austral. Mus. 8 (4): 667. Type species, *Terebra albida* Gray, 1834, recent, Australia. "Shell ivory white, seldom brownish white, sometimes with a row of small brown spots; shell rather wide; whorls flat; whorls smooth except for sinuous, oblique accremental striae sometimes gathered into groups to form obsolete, very flat, low angular riblets; protoconch of 1½ whorls."

Oxymeris Dall, 1903, Proc. U. S. Nat. Mus. 26, no. 1342: 951 [footnote]. Type species, *Terebra maculata* Lamarck = *T. maculata* (Linnaeus, 1758), recent, Indo-Pacific. The name was proposed in this fashion: "For the section of *Terebra* commonly called *Acus*, also a preoccupied name, I now suggest the designation of *Oxymeris*." He did not mention the author of the *Acus* he was replacing. In 1908, (Nautilus 21 (11): 124), Dall diagnosed *Oxymeris* as having rapidly enlarging whorls. In 1923 Bartsch diagnosed it as "having a subsutural groove present only on early whorls."

Panaterebra Olsson, 1967, Paleo. Research Inst. p. 14. Type species, *Terebra robusta* Hinds, 1844, recent, Eastern Pacific. "Like *Paraterebra* in general appearance, shell large, robust; sculpture of three, axially ribbed or noded, encircling bands but restricted to a relatively short stretch, the sculpture becomes obsolete on the middle or final sections of the shell; columella with a single fold."

Paraterebra Woodring, 1928, Carnegie Inst. Pub. no. 395: 135. Type species, *Terebra texana* Dall, 1898 = *T. taurina* (Lightfoot, 1786), recent, Western Atlantic. "Shell large, moderately slender; aperture relatively narrow; outer lip inflected forward at about a third of distance from suture to base of anterior canal; anterior canal long, bearing wide, deep notch; siphonal fasciole wide, striate, with keel; columella with parietal callus; sculpture of early whorls consisting of a sutural band, below which lies a narrower band, and of fine axial ribs curved forward; sculpture partly disappearing on later whorls."

Partecosta Dance & Eames, 1966, Proc. Malac. Soc. Lond. 37, pt. 1, p. 42. Type species, *Strioterebrum (Partecosta) wilkinsi* Dance & Eames, 1966 = *Terebra fuscobasis* E. A. Smith, 1877 recent, Persian Gulf. "Shell terebriform; protoconch smooth initially, becoming ribbed later; early postnuclear whorls knobbly, later whorls flatter and resembling those of *Hastula;* there is a subsutural band and groove." Subgenus of *Strioterebrum.*

Parviterebra Pilsbry, 1904, Proc. Acad. Nat. Sci. Philadelphia, 56:5. Type species, *Parviterebra paucivolvis* Pilsbry, 1904, recent, Japan. "Small number of whorls; absence of any subsutural groove or band; long tapering body whorl without a siphonal fasciole at base of the columella." This group is closely allied to *Mazatlania* Dall 1900, and is now placed in the family Columbellidae.

Perirhoe Dall, 1908, Nautilus 21 (11): 125. Type species, *Terebra circumcincta* Deshayes, 1857, recent, Indo-Pacific. "Shell short, small; axial sculpture obsolete, spiral emphatic; spiral lines not punctate." Subgenus of *Terebra.*

Pervicacia Iredale, 1924, Proc. Linn. Soc. N.S.W., vol. 49, pt. 3: 262. Type species, *Terebra ustulata* Deshayes, 1857, recent, Australia. "There is only the basal twist of the columella to represent a fold."

Punctoterebra Bartsch, 1923, Nautilus 37 (2): 63. Type species, *Terebra nitida* Hinds, 1844, recent, Indo-Pacific. "Spiral sculpture consisting of subsutural groove only; axial ribs present on all whorls; axial ribs strong; subsutural groove cutting intercostal spaces only." Subgenus of *Terebra.*

Strioterebrum Sacco, 1891, Moll. Terr. Terz. del Piemonte Liguria 10: 33. Type species, *Terebra basteroti* Nyst, 1843, fossil, Belgium. "with axial ribs and spiral cords and a deep subsutural groove." This genus has been assigned to a number of recent species.

Subula Schumacher, 1817, Ess. Nouv. Syst. p. 233. Type species, *Terebra dimidiata* (Linnaeus, 1758), recent, Indo-Pacific.

Terebra Bruguière, 1789, Encyclop. Méthod. Hist. Nat. Vers., 1:15. Type species, *Buccinum subulatum* Linnaeus, 1767. Recent, Indo-Pacific.

Terebrina Bartsch, 1923, Nautilus 38 92): 63. Type species, *Terebra cingulifera* Lamarck, 1822, recent, Indo-Pacific. "Subsutural cords nodulose on early whorls only; spiral lines punctate."

Terebrum Montfort, 1810, Conch. Syst. 2: 331. Type species, *Terebrum tigreúm* Montfort, 1810 = *T. subulata* (Linnaeus, 1767), recent, Indo-Pacific. "Eyes at the base of the short tentacles and shell which is slender and shiny with square or round black or brown regularly placed spots."

Terenolla Iredale, 1929, Mem. Qld. Mus., 9 (3): 282. Type species, *Terebra pygmaea* Hinds, 1844, recent, Indo-Pacific.

Triplostephanus Dall, 1908, Nautilus 21 (11): 124. Type species, *Terebra triseriata* Gray, 1834, recent, Indo-Pacific. "Shell elongate; whorls medially constricted; whorls nodulose at both margins."

SPECIES EXCLUDED FROM THE TEREBRIDAE

The following species were originally described as *Terebra* but have been subsequently shown to belong to other families:

Terebra acus Bosc, 1801, Hist. Nat. Coq. 5: 16. This is *Pyramidella acus* (Gmelin, 1791).

Terebra (Euryta) angasi Tryon, 1885, Man. Conch. 7: 38, pl. 12, figs. 25, 26. This is a synonym of *Parviterebra brazieri* (Angas, 1875), family Columbellidae.

Terebra arguta Gould, 1853, Boston Journ. Nat. Hist. 6: 380, fig. 19. This is a synonym of *Mazatlania fulgurata* (Philippi, 1846), family Columbellidae, according to Gould himself.

Terebra buccinoidea Blainville 1824, Dict. Sci. Nat. 32: 206. This is a synonym of *Bullia vittata* (Linnaeus, 1767), family Nassariidae.

Terebra buccinulum Deshayes, 1857, Journ. Conchyl. 6: 92, pl. 5, fig. 12. This is a member of the genus *Bullia*, family Nassariidae. The locality of "East coast of Australia" is erroneous.

Terebra cosentini Philippi, 1836, Enum. Moll. Sicilae 1:227, pl. 11, fig. 29. This is a species of *Mazatlania* from the west and east coasts of America (see Bouchet and Gofas, 1983).

Terebra costata Borson, 1825, Mem. Real. Acad. Sci. Torino, 29: 309. This is a non-terebrid fossil, probably a buccinid.

(?) Terebra dyscritos Verco, 1906, Trans. Rep. Royal Soc. S. Australia 30: 149, pl. 4, figs. 3-5. This species is now assigned to *Exomilus* Hedley, family Turridae.

Terebra fluminea Bosc, 1801, Hist. Nat. Coq. 5: 17. This is *Buccinum flumineum* Gmelin, 1791, a fresh water species.

Terebra fluviatilis Bosc, 1801, Hist. Nat. Coq. 5: 21. This is a fresh or brackish water species.

Terebra fulgurata Philippi, 1846, Zeit. f. Malakozool. 3: 53. This is a synonym of *Mazatlania fulgurata* (Philippi, 1846), family Columbellidae.

Terebra gracilis Lea, 1833, Contrib. Geol. p. 166, pl. 5, fig. 171. This is a fossil species of Buccinidae.

Terebra granulosa Lamarck, 1822, Hist. Nat. Anim. S. vert. 7: 291. This is the West African *Dorsanum (Fluviodorsum) granulosum*, family Nassariidae.

Terebra hedleyi Tate, 1901, Proc. Linn. Soc. New S. Wales 26 (2): 214. This is a synonym of *Cingulina brazieri* (Angas, 1877), family Pyramidellidae.

Terebra lineolata G. B. Sowerby, 1825, Cat. shells coll. Tankerville, App. p. xxiii. This is a synonym of *Bullia tranquebarica* (Röding, 1798), family Nassariidae.

Terebra lividula Bosc, 1801, Hist. Nat. Coq. 5: 20, a generic re-assignment for *Buccinum lividulum* Gmelin, 1791, which is a member of the family Cerithiidae.

Terebra nodosoplicata Dunker, 1853, Zeit. f. Malakozool.: 10 (7): 110. This is a synonym of *Mazatlania cosentini* (Philippi, 1836), family Columbellidae.

Terebra obesa Hinds, 1844, Proc. Zool. Soc. Lond., pt. 10: 158. The outer lip of the aperture of the holotype is lirate, showing it does not belong in the Terebridae.

Terebra phallus Bosc, 1801, Hist. Nat. Coq. 5: 17. This is a synonym of *Clionella sinuata* (Born, 1778), family Turridae.

Terebra polita Gray, 1834, Proc. Zool. Soc. Lond. pt. 2: 63, a generic re-assignment of *Buccinum politum* Lamarck, 1822, which is a synonym of *Dorsanum miran* (Bruguière, 1789), family Nassariidae.

Terebra rambhanensi Preston, 1914, Rec. Ind. Mus. Calcutta 10: 279, fig. 5. This is a *Turbonilla*, family Pyramidellidae.

Terebra radiata Bosc, 1801, Hist. Nat. Coq. 5: 20, a generic re-assignment of the non-terebrid *Buccinum radiatum* Gmelin, 1791.

Terebra speciosa Bean in Thorpe, 1844, Brit. Mar. Conch., App. p. 267. This is an *Odostomia*, family Pyramidellidae.

Terebra tahitensis Gray, 1834, Proc. Zool. Soc. Lond. pt. 2: 63. This is a synonym of the problematical *Buccinum othaeitense* Bruguière, 1789, which is tentatively considered a *Bullia*, family Nassariidae.

Terebra thyraea Melvill, 1897, Mem. Proc. Manchester Lit. Phil. Soc. 41 (3): 10, pl. 6, fig. 13. This is a member of *Parviterebra* Pilsbry, family Columbellidae.

Terebra tuberculata Bosc, 1801, Hist. Nat. Coq. 5: 18, a generic re-assignment of *Buccinum tuberculatum* Gmelin, 1791.

Terebra virginica Bosc, 1801, Hist. Nat. Coq. 5: 21, is a generic re-assignment of the fresh water, pleurocerid species, *Buccinum virginicum* Gmelin, 1791.

SPECIES OF DUBIOUS STATUS IN THE TEREBRIDAE

This list contains species which were based on inadequate or ambiguous descriptions, or on very poor illustrations whose type-specimens are no longer extant.

Terebra (Acus) limatula acrior Dall, 1889, Bull. Mus. Comp. Zool. Harvard 18 (2): 66. The type is lost and the description was too brief for identification. In 1901 Dall published a figure of this species, but it was of a shell with most of the body whorl missing. The figure somewhat resembles that of *T. nassula* Dall.

Turritella aequalis Say, 1827, Journ. Acad. Nat. Sci. Philadelphia ser. 1, vol. 2 (2): 208. This species was not figured, and no type exists. Say described other *Terebra* as *Turritella*, but this shell, 1/5th inch in length could be Pyramidellidae.

Terebra ammiralis Fischer von Waldheim, 1807, Mus. Demidoff: 180. The type is lost, and the species was not figured.

Terebra arguta Gould, 1860, Proc. Boston Soc. Nat. Hist. 7: 330. This species from the Cape Verde Islands is a primary homonym of *T. arguta* Gould, 1853, and at the same time is a *nomen dubium* since no type is extant and the species has not been illustrated.

Terebra approximata Deshayes, 1859, Proc. Zool. Soc. Lond. pt. 27: 299. The type is lost, and the species was not figured.

Terebra belcheri Philippi, 1851, Zeit. f. Malakozool. 8 (8): 123. The species was not figured. Philippi's types are supposed to be in the National Museum in Santiago, Chile, but at this time are not accessible to research workers.

Terebra bellanodosa Grabau & King, 1928, Shells of Peitaiho: 214, pl. 9, fig. 80. The original description is very short, and the illustration is too small and may depict either *T. bifrons* Hinds or *T. pustulata* E. A. Smith. We have no other information on the species, and the whereabouts of the type is not known to us.

Terebra bicincta Hinds, 1844, Proc. Zool. Soc. Lond. 11: 150; 1844 Hinds in Sowerby, Thes. Conchyl. 1:175, pl. 44, fig. 72. The type is lost, and the locality is unknown. The figure somewhat resembles the Panamic *T. glauca* Hinds or the Indo-Pacific *T. pustulosa* E. A. Smith, except that both have brownish shells. The figure of *T. bicincta* is white, and Hinds stated that the artist added an extra row of pustules.

Terebra bipartita Gould, 1861, Proc. Boston Soc. Nat. Hist. 7: 330. The species was never figured although it has appeared on two unillustrated lists since its original description.

Terebra cancellata Gray, 1834, Proc. Zool. Soc. Lond. pt. 2: 62. (non Quoy & Gaimard 1833): The type is lost, the species was not figured, and the name is a primary homonym.

Terebra chinensis Deshayes, 1859, Proc. Zool. Soc. Lond. pt. 27: 309. The type is lost, and the species was not figured. Reeve (1860) considered it conspecific with *T. cingulifera* Lamarck.

Terebra costata Küster, 1839, Syst. Conchyl. Cab. Martini & Chemnitz, ed. 2, 5 (2): 26, pl. 6, figs. 4,5 [non Menke, 1828; nec Lea, 1833] The type is lost, and the name is a primary homonym.

Buccinum cuspidatum Gmelin, 1791, Syst. Nat. ed. 13:3505. The cited figure in Seba is of an indeterminable terebrid, probably belonging to the genus *Terebra s. s.* It is a *nomen dubium.*

Terebra decussata Philippi, 1851, Zeit. f. Malakozool. 8 (8): 124. The species was not figured, and Philippi's types are unavailable for examination.

Terebra difficilis Deshayes, 1859, Proc. Zool. Soc. Lond. pt. 27: 304. The type is lost, the species was not figured, and the type locality is unknown. Reeve (1860) said the species is similar to *T. larvaeformis* and *T. souleyeti*, while Tryon (1885) considered it conspecific with *cancellata* Quoy & Gaimard.

Terebra dispar Deshayes, 1859, Proc. Zool. Soc. Lond. pt. 27: 284. The type is lost and the species was not figured. Reeve (1860) stated that it is similar to *micans*, and Tryon thought it was a synonym *cinerea* Born.

Terebra dunkeri Deshayes, 1859, Proc. Zool. Soc. Lond. pt. 27: 285. This is a substitute name for *T. eburnea* Dunker.

Terebra eburnea Dunker, 1853, Zeit. f. Malakozool. 10:96. (non Hinds, 1844). The type is lost, and the species was not figured. The type locality is Japan.

Terebra elegans Küster, 1841 Syst. Conch. Cab. Martini & Chemnitz, ed. 2, 5 (2): 31, pl. 6, fig. 13. The figure looks like *H. strigilata* Linnaeus, *H. acumen* Deshayes, or *H. parva* Baird in Brenchley, but the figure is so small it is impossible to distinguish. Patel (1888) said it is a *Bullia* rather than a *Terebra.*

Buccinum ferrugineum Born, 1778, Index rer. Nat. Mus. Caes. Vind. p. 263. The type is lost. Tomlin (1944) interpreted it as *T. festiva*, Tryon (1885) as *T. dimidiata.*

Terebra fuscomaculata Sowerby, 1825, Cat. Shells Coll. Tankerville, App. xxiii. The type is lost, and the species was never figured. The type locality is unknown.

Buccinum gemicum Linnaeus, 1771, Mantissa plantarum Reg. Anim. appendix. p. 550. Gmelin (1791), Dillwyn (1817), and Dodge (1956) have declared this species to be indeterminable.

Buccinum geminum L., of Gmelin, 1791 and others is *gemicum* (see above).

Acus limatulus Dall, 1889, Bull. Mus. Comp. Zool. Harvard 18 (2): 63. The type is lost, and the species was not figured. The description alone is too brief for identification.

Acus protextus lutescens Dall, 1889, Bull. Mus. Comp. Zool. Harvard 18 (2): 63 (non *lutescens* E. A. Smith, 1873). No figure was published, and the type is lost.

Terebra metcalfi Reeve, 1860, Conch. Icon. 12: pl. 26, fig. 139. The type is lost. Reeve's figure looks as though it might belong to the genus *Mazatlania.* His type locality is the West Indies, but it resembles no terebrid species from the Western Atlantic.

Buccinum monile Linnaeus, 1771, Mantissa plantarum Reg. Anim. appendix, p. 550. Dodge (1956), who treated Linnaeus' *Buccinum* species in detail, and several authors before him, concluded that *B. monile* Linnaeus is a *Terebra* of unknown identity. A specimen in the Linnaean collection of the Linnean Society, London, contains a *Terebra* species 36.7 × 7.2 mm, which appears to be a very faded *T. undulata* Gray. This particular specimen is undocumented and was segregated by Hanley from an unmarked box. Since the origin of this probable type specimen is unknown, it is better to consider it a *nomen dubium.*

Terebra muricina Bosc, 1801, Hist. Nat. Coq. 5: 17: No figure was cited, no reference to a figure in an earlier

---

Color Plate F

1: *Terebra connelli* Bratcher & Cernohorsky; S. Africa; holotype; 97.6 mm. (sp. **267**).
2: *Duplicaria dussumierii* (Kiener); Borneo; 85.9 mm. (sp. **243**).
3: *Terebra punctatostriata* Gray; Papua New Guinea; 67.3 mm. (sp. **24**).
4: *Terebra alveolata* Hinds; Taiwan; 55.3 mm. (sp. **78**).
5: *Terebra alveolata* Hinds; Taiwan; 50.1 mm. (sp. **78**).
6: *Terebra turrita* (E. A. Smith); Taiwan; 58.8 mm. (sp. **99**).
7: *Terebra lima* Deshayes; Japan; 79.3 mm. (sp. **42**).
8: *Duplicaria bernardii* (Deshayes); Australia; 45.3 mm. (sp. **239**).
9: *Duplicaria australis* (E. A. Smith); W. Australia; 56.0 mm. (sp. **246**).
10: *Terebra boucheti* Bratcher, paratype; Philippines; 46.3 mm. (sp. **46**).
11: *Duplicaria duplicata* (Linnaeus); India; 68.9 mm. (sp. **237**).
12: *Duplicaria duplicata* (Linnaeus); Gulf of Oman; 70.5 mm. (sp. **237**).
13: *Duplicaria duplicata* (Linnaeus); West Australia; 72.4 mm. (sp. **237**).

Color Plate F. Indo-Pacific *Terebra* and *Duplicaria*.

publication was given, and the brief description is insufficient for an identification.

Terebra nectarea Melvill & Standen 1895, Journ. Conch. 8 (4): 91, pl. 2, fig. 2. The type could not be located at the Manchester University Museum, and the figure makes placing it in the Terebridae doubtful.

Buccinum obliquum Gmelin, 1791, Syst. Nat. ed. 13: 3404 (ref. Rumphius, pl. 30, fig. F.) The cited figure in Rumphius is indeterminable, and there is no type specimen.

Terebra oblongoguttata Fischer v. Waldheim, 1807, Mus. Demidoff 3: 181 (ref. Born, 1780, pl. 10, fig. 10). It is impossible to determine from the cited figure whether this is *Hastula strigilata* or *H. matheroniana*.

Terebra orizoides Lesson, 1842, Actes Soc. Linn. Bordeaux 7: 203. The type is lost, and the species has not been illustrated.

Terebra perlata Lesson, 1842, Actes Soc. Linn. Bordeaux 7: 203. The type is lost, and the species has not been illustrated.

Terebra prolixa Gould, 1860, Proc. Boston Soc. Nat. Hist 7: 330. The species was not figured, and no type has been found. The original description compares it with *T. amoena* Deshayes. The type locality is off the China coast.

Terebra propinqua Pease, 1869, Amer. Journ. Conch. 5: 66. The type from Hawaii is lost and the species was not figured. Alison Kay (1979, p. 395) considered it a synonym of *columellaris* Hinds, 1844.

Buccinum proximatum Linnaeus, 1771, Mantissa Plantarum Reg. Anim., appendix p. 550. The description was brief, the species was not illustrated, no locality was given, and no type exists. Gmelin (1791) and Dillwyn (1817) both listed it, but merely copied or translated the Linnaean description. Dillwyn said, "No subsequent author has ascertained this species." Dodge, (1956) said he could add only that it is undoubtedly a *Terebra*.

Buccinum pugio Gmelin 1791, Syst. Nat., ed. p. 1305 (ref. Argenville, pl. 11, fig. Q). The figure cited is indeterminable and in shape resembles *T. felina* Dillwyn, in other ways it looks like *H. parva* Baird in Brenchley.

Terebra punctata Gray, 1834, Proc. Zool. Soc. Lond. pt. 2: 61. The type is lost, and the species was never figured. Hinds (1844) and Tryon (1885) said it is conspecific with *T. corrugata* Lamarck. The type locality is unknown.

Terebra radula Gravenhorst, 1807, Vergl. Uebers. Zool. Syst. p. 27. Original publication not seen.

Terebra rosacea Pease, 1869, Amer. Journ. Conch. 5: 65. The type is lost, and the species was never figured. The description is interpreted by Kay (1979) as a different species from that which we and Weaver (1960) interpret it to be. The type locality is Oahu, Hawaii.

Terebra rustica Hinds, 1844, Proc. Zool. Soc. Lond. 11: 158. This was described from the Metcalfe collection, the whereabouts of which are unknown. The type locality is unknown and the figure indeterminable. Tryon (1885) listed it as *T. fulgurata* Philippi, which has since been placed in the Columbellidae, but the aperture appears to be too flaring for that genus. It somewhat resembles the form of *T. armillata* Hinds that Carpenter described as *T. albocincta*.

Terebra spicatus Gray in Dieffenbach, 1843, Travels in New Zealand 2: 234 (refers to *Limax spicatus* Martyn, Univ. Conch. pl. 121, fig. 1). The species is not identifiable from the figure but belongs to *Hastula (Impages)*.

Terebra subdivisa Philippi, 1851, Zeit f. Malakozool. 8 (5): 96. The species was not figured, and Philippi's types are unavailable for examination. The type locality is unknown.

Terebra subnodosa Carpenter, 1857, Cat. Mazatlan Shells p. 386. The holotype is so badly damaged that it is indeterminable. Carpenter mentioned that the spire is bent at an angle through fracture. There is another fracture as serious at the penultimate whorl, and the entire outer lip is missing. Carpenter said the specimen differs from his *T. albocincta* (= *T. armillata* Hinds) by having raised instead of indented sculpture and a subnodulous periphery.

Terebra tenuisculpta Sowerby, 1870, Proc. Zool. Soc. Lond. p. 252. This species was never figured, the type is lost, and the description is very brief. The type locality is China Seas.

Terebra thielei Barnard, 1958, Ann. S. African Mus. 44 (4): 80. Although there are four specimens in the type lot, it is impossible to determine the species. Two of the specimens are fragments of the apex portions, the largest being the first four and a half whorls of the teleoconch. Of the other two, the larger and more complete has only four teleoconch whorls. None had developed the mature sculpture, aperture shape, or outline, and all are in such poor condition that there is no indication of their color. It is possible that they are specimens of *kowiensis* Turton or *planecosta* Barnard.

Terebra tuberosa Hinds, 1844, Proc. Zool. Soc. Lond. pt. 11: 152. 1844, Hinds in Sowerby, Thes. Concyl. 1: 183, pl. 45, fig. 99. The type is lost, and the figure suggests this is possibly a juvenile *Cerithium*.

Terebra violascens Hinds, 1844, Proc. Zool. Soc. Lond. pt. 11: 154; 1844, Hinds in Sowerby, Thes. Conchyl. 1:177, pl. 45, fig. 98. The type is lost, and the figure is impossible to interpret.

Genus *Terebra* Bruguière, 1789

Type species, *Terebra subulata* (Linnaeus, 1758).

Although a number of generic and subgeneric divisions have been proposed, they are artificial and of little use in a family with shell characters as variable as those found in the Terebridae. As it is impossible to devise additional groupings on shell characters alone that will not overlap, we are including here all terebrids with shell characters which do not have the requirements for our concept of the genera *Hastula, Duplicaria* or *Terenolla*.

The majority of the species we are listing under *Terebra s.s.* have no radula. Some, like *T. succincta* have barbed, harpoon-shaped radular teeth and a poison sac like most of those of the genus *Hastula* although their shell characteristics differ greatly. The radula of most have never been observed nor have the soft parts been available for radula study. Until much more work is done on this problem, we place the following in the genus *Terebra*.

INDO-PACIFIC *TEREBRA*

1. *Terebra areolata* (Link, 1807)

(Pl. 1, figs. 1a-e)
(Color pl. B, fig. 11)

1807 *Vertagus areolatus* Link, Beschr. Nat.-Samml. Univ. Rostock Abt. 3:128 (Ref. Chemnitz 1780 vol. 4, pl. 154, fig. 1443).

1816 *Terebra subulata* Lamarck, Tabl. Encycl. Meth. p. 3, pl. 402, figs.2a-b [non *Buccinum subulatum* Linnaeus, 1767]. Okutani, Moll. Japan p. 154, fig. 9.

1822 *Terebra muscaria* Lamarck, Hist. Nat. Anim. S. Vert. 7:285 1860 Reeve, Conch. Icon, 12: pl. 3, sp. 9; 1917 Hirase, Terebridae Jap. Emp. p. 23, pl. 1, fig. 5; 1953 Risbec, Bull. Mus. Nat. Hist. Paris, 2nd ser. 25:576, anatomy.

1935 *Terebra (Subula) muscaria* Lamarck, Dautzenberg, Mem. Mus. Roy. d'Hist. Nat. Belgique 2(17):31, pl. 2, fig. 1.

1952 *Terebra areolata* (Link), Kuroda & Habe, Rec. Moll. Japan. p. 89; 1960 Weaver, Hawaiian Mar. Moll. 1(3):10, pl. 3, figs. center; 1967 Cernohorsky, Mar. Shells Pacific 1:196, pl. 49, fig. 353; 1975 Salvat & Rives, Coq. Polynésie p. 359, fig. 379; 1979 Kay, Hawaiian Mar. Shells p. 393, fig. 127c; 1982 Abbott & Dance, Compendium of Sea Shells p. 273; 1984 Trondle, Xenophora 15:8, fig. 5; 1984 Aubry, Terebridae pl. 5; 1984 Salvat & Rives, Shells of Tahiti p. 134, pl. 34, fig. 5.

1961 *Subula areolata* (Link), Oyama & Takemura, Moll. Shells, Resources Exploitation Inst. 5: *Subula* fig. 4, protoconch; 1967 Habe & Kosuge, Stand. Book. Jap. Shells Col. 3:101, pl. 40, fig. 4.

1966 *Terebra (Subula) areolata* (Link), Cernohorsky & Jennings, Veliger 9(1):43, pl. 4, fig. 3.

Description: Shell to 167 mm; color cream to orange-beige, with 3 rows of squarish brown spots per whorl, 4 on body whorl; outline of whorls slightly convex with somewhat indented subsutural band; protoconch $3^{1/2}$ conical whorls; subsutural band flat, defined by groove which becomes progressively more shallow; early sculpture may be very strong with well developed nodes on subsutural band except for subsutural band and microscopic axial striae; aperture quadrate; columella curved; operculum opaque brown.

Type locality: None. The Island of Ticao, Philippines, is here selected as the type locality; *T. muscaria* "Indian Ocean."

Distribution: Indo-Pacific, from the Red Sea to Polynesia and the Hawaiian Islands; 3 to 50 m.

Type: The type figure from Chemnitz (1780) is designated as the illustrated lectotype of *T. areolata* (Link); *T. muscaria* holotype (pl. 1, fig. 1a) MHNG no. 1102: 93.2 mm.

Discussion: There has been more work done on this species than on most terebrids. J. D. Taylor (unpubl. Ph. D. thesis) reported that in Hawaii the veligers are found in plankton during July and August. They hatch when one whorl is complete, and they add $1^{1/2}$ more whorls while planktonic. B. A. Miller (1966) noted that in Enewetak, Marshall Islands, they feed on the enteropneust, *Ptychodina flava*. Risbec (1953) reported that they have neither radula nor poison gland.

T. achates (**3**) and *T. consobrina* (**4**) may be separated from this species by having shells with spiral rows of punctations and 2 rows of spots per whorl, 3 on body whorl. Refer to *T. subulata* (**2**) for comparison.

2. *Terebra subulata* (Linnaeus, 1767)

(Pl. 1, figs. 2a-b)
(Color pl. B, fig. 10)

1767 *Buccinum subulatum* Linnaeus, Syst. Nat. ed. 12:1205.

1810 *Terebrum tigreum* Montfort, Conchyl. Syst. 2:430, fig. 2.

1811 *Terebra fusca* Perry, Conchology or Nat. Hist. Shells, London, p. 15, pl. 16, fig. 3.

1859 *Terebra (Subula) tigrina* Gmelin, Chenu, Man. Conchyl. 1:219, fig. 1207 [non *Buccinum tigrinum* Gmelin, 1791, p. 3475, a *Cominella*].

1928 *Terebra subulata* (Linnaeus), Hirase, Terebridae Jap. Emp. p. 21, pl. 1, fig. 4; 1935 Dautzenberg, Mem. Mus. Roy. d'Hist. Belgique 2(17):8, pl. 1, fig. 2; 1960 Weaver, Hawaiian Mar. Moll. 1(7):26, pl. 7, fig. top left; 1966 Cernohorsky & Jennings, Veliger 9(1):40, pl. 4, fig. 2; 1975 Okutani, Moll. Japan p. 154, fig. 9; 1979 Kay, Hawaiaan Mar. Shells p. 402, fig. 127e; 1982 Abbott & Dance, Compendium of Seashells p. 273; 1984 Trondle, Xenophora, 15:13, fig. 28; 1984 Aubry, Terebridae pl. 8; 1984 Salvat & Rives, Shells of Tahiti p. 134, pl. 34, fig. 4.

1983 *Terebra subulaia* [sic] Linnaeus, Habe & Kosuge, Common Shells Japan P. 40, fig. 24.

Description: Shell to 168 mm; color cream with 2 rows of squarish brown spots per whorl, 3 on body whorl; outline of whorls straight; protoconch of 3 conical whorls; subsutural band with a row of small nodes on early whorls only, band marked by weak groove, sometimes followed by a second row of minute nodes, both fading completely; axial and spiral sculpture absent in later whorls except for microscopic growth striae, and rarely very weak, almost microscopic spiral punctations; aperture quadrate; columella recurved.

Type locality: *T. subulata*, "India"; *T. fusca*, "W. Indies" (error).

Distribution: Tropical Indo-Pacific; intertidal to 10 m.

Type: *T. subulata* holotype, the Linnean Society London, 76.5 × 14.9 mm; *T. fusca*, type lost.

Discussion: *T. tigreum* Montfort, 1810 (=*T. subulata*) is the type species of the genus *Terebrum* Montfort, 1810, described as having eyes at the base of the short tentacles and shell which is slender and shiny with square or round, black or brown, regularly placed spots. Lamarck was the first to assign *subulatum* Linnaeus to the genus *Terebra* Bruguière, thus making it the type of the genus.

This species has harpoon-shaped radular teeth and a poison sac (Mills, 1979). Although it is considered not to have punctate sculpture, some large specimens, including the holotype, have several rows of extremely obsolete punctations. It may be separated easily from the other large brown-spotted species, such as *T. areolata* (**1**)), which has 3 rows of spots per whorl, 4 on body whorl, has ribbed early whorls, and retains the subsutural groove into maturity. *T. subulata* has only 2 rows of spots per whorl, 3 on the body whorl; no ribs on early whorls; and no subsutural groove on mature whorls. *T. achates* (**3**)) and *T. consobrina* (**4**) have spiral rows of well-marked punctations below the subsutural band. The early whorls of *T. achates* have 2 rows of heavy, well-developed round nodes, followed by about 4 rows of deep punctations.

3. *Terebra achates* Weaver, 1960

(Pl. 1, figs. 3a-c)
(Color pl. B, fig. 9)

1960 *Terebra achates* Weaver, Hawaiian Mar. Moll. 1(1):2, pl. 1, fig. center; 1979 Kay, Hawaiian Mar. Shells p. 391, fig. 127d; 1982 Abbott & Dance, Compendium of Seashells p. 273; 1984 Aubry, Terebridae pl. 1.

Description: Shell to 151 mm; color, white with 2 spiral rows of squarish brown spots, 3 on body whorl; outline of whorls flat with convex subsutural band; protoconch of 2 bulbous glassy whorls; subsutural band noded, followed by a second band, the two separated by a series of punctations which become interrupted lines on later whorls, the 2 bands occupying half the whorl in early whorls; axial sculpture absent; spiral sculpture of rows of punctations, the punctations coalescing into punctate spiral lines on extremely large individuals and the nodes becoming less prominent; body whorl smooth anterior to periphery except for axial striae; aperture semi-elongate; columella recurved, with moderate parietal callus on large specimens.

Type locality: "Punuluu, Oahu, Hawaiian Islands."

Distribution: This species is believed to be endemic to the Hawaiian Island chain in 3 to 116 m, the largest individuals having been found on Midway Island. However, three specimens, all dead collected from the Panama area, are in various United States museum collections.

Type: Holotype in the Bernice P. Bishop Museum, Honolulu, no. 212863; 104 × 18 mm.

Discussion: Most specimens under 40 mm have been collected in water from 20 m. or deeper, while most of the larger specimens have been found at a depth of about 6 m. or less, indicating that they probably breed in deep water. They are sometimes found living with *T. areolata*. Refer to *T. subulata* (**2**) for comparison.

Plate 1

FIGS. **1a-e,** *Terebra areolata* (Link). **1a,** Holotype of *T. muscaria* Lamarck, Museum d'Histoire Naturelle, Geneva, No. 1102; 93.2 mm. **1b,** Fiji Is.; 113.2 mm. **1c,** Specimen with one row of spots missing. **1d,** Early whorls in juvenile specimen. **1.e,** Middle whorl. **2a,b,** *T. subulata* (Linnaeus). **2a,** Fiji Is.; 112.0 mm. **2b,** Middle whorl. **3a-c,** *T. achates* Weaver. **3a,** Holotype Bishop Museum, Honolulu, No. 212863; 104.0 mm. **3b,** Last two whorls. **3c,** Middle whorl.

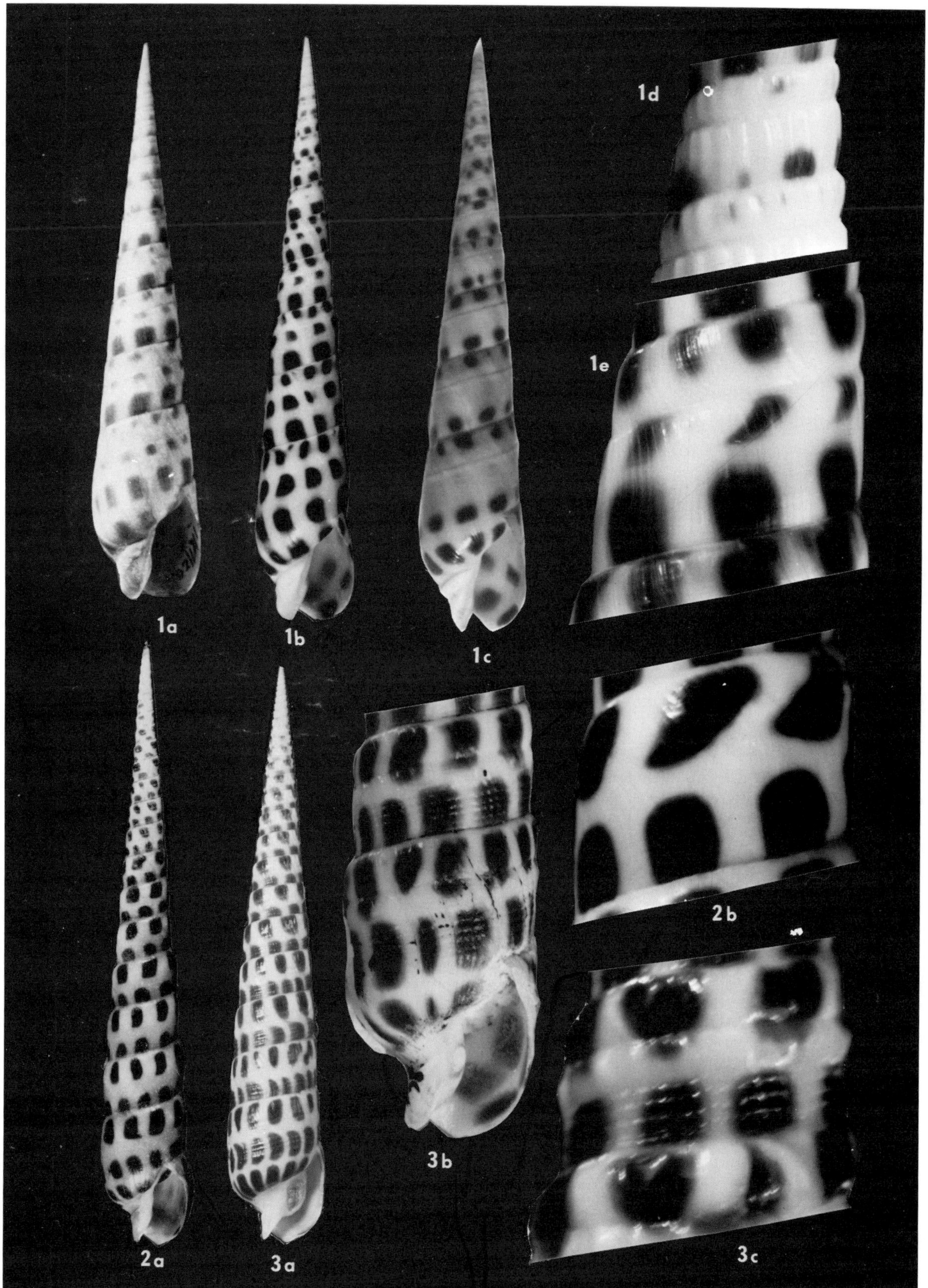

Plate 1. Indo-Pacific *Terebra*.

4. *Terebra consobrina* Deshayes, 1857

(Pl. 2, figs. 4a-b)
(Color pl. B, fig. 17)

1857 *Terebra consobrina* Deshayes, Journ. Conchyl. 6:72, pl. 3, fig. 3; 1860 Reeve, Conch. Icon. 12: pl. 6, sp. 23; 1984 Aubry, Terebridae pl. 11.

1967 *Terebra (Dimidiacus) consobrina* Deshayes, Bratcher & Burch, Veliger 10(1):9; 1970 Mienis, Argamon 1(2):38, 40, fig. 3.

Description: Shell to 120 mm; color cream, ornamented with 2 rows of squarish brown spots, 3 on body whorl; outline of whorls straight, very slightly turreted, subsutural band unnoded, divided in center by extremely shallow, usually punctate groove, band defined by almost obsolete row of punctations, which becomes progressively weaker; remainder of whorl sculptured with 4 to 6 spiral rows of punctations; aperture quadrate; columella short, recurved.
Type locality: "Red Sea."
Distribution: From the Red Sea to Madagascar, Indian Ocean.
Type: Lectotype BM(NH) no. 197956; 89.3 × 12.0 mm.
Discussion: The most outstanding feature of this species, which separates it from the other large spotted species, is its subsutural band, which is divided by a shallow punctate groove. This species is found mainly in the Red Sea and has frequently been confused with the somewhat similar *T. subulata* (**2**), a species not found in the Red Sea.

5. *Terebra tessellata* Gray, 1834

(Pl. 2, figs. 5a-e)
(Color pl. B, fig. 16)

1834 *Terebra tessellata* Gray, Proc. Zool. Soc. Lond. pt. 2:61; 1844 Hinds in Sowerby, Thes. Conch. 1:166, pl. 45, fig. 124; 1860 Reeve, Conch. Icon. 12: pl. 19, sp. 88.

1867 *Terebra decorata* Deshayes, Journ. Conchyl. 6:75, pl. 4, fig. 13.

Description: Shell to 60 mm; color cream with 2 spiral rows of brown spots per whorl plus a row on the subsutural band and an additional row below the periphery of the body whorl; outline of whorls straight; subsutural band convex, narrow, noded, defined by a deep groove; anterior to this band is a row of weak, small nodes which become progressively weaker, usually disappearing completely on later mature whorls; remainder of whorls smooth except for axial striae; aperture quadrate; columella recurved; siphonal fasciole with a very sharp keel.
Type locality: *T. tessellata*, none. Aden is here designated as the type locality; *T. decorata* was from "Padang," Sumatra.
Distribution: Indo-Pacific, from Aden to Indonesia.
Type: *T. tessellata* holotype BM(NH) no. 1979120, 29.0 mm; *T. decorata* holotype (pl. 2, fig. 5c) BM(NH) no. 197958, 39.9 × 9.0 mm.
Discussion: The holotype is a juvenile specimen which still retains the nodes anterior to the subsutural band and has just begun to develop the color pattern typical of this species.

Two other Indo-Pacific species also have four rows of brown spots per whorl. *T. areolata* (**1**) has a flat subsutural band and is axially ribbed on early whorls while *T. tessellata* has a narrow, convex band and lacks the axial ribs. *T. ornata* (**139**) a Panamic species, has a broader shell with a wider subsutural band.

6. *Terebra ligata* Hinds, 1844

(Pl. 2, figs. 6)

1844 *Terebra ligata* Hinds, Proc. Zool Soc. Lond. 11:153; 1844 Hinds in Sowerby, Thes. Conch. 1:166, pl. 45, figs. 117-118.

Description: Shell to 31 mm; color whitish with small, unevenly spaced fulvous brown spots on the subsutural band, larger ones on the remainder of the whorl; outline of whorls concave; protoconch of 1½ narrow whorls; subsutural band of double row of nodes; axial sculpture of almost obsolete threads; spiral sculpture of 3 or 4 rough cords; aperture short, quadrate; columella recurved.

Plate 2
FIGS. **4a,b,** *Terebra consobrina* Deshayes. **4a,** Lectotype British Museum (N.H.) No. 197956; 89.3 mm. **4b,** Middle whorls. **5a-e,** *T. tessellata* Gray. **5a,** Immature holotype, British Museum (N.H.) No. 1979120; 29.0 mm. **5b,** Illustration of type from Hinds, 1844, pl.45, fig. 124. **5c,** Type-figure of *T. decorata* Deshayes, from Deshayes, 1867, pl.4, fig. 13. **5d,** Djibouti, French Somalia; 62.0 mm. **5e,** Middle whorls. **6,** *T. ligata* Hinds. Holotype British Museum (N.H.) No. 1979128; 31.1 mm. **7a-c,** *T. eximia* Deshayes. **7a,** Lectotype Museum National Histoire Naturelle, Paris; 46.2 mm. **7b,** Madras, India; 74.9 mm. **7c,** Middle whorls of lectotype.

Plate 2. Indo-Pacific *Terebra*.

Type locality: "Marquesas Islands."
Distribution: Indo-Pacific, known from type locality only.
Type: Holotype BM(NH) no. 1979128; 31. 1 mm.
Discussion: This species is unknown to us except for the holotype, which is obviously a juvenile and may not have developed its mature sculpture. We are listing it tentatively as valid. Tryon considered it a synonym of *T. tessellata* (**5**) but that species has a straight instead of concave outline and has a single rather than a double row of nodes on the subsutural band.

7. *Terebra eximia* Deshayes, 1859

(Pl. 2, figs. 7a-c)

1859 *Terebra eximia* Deshayes, Proc. Zool. Soc. Lond. pt. 27:314; 1860 Reeve, Conch. Icon. 12: pl. 21, sp. 106; 1977 Bratcher, Nautilus 91(2):40, fig. 5; 1984 Aubry, Terebridae pl. 15.

Description: Shell to 75 mm; color white with fulvous spots on subsutural band, at periphery of body whorl, and occasionally posterior to suture; outline of whorls flat; protoconch of 1½ brown mamillate whorls; noded subsutural band convex in early whorls, becoming flat and smooth or with slanting obsolete ribs later; remainder of whorl with spiral rows of punctations between unevenly spaced rows of raised spiral cords; weak nodes on cords in middle of whorls; body whorl short; aperture unknown (broken on holotype); columella extremely sinuous and recurved.
Type locality: None. Madras, India is here designated as the type locality.
Distribution: Indo-Pacific, from the Red Sea to India and China.
Type: Lectotype MNHNP; 46.2 × 8.0 mm.
Discussion: Deshayes published the measurements of this species as 92 × 8 mm, which must have been a typographical error, as the width of the lectotype matches his measurements exactly. This rare species, with its rather broad subsutural band, followed by about three constricting rows of deep punctations is not likely to be confused with any other species in the family.

8. *Terebra guttata* (Röding, 1798)

(Pl. 3, figs. 8a-b)

1778 *Buccinum candidum* Born, Index rer. Nat. Mus. Caes. Vindob., p. 257; 1780 Born, Mus. Caes. Vindob. p. 263, pl. 10, fig. 8. [*nomen oblitum*].

1798 *Epitonium guttatum* Röding, Mus. Bolten p. 94 (ref. Chemnitz, 1780, Neues syst. Conchyl. Cab. 4, pl. 153, fig. 1442 [originally spelled "*guttacum*"].

1817 *Buccinum oculatum* Dillwyn, Des. Cat. Rec. Shells 2:642, no. 133.

1822 *Terebra oculata* Lamarck, Hist. Nat. Anim. S. Vert. 7:286.

1834 *Terebra laevis* Gray, Proc. Zool. Soc. Lond. 2:61.

1858 *T. nebulosa* Lorois, Journ. Conchyl. 7: 90, pl. 1, fig. 4 [non Sowerby, 1825].

1859 *T. loroisi* Deshayes, Proc. Zool. Soc. Lond. pt. 27:313 [*nom nov. pro T. nebulosa* Lorois].

1869 *Terebra sculptilis* Pease, American. Journ. Conch. 5(2):64.

1915 *Epitonium guttatum* Röding, Dall, Index. Mus. Bolten p. 30 (valid emendation of the misspelling *guttacum*).

1961 *Terebra guttata* (Röding), Oyama, Moll. Shells, Resources Exploitation Inst. 5, *Terebra* pl., fig. 7; 1962 Kira, Shells W. Pacific Col. I:111, pl. 39, fig. 20; 1966 Habe & Kosuge, Shells World Col. 2:100, pl. 39, fig. 20; 1967 Cernohorsky, Marine Shells Pacific I:202, pl. 50, fig. 370; 1972 Hinton, Shells New Guinea & Cent. Indo-Pacific p. 48, pl. 24, fig. 6; 1975 Salvat & Rives, Coq. Polynésie p. 358, fig. 375; 1977 Mills, Veliger 19(3):263, figs. 9-15, radula; 1979 Kay, Hawaiian Mar. Shells p. 399, fig. 127b; 1982 Abbott & Dance, Compendium of Sea shells p. 273; 1984 Trondle, Xenophora 15:11, fig. 19; 1984 Aubry, Terebridae pl. 5; 1984 Salvat & Rives, Shells of Tahiti p. 134, pl. 34, fig. 3.

1964 *Terebra (Terebra) guttata* (Röding), Shikama, Sel. Shells World Col. 1:125, pl. 102, fig. 7; 1967 Cernohorsky & Jennings, Veliger 9(1):40, pl. 4, fig. 5.

Description: Shell to 175 mm; color early whorls of yellowish beige or tan, usually with white subsutural band, later whorls tan with oblong white spots below suture; outline of whorls concave; protoconch of 1½ whorls; subsutural band of white nodes separated from a second row of smaller nodes by a moderately deep subsutural groove, band and groove fading completely as oblong white spots appear, leaving only a subsutural swelling; remainder of whorl smooth except for fine axial growth striae; aperture moderately elongate with pinkish beige or yellowish beige interior; columella recurved.

Plate 3
FIGS. **8a,b,** *Terebra guttata* Röding. **8a,** Illustrated lectotype from Chemnitz, 1780, pl.153, fig. 1442. **8b,** Fiji Is.; 98.0 mm. **9,** *T. caledonica* Sowerby. Holotype British Museum (N.H.) No. 1909.10.19.103.; 46.8 mm. **10,** *T. dimidiata* (Linnaeus). Fiji Is.; 110.0 mm. **11a-f,** *T. crenulata* (Linnaeus). **11a,** Lectotype of *T. fimbriata* Deshayes, Museum National Histoire Naturelle, Paris; 87.0 mm. **11b,** Guadalcanal, Solomon Is.; 36.9mm. **11c,** Fiji Is.; 83.0 mm. Smooth form. **11d,** Holotype of var. *booleyi* Melvill & Sykes, British Museum (N.H.) No. 1898.4.40.1.; 68.7 mm. **11e,** Middle whorls of holotype of *T. interlineata* Deshayes. **11f,** Middle whorls.

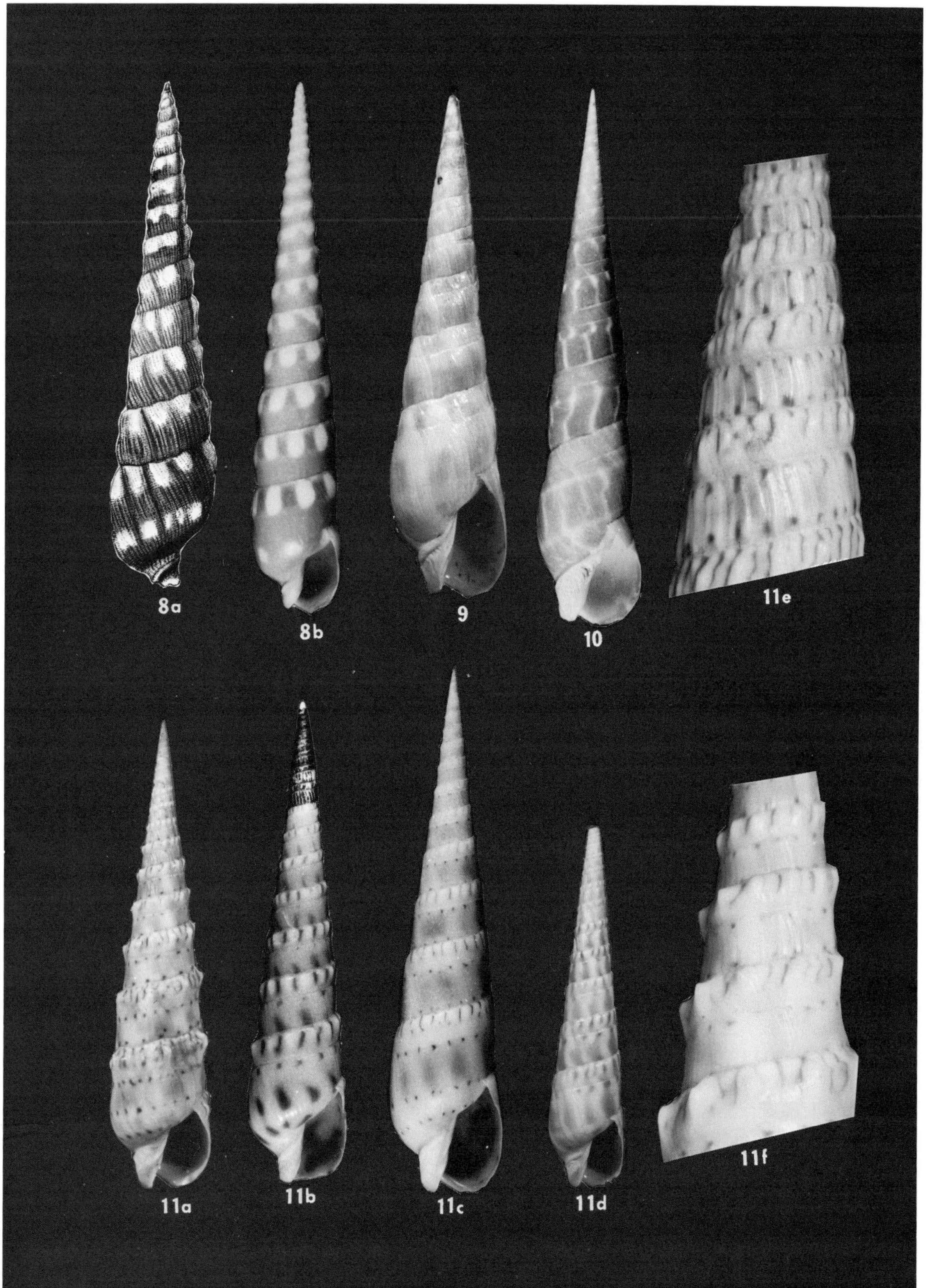

Plate 3. Indo-Pacific *Terebra*.

Type locality: *T. guttata*, none. East Indies indicated by Chemnitz, 1780 is here designated as the type locality; *B. candidum* none; *B. oculatum*, "*East Indian Seas;*" *T. laevis* none; *T. nebulosa* Lorois, none; *T. sculptilis*, "Oahu, Hawaii."
Distribution: Tropical Indo-Pacific; intertidal to 116 m.
Type: *T. guttata*, the specimen depicted in figure 1442, on plate 153 of Chemnitz, 1780, is here selected as the illustrated lectotype; *B. candidum*, type NHMV, 117.6 × 20.1 mm; *B. oculatum* no type; *T. laevis* holotype BM(NH) no. 1874.11.10.10, 34.3 mm (juvenile); *T. nebulosa* Lorois type probably in MNHNP but not segregated [Dr. P. Bouchet in litt.], original measurements length 46 × 89 mm, obviously a misprint for 8.9 mm; *T. sculptilis* lectotype MCZ no. 248804, 24.1 × 4.3 mm.
Discussion: Although the holotype shows *Buccinum candidum* Born to be the same species as *T. guttata*, the name *candidum* has not been applied once as the name of a valid species during the last 50 years and must be considered a *nomen oblitum*.

This is one of the species known to have harpoon-shaped radular teeth and a poison sac (Mills, 1977).

9. *Terebra caledonica* G. B. Sowerby, 1909

(Pl. 3, fig. 9)

1909 *Terebra caledonica* Sowerby, Proc. Malac. Soc. Lond. 8(4):198, text fig.

Description: Shell to 47 mm; color shiny grayish tan with a row of small pale brown squares on the subsutural band and a single row of very large ones on remainder of whorl, plus a row of small squares on base of body whorl; outline of whorls slightly convex; subsutural band marked by a shallow, finely punctate, subsutural groove; no other sculpture; body whorl long, inflated; aperture moderately quadrate; columella recurved.
Type locality: "Isle of Pines, New Caledonia."
Distribution: Indo-Pacific, type locality only.
Type: Holotype BM(NH) no. 1909.10.19.103; 46.8 × 11.0 mm.
Discussion: This species requires further research as it has not been recognized since the date of the original description and is known only from the type specimen.

10. *Terebra dimidiata* (Linnaeus, 1758)

(Pl. 3, fig. 10)
(Color pl. C, figs. 13–15)

1758 *Buccinum dimidiatum* Linnaeus, Syst. Nat. ed. 10:742; 1956 Dodge, Bull. Amer. Mus. Nat. Hist. 111(3):232.
1811 *Terebra carnea* Perry, Conchology pl. 16, fig. 1.
1853 *Acus dimidiata* H. & A. Adams, Genera Rec. Moll. 1(8):224.
1857 *Terebra splendens* Deshayes, Journ. Conchyl. 6:73, pl. 5, fig. 11.
1895 *Terebra (Subula) dimidiata* Linnaeus, Melvill & Standen, Journ. Conch. 8:90; 1964 Shikama, Sel. Shells World Col. 1:126, pl. 102, fig. 1; 1966 Cernohorsky & Jennings, Veliger 9(1):43, pl. 4, figs. 4,4a.
1935 *Terebra (Subula) dimidiata* var. *pallida* Dautzenberg, Mem. Mus. Hist. Nat. Belgique 2(17):23 [non Deshayes, 1857]. (Refers to Chemnitz, vol. 11, figs. 1817, 1818).
1935 *Terebra (Subula) dimidiata* var. *circumvoluta* "Chemnitz" Dautzenberg, Mem. Mus. Hist. Nat. Belgique, 2(17):22, pl. 1, fig. 4. (Refers to Chemnitz, vol. 4, fig. 1444).
1960 *Terebra dimidiata* (Linnaeus), Weaver, Hawaiian Mar. Moll. 1(6):22, pl. 6, fig. center; 1967 Cernohorsky, Mar. Shells Pacific 1:199, pl. 50, fig. 362; 1972 Hinton, Shells New Guinea & Cent. Indo-Pacific p. 48, pl. 24, fig. 5; 1975 Salvat & Rives, Coq. Polynésie p. 359, fig. 381; 1979 Kay, Hawaiian Mar. Shells p. 397, fig. 127a; 1982 Signor, Nautilus 25 (1) 79, figs. 1-6; 1982 Abbott & Dance, Compendium of Seashells p. 273; 1984 Trondle Xenophora 15:10, fig. 14; 1984 Aubry, Terebridae pl. 1; 1984 Salvat & Rives, Shells of Tahiti p. 134, pl. 34, fig. 1.
1961 *Subula dimidiata* (Linnaeus), Oyama, Venus 21(2):177, text. fig. 3 (protoconch); 1962 Kira, Shells W. Pacific Color 1: 111, pl. 39, fig. 17; 1967 Habe & Kosuge, Stand Book Jap. Shells Color 3:102, pl. 40, fig. 23.

Description: Shell to 166 mm; color, yellow or pinkish orange with ivory flammules, or ivory with pinkish orange subsutural band and squarish spots; outline of whorls flat with slightly indented subsutural band; protoconch of 3½ conical whorls; subsutural band faintly noded on early whorls, flat later, defined by a groove which is punctate early and smooth later; axial ribs numerous, curved, closely spaced on early whorls; later whorls smooth except for subsutural groove; aperture quadrate; columella almost straight.
Type locality: *T. dimidiata*, "African Ocean." Dar-es-Salaam, Tanzania, is here designated as the type locality; *carnea*, none; *T. splendens*, "China."
Distribution: Indo-Pacific, from the Red Sea

through the tropical Indo-Pacific to Polynesia and Hawaii.

Type: *T. dimidiata* holotype in the Linnean collection, London, 54.3 × 10.1 mm; *T. splendens* lectotype MNHNP, 55.5 mm.

Discussion: This species has neither radula nor poison gland (Rudman, 1969). It exhibits two distinct color patterns, one almost the reverse of the other, though the sculpture and shape remain constant. It is the type species of the genus *Subula* Schumacher, 1817. *T. consors* **(14)** may be separated from this species by its lack of an indented subsutural band, and by its narrower apical angle, and pale amber spots beginning in the subsutural band and continuing to the following suture. The spots of *T. dimidiata* are on the anterior portion of the whorl.

(11) *Terebra crenulata* (Linnaeus, 1758)

(Pl. 3, figs. 11a-f)
(Color pl. C, figs. 11, 12)

1758 *Buccinum crenulatum* Linnaeus, Syst. Nat. ed. 10:741; 1956 Dodge, Bull. American Mus. Nat. Hist. 111(3):221.

1791 *Buccinum varicosum* Gmelin, Syst. Nat. ed. 13:3505 (Ref. Seba, 3: pl. 561, fig. 17).

1811 *Terebra maculata* Perry, Conchology, pl. 16, fig. 2, [non *Buccinum maculatum* Linnaeus, 1758].

1845 *Buccinum luteolum* "Martyn", Chenu, Univ. Conch. p. 25, pl. 33, fig. 1a.

1857 *Terebra fimbriata* Deshayes, Journ. Conchyl. 6:71, pl. 5, fig. 1.

1859 *Terebra interlineata* Deshayes, Proc. Zool. Soc. Lond. pt. 27: 277.

1898 *Terebra crenulata* var. *booleyi* Melvill & Sykes, Proc. Malac. Soc. Lond. 3:42, pl. 3, fig. 5.

1923 *Terebra (Oxomeris) maculata* "Lamarck" Bartsch, Nautilus 37: 63, naming *T. maculata* as type species.

1935 *Terebra crenulata* and variety *fimbriata* Deshayes, Dautzenberg, Mem. Mus. Roy. d'Hist. Nat. Belgique, vol. 2, fasc. 17, pl. 1, figs. 5, 6.

1960 *Terebra crenulata* (Linnaeus), Weaver, Hawaiian Mar. Moll. 1 (4):14, pl. 4, figs. center row; 1971 Bratcher & Burch, Proc. Cal. Acad. Sci. 37(21):555, fig. 26; 1971 Bratcher & Burch in Keen, Seashells Trop. W. America ed. 2:674, fig. 1532; 1972 Hinton, Shells New Guinea & Cent. Indo-Pacific p. 48, pl. 24, figs. 1-3; 1975 Salvat & Rives, Coq. Polynésie p. 358, fig. 377; 11979 Kay, Hawaiian Mar. Shells p. 397, fig. 129h; 1982 Abbott & Dance, Compendium of Seashells p. 274; 1984 Trondle, Xenophora 15:9, fig. 12, 13; 1984 Aubry, Terebridae pl. 1; 1984 Salvat & Rives, Shells of Tahiti p. 134, pl. 34, fig. 6.

1961 *Subula (Oxymeris) crenulata* (Linnaeus), Oyama & Takemura, Moll. Shells, Resources Exploitation Inst. 5: *Subula* figs. 10, 11.

1962 *Subula crenulata* (Linnaeus), Kira, Shells W. Pacific Color 1:111, pl. 39, fig. 16.

1964 *Terebra (Oxymeris) crenulata* (Linnaeus), Shikama, Sel. Shells World Color 1:126, pl. 102, fig. 2; 1966 Cernohorsky & Jennings, Veliger 9(1):41, pl. 4, figs. 6, 6a.

Description: Shell to 140 mm; color flesh to beige, usually blended with fawn or gray zones, ornamented with 2 row of small brown dots, one anterior to the subsutural band, the other posterior to the suture, also with short wavy brown axial lines immediately anterior to the suture; outline of whorls straight, with the subsutural area narrowly convex or angular; protoconch of $3^{1}/_{2}$ to $4^{1}/_{2}$ conical whorls; subsutural band with rounded nodes or narrow axial ribs, band defined by subsutural groove, and deep punctations between nodes, groove becoming progressively obsolete; axial ribs numerous, sharp, on early whorls, fading completely; later whorls smooth except for nodes below suture; nodes small and inconspicuous or large and triangular; aperture quadrate; columella almost straight.

Type locality: *T. crenulata*, none originally given. The Moluccas Islands, Indonesia, are here designated as the type locality; *B. varicosum*, none; *maculata* Perry, "Brazil and West Indies," error; *T. fimbriata* none; *T. interlineata*, "Hawaii;" *T. crenulata booleyi*, "Andaman Is."

Distribution: Indo-Pacific, from the Red Sea throughout the tropical Indo-Pacific to the Revillagigedo Ids., Mexico; intertidal to 73.

Type: *T. crenulata* lectotype Linnean Society, London, 51.1 × 12.7 mm; *T. fimbriata* lectotype (pl. 3, fig. 11a) MNHNP, 87.0 × 190. mm; *T. interlineata* holotype (pl. 3, fig. 11f) BM(NH) no. 197998, 61.2 × 13.0 mm; *T. crenulata booleyi* holotype (pl. 3, fig. 11d) BM(NH) no. 1848.4.40.1, 68.7 mm.

Discussion: This species is so variable in outline and pattern that it is easy to understand why it was considered a different species or subspecies by a number of authors, although the various forms clearly intergrade into one another when a large number of specimens are available for examination. *T. crenulata* is netted in the plankton from April through September in Hawaii. The larvae settle when their length is approximately 1 mm. Upon metamorphosis the protoconch becomes opaque cream color. (J. B. Taylor, unpubl. Ph.D. thesis).

12. *Terebra succinea* Hinds, 1844

(Pl. 4, figs. 12a-e)
(Color pl. C, fig. 8)

1844 *Terebra succinea* Hinds, Proc. Zool. Soc. Lond. pt. 11:149; 1844 Hinds in Sowerby, Thes. Conchyl. 1:151, pl. 42, fig. 40; 1967 Cernohorsky, Marine Shells Pacific 1:206, pl. 52, fig. 386; 1984 Aubry, Terebridae pl. 10.

1966 *Terebra (Perirhoe) succinea* Hinds, Cernohorsky & Jennings, Veliger 9(1):52, pl. 4, fig. 0.

Description: Shell to 118 mm; color yellowish peach with a few tiny pinpoint brown spots scattered anterior to the subsutural band and posterior to the suture and at the periphery of the body whorl; outline of whorls flat; subsutural band slightly convex, sometimes divided in center by very faint groove, defined by narrow groove, becoming almost obsolete on later whorls, band may fade to a slight swelling; sculpture extremely variable; axial sculpture on some specimens of curved, sharp ribs, on others they are curved striae; spiral sculpture may be almost obsolete or of well-developed, unevenly spaced cords which may or may not cross ribs; body whorl with almost obsolete cord at periphery; aperture quadrate; columella recurved, with 2 heavy plications.
Type locality: Nagada, Papua New Guinea, is here designated as the type locality.
Distribution: From the Ryukyu Is. to Papua New Guinea and the Fiji Is.; intertidal to 41 m.
Type: Lectotype BM(NH) no. 1968172, here designated; 113.5 mm.
Discussion: The early whorls of this rare species tend to be eroded on the top third of the dorsum, indicating they spend much time exposed. Three specimens collected in Okinawa were found exposed, crawling on sand and broken coral during the day. This is one of the species known to have harpoon-shaped radular teeth and poison sac.

13. *Terebra argus* Hinds, 1844

(Pl. 4, figs. 13a-d)
(Color pl. C, fig. 7)

1839 *Terebra nebulosa* Kiener, Spec. Gen. Icon. Coq. Viv. 9:23, pl. 10, fig. 22 [non Sowerby, 1825].

1844 *Terebra argus* Hinds, Proc. Zool. Soc. Lond. pt. 11:160 [nom. nov. pro *T. nebulosa* Kiener, 1839]; 1844 Hinds in Sowerby, Thes. Conch. 1: × 154, pl. 43, fig. 64; 1917 Hirase, Terebridae Jap. Emp. p. 24, pl. 7, fig. 107; 1972 Hinton, Shells New Guinea & Cent. Indo-Pacific p. 48, pl. 49, fig. 14; 1975 Salvat & Rives, Coq. Polynesie p. 359, fig. 380; 1982 Abbott & Dance, Compendium of Seashells p. 273, fig.; 1984 Trondle, Xenophora 15:8, fig. 6; 1984 Aubry, Terebridae pl. 2; 1984 Salvat & Rives, Shells of Tahiti p. 134, pl. 34, fig. 7.

1921 *Terebra argus brachygyra* Pilsbry, Proc. Acad Nat. Sci. Philadelphia 72:304; 1979 Kay, Hawaiian Mar. Shells p. 393, fig. 129a.

1961 *Subula argus* (Hinds), Oyama, Venus 21 (2):186; 1962 Kira, Shells W. Pacific Col. 2:209, pl. 71, fig. 15; 1966 Habe & Kosuge, Shells World Col. 1:98, pl. 39, fig. 5.

1966 *Terebra (Subula) argus argus* Hinds, Cernohorsky & Jennings, Veliger 9(1):43, pl. 4, fig. 7.

1967 *Terebra argus argus* Hinds, Cernohorsky, Mar. Shells Pacific 1:196, pl. 49, fig. 354.

Description: Shell to 108 mm; color ivory, usually with 3 rows of faint nebulous yellow spots per whorl, 4 on body whorl; outline of whorls straight; protoconch $3^{1}/_{2}$ whorls, the last one being unusually long; subsutural band convex, defined by a shallow punctate groove; early whorls of teleoconch elongate, sometimes crenulate, later whorls may have obsolete axial ribs or may be smooth with only axial striae; aperture quadrate, columella recurved.
Type locality: *T. argus*, "Tahiti, Society Islands and Nukuhiva, Marquesas Islands." Tahiti, French Polynesia is here restricted as the type locality; *T. argus brachygyra*, "off Honolulu, Hawaii," 6-15 m.
Distribution: From the Red Sea to French Polynesia and the Hawaiian Ids., 3-116 m.
Type: *T. argus* lectotype BM(NH) no. 1968253; 88.7 mm; *T. argus brachygyra* holotype (pl. 4, fig. 13c) ANSP no. 46913, 40.0 × 8.0 mm.
Discussion: This species was described as *Terebra nebulosa* by Kiener, a name which was preoccupied but which was apt because of the extremely nebulous yellow spots. In some individuals the spots are scarcely noticeable. The early whorls are without spots, which often show up first as faint yellowish lines, becoming squarish spots on subsequent whorls. Although this is not a particularly slender shell, the first four of five whorls are long and slender. Specimens from Hawaii tend to be more coarsely sculptured and

---

Plate 4
FIGS. **12a-e,** *Terebra succinea* Hinds. **12a,** Lectotype British Museum (N.H.) No. 1968172; 113.5 mm. **12b,c,** Fiji Is.; 117.4 mm and 78.0 mm. **12d,** Middle whorls in lectotype. **12e,** Middle whorls in paralectotype. **13a-d,** *T. argus* Hinds. **13a,** Lectotype British Museum (N.H.) No. 1968253; 88.7 mm. **13b,** Tahiti, Society Is.; 57.0 mm. **13c,** Middle whorls in holotype of *T. argus brachygyra* Pilsbry. **13d,** Middle whorls.

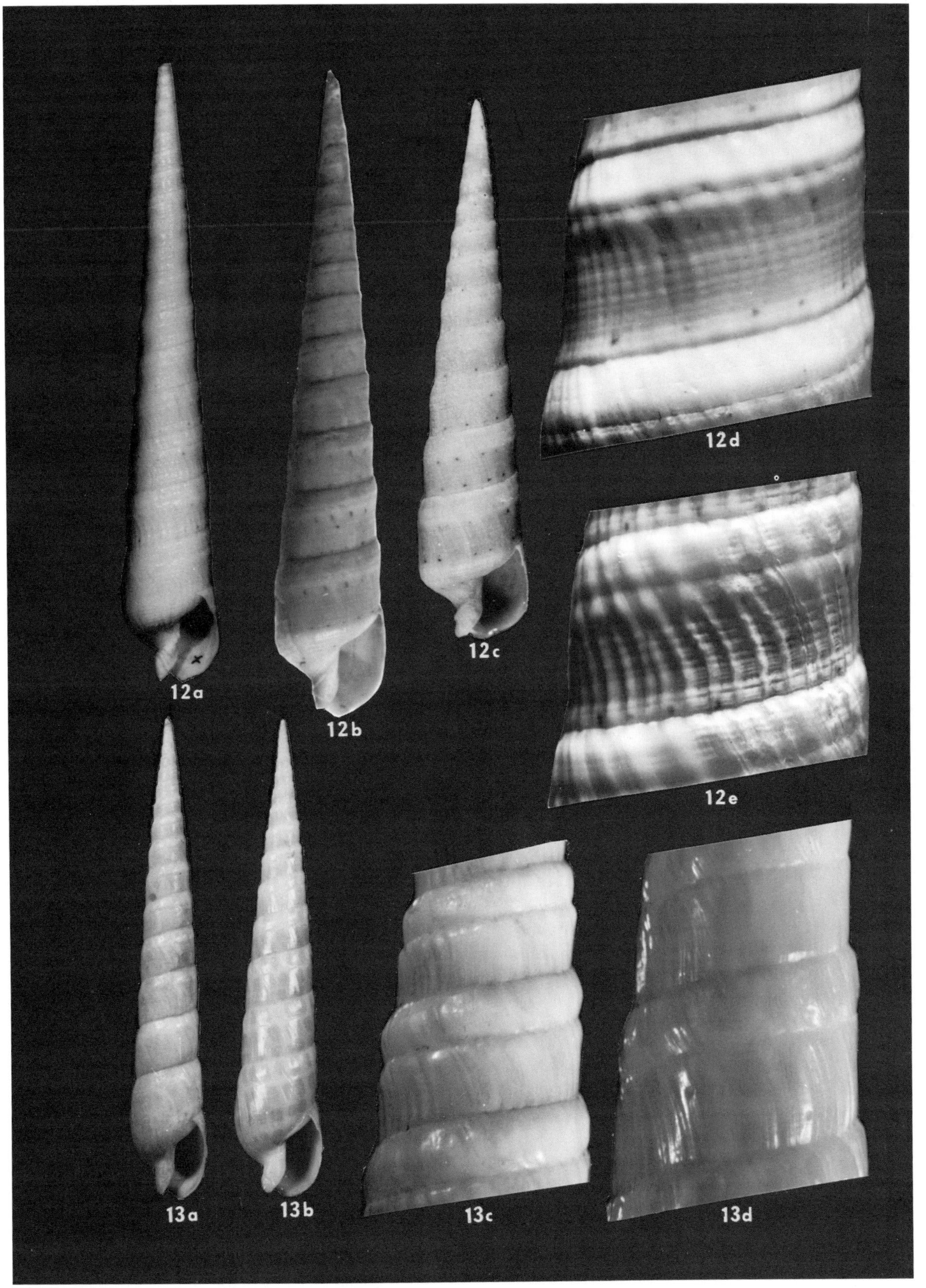

Plate 4. Indo-Pacific *Terebra*.

generally smaller. The Hawaiian form is found occasionally elsewhere, but the smoother form appears not to exist in Hawaii.

14. *Terebra consors* Hinds, 1844

(Pl. 5, fig. 14)

1844 *Terebra consors* Hinds, Proc. Zool. Soc. Lond. pt. 11:150; 1844 Hinds in Sowerby, Thes. Conch. 1:154, pl. 42, fig. 26; 1860 Reeve, Conch, Icon. 12: pl. 1, sp. 2; 1984 Aubry, Terebridae pl. 5.

1857 *Terebra glabra* Deshayes, Journ. Conchyl. 6:101, pl. 5, fig. 13.

1857 *Terebra virginea* Deshayes, Journ. Conchyl. 6:83, pl. 4, fig. 12.

Description: Shell to 71 mm; color white to cream with large rectangular peach to brown nebulous markings; outline of whorls flat; subsutural band, sometimes slightly crenulate, marked by shallow, faintly punctate groove; sculpture of microscopic axial growth striae only; body whorl with light band at periphery and square spots below; aperture quadrate; columella straight.

Type locality: *T. consors*, "Tahiti, French Polynesia;" *T. glabra*, "China Seas;" *T. virginea*, "Zanzibar."

Distribution: From Zanzibar to French Polynesia.

Type: *T. consors* lectotype BM(NH) no. 1968173; 70.9 mm; *T. glabra* lectotype MNHNP, 70.0 × 13.0 mm; *T. virginea* holotype BM (NH) no. 197953, 52.3 × 11.0.

Discussion: The rectangular nebulous markings of this species often extend into the subsutural band and may be very light in color. Although the species has quite a wide distribution, it is not common anywhere. Refer to *T. dimidiata* (**10**) for comparison.

15. *Terebra maculata* (Linnaeus, 1758)

(Pl. 5, figs. 15a-c)

1758 *Buccinum maculatum* Linnaeus, Syst. Nat. ed. 10:741; 1956 Dodge, Bull. Amer. Mus. Nat. Hist. III (3):218.

1839 *Terebra maculata* (Linnaeus), Kiener, Spec. Gen. Icon. Coq. Viv. 9:4, pl. 1, fig. 1; 1860 Reeve, Conch Icon. 12, pl. 1, fig. 4; 1917 Hirase, Terebridae Jap. Emp. p. 25, pl. 1, fig. 2.; 1960 Weaver, Hawaiian Mar. Moll. 1(7):26, pl. 7, fig. center; 1967 Cernohorsky, Mar. Shells Pacific 1:204, pl. 51, fig. 376; 1969 Rudman, Veliger 12(1):59, anatomy; 1971 Bratcher & Burch, Proc. Calif. Acad. Sci. ser. 4:37(21):544, figs. 12, 13; 1971 Bratcher & Burch in Keen, Seashells Trop. W. America 2nd ed.:680, fig. 1546; 1972 Hinton, Shells of New Guinea & Cent. Indo-Pacific p. 48, pl. 24, fig. 4; 1975 Salvat & Rives, Coq. Polynesie p. 358, fig. 378; 1979 Kay, Hawaiian Mar. Shells p. 399, fig. 129j; 1982 Abbott & Dance Compendium of Seashells p. 273; 1984 Trondle, Xenophora 15:11, fig. 22; 1984 Aubry, Terebridae pl. 1.

1840 *Terebra maculosa* Pfeiffer, Krit. Reg. Martini-Chemnitz Conch. Cab. p. 41.

1935 *Terebra maculata confluens* Dautzenberg, Mem. Mus. Roy. Hist. Nat. Belgique 1 (17):31.

1939 *Terebra (Subula) roosevelti* Bartch & Rehder, Smithsonian Misc. Coll. 98(10):1, pl. 1, fig. 6; 1982 Abbott & Dance, Compendium of Seashells p. 277, holotype fig'd.

1961 *Subula (Oxymeris) maculata* (Linnaeus) Oyama & Takemura, Moll. Shells, Resources Exploitation Inst. 5, *Subula* figs. 7-9; 1962 Kira, Shells W. Pacific Color 1:111, pl. 39, fig. 21.

1964 *Terebra (Oxymeris) maculata* (Linnaeus), Shikama, Select. Shells World Colorl:26, pl. 102, fig. 6; 1966 Cernohorsky & Jennings, Veliger 10(1):42, pl. 4, fig. 1.

1967 *Oxymeris maculatua* (Linnaeus), Habe & Kosuge, Stand. Book Jap. Shells Col. 3: 102, pl. 40, fig. 21; 1975 Okutani, Moll. Japan p. 155, fig. 5; 1982 Habe & Kosuge, Common Shells Japan Color pl. 40 fig. 21.

1970 *Terebra maculata roosevelti* Bartsch & Rehder, Bratcher, Hawaiian Shell News 18(3):6, text fig.; 1971 Bratcher & Burch, Proc. Calif. Acad. Sci. ser. 4, 37(21):544, figs. 10,11; Bratcher & Burch in Keen, Seashells Trop. W. America 2nd ed. 680, fig. 1550.

Description: Shell to 274 mm, sturdy; color cream with a broad band of blackish brown streaks and maculations anterior to the suture and a narrower band mid-whorl or posterior to the suture, body whorl with spiral rows of fawn colored tessellations; outline of whorls almost straight, broad apical angle; protoconch of 3 conical whorls; subsutural band flat, defined by groove on early whorls before fading completely, band indicated only by color on mature whorls; axial ribs numerous, sharp, evenly-spaced on early whorls, fading to very fine microscopic growth striae giving the appearance of completely smooth whorls; spiral sculpture absent;

Plate 5

FIG. **14**, *Terebra consors* Hinds. Seychelles Is.; 33.0 mm. FIGS. **15a-c**, *T. maculata* (Linnaeus). **15a**, Eilat, Gulf of Aqaba, Israel; 134.5 mm. **15b**, Holotype of *T. roosevelti* Bartsch & Rehder, National Museum Natural History, Washington No. USNM-472534; 43.2 mm. **15c**, Teratological specimen from Iba Zambate, Philippines; 48.2 mm. **16a,b**, *T. chlorata* Lamarck. **16a**, Fiji Is.; 53.0 mm. **16b**, Middle whorls. **17a-c**, *T. felina* (Dillwyn). Illustrated lectotype from Chemnitz, 1780, pl.154, FIG. 1448. **17b**, Lectotype of *T. suffusa* Pease, Museum of Comparative Zoology, Harvard No. 248802; 29.3 mm. **17c**, Fiji Is.; 63.0 mm.

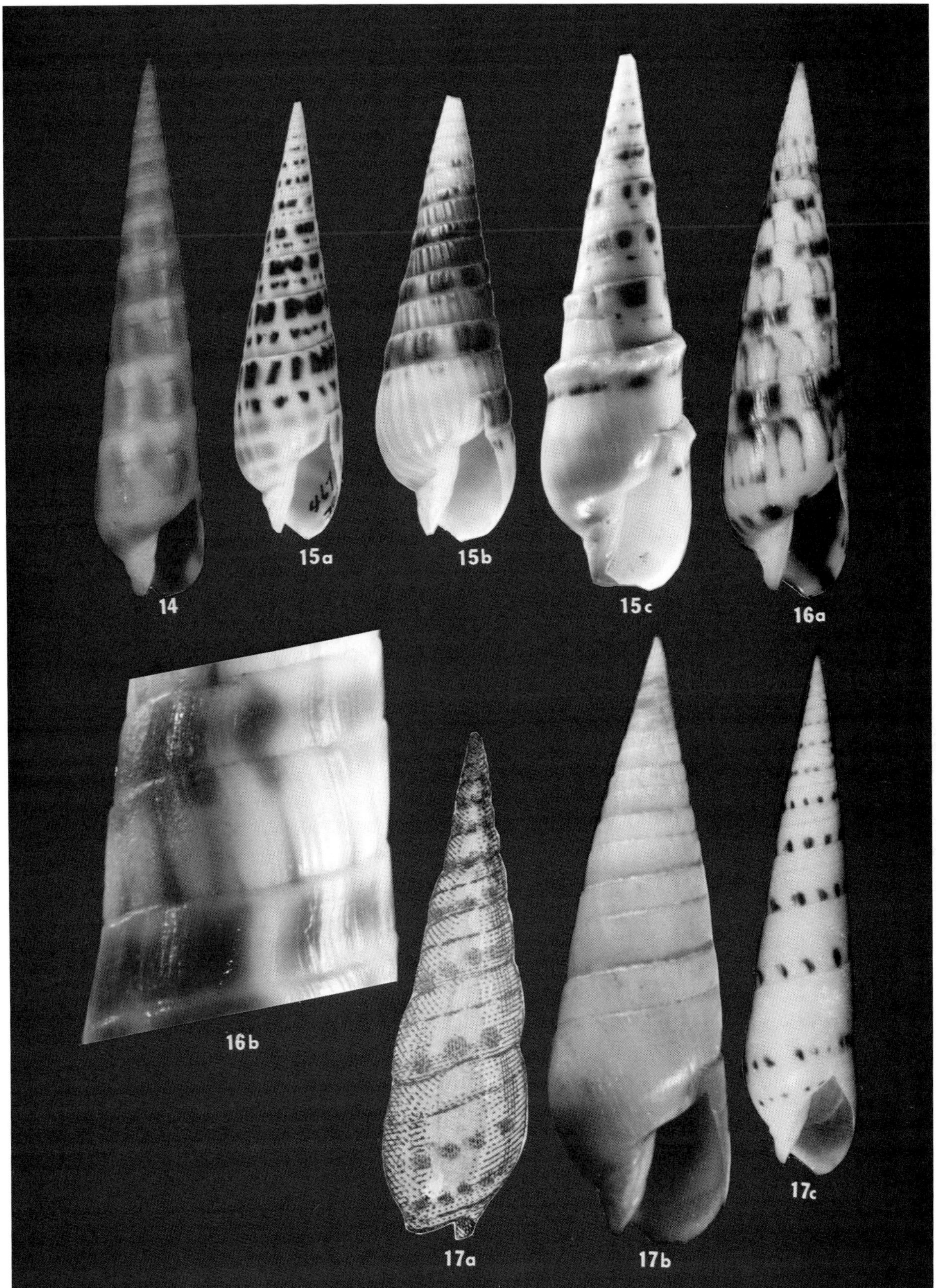

Plate 5. Indo-Pacific *Terebra*.

body whorl elongate; aperture moderately quadrate; columella recurved; siphonal fasciole with heavy, almost rib-like striations, posterior keel.
Type locality: *T. maculata*, "African Ocean;" *T. maculata confluens*, "New Caledonia and Tahiti"; *T. maculata roosevelti*, "Soccorro Id., Mexico, 7-8 fms."
Distribution: From the Red Sea throughout the tropical Indo-Pacific to tropical eastern Pacific, rare in the eastern Pacific; subtidal to 210 m.
Type: *T. maculata* holotype Linnean coll., Linnean Soc. London; *T. maculata confluens* type IRHN Brussels; *T. maculata roosevelti* holotype (pl. 5, fig. 15b) USNM no. 472534, 43.2 × 12.1 mm.
Discussion: This is the type species of the genus *Oxymeris* Dall, 1903, a replacement name for *Acus* H. & A. Adams, non Edwards. Dall diagnosed the genus as "Having rapidly enlarging whorls." Bartsch later diagnosed the genus as "Having a subsutural groove present only on early whorls." This largest and heaviest of the terebrids has commonly been used as a tool in many of the South Pacific cultures. It is a common species throughout most of its wide range and has neither radula nor poison gland. The holotype of *T. (Subula) roosevelti* Bartsch & Rehder (=*T. maculata*) is a juvenile specimen and matches well with juveniles found in Hawaii.

This species has planktonic development, hatching when one whorl is complete and adding two or more while in the planktonic stage (J. B. Taylor, unpbl. Ph.D. Thesis). It possesses eyes on small stalks. There is no poison gland (Rudman, 1969). The animal feeds on the polychaete worm, *Dasybranchus caducus* (B. A. Miller, 1970).

16. *Terebra chlorata* Lamarck, 1822

(Pl. 5, figs. 16a-b)
(Color pl. B, fig. 3)

1822 *Terebra chlorata* Lamarck, Hist. Nat. Anim. sans Vert. 9:288; 1839 Kiener, Spéc. Gén. Icon. Coq. Viv. 9:8, pl. 4, fig. 8, 8a; 1960 Weaver, Hawaiian Mar. Moll. 1(2):6, pl. 2, figs. center; 1967 Cernohorsky Mar. Shells Pacific 1: 198, pl. 49, fig. 357; 1972 Hinton, Shells New Guinea & Cent. Indo-Pacific p. 48, pl. 24, fig. 11; 1979 Kay, Hawaiian Mar. Shells p. 395, fig. 129i; 1982 Abbott & Dance, Compendium of Seashells p. 273; Trondle, Xenophora 125:9, fig. 9; 1984 Aubry Terebridae pl. 5; 1984 Salvat and Rivês, Shells of Tahiti p. 134, pl. 34, fig. 8.

1834 *Terebra knorrii* Gray, Proc. Zool. Soc. Lond. pt. 2:59.

1961 *Subula (Oxymeris) chlorata* (Lamarck), Oyama & Takemura, Moll. Shells, Resources Exploitation Inst. 5: *Subula* fig. 4; 1962 Kira, Shells W. Pacific Color 1:209, pl. 71, fig. 17.

1966 *Terebra (Oxymeris) chlorata* Lamarck, Cernohorsky & Jennings, Veliger 9(1):41, pl. 4, fig. 8.

Description: Shell to 101 mm; color ivory with brown or rust colored splotches on subsutural band and flammules of same color on remainder of whorl; outline of whorls straight; protoconch of 1½ rounded whorls; subsutural band flat, unnoded, marked by suture and moderately deep subsutural groove; band and remainder of whorl ribbed in early whorls, both becoming smooth later with only microscopic axial growth striae evident; aperture quadrate; columella straight; the operculum is reddish brown, transluscent and almost smooth.
Type locality: *T. chlorata*, none. Laminusa, Siasi, Sulu, Philippines, is here designated as the type locality; *T. knorrii*, none given.
Distribution: From East Africa to Pitcairn Island and Hawaii.
Type: *T. chlorata* lectotype MHN Geneva no. 1101/25, 50.2 × 12.2 mm; *T. knorrii* holotype BM(NH) no. 1874.11.10.2, 72.7 mm.
Discussion: This species has a long labial tube and a broad fleshy foot specialized for burrowing in the sand. When it contacts its prey, a polychaete worm, in its burrow in loosely compacted sand, the foot digs down and pulls the shell into the sand at a 90 degree angle. At the same time the long slender labial tube everts, enters and extends into the burrow until it grasps the prey (Miller, 1971). See *T. felina* **(17)** for comparison with that species and with *T. troendlei* **(18)**.

17 *Terebra felina* (Dillwyn, 1817)

(Pl. 5, figs. 17a-c)
(Color pl. 3, fig. 2)

1791 *Buccinum tigrinum* Gmelin, Syst. Nat. 13th ed.: 3502 [non Gmelin, 1791, p. 3475, a *Cominella* in the family Buccinidae].

1817 *Buccinum felinum* Dillwyn, Descr. Cat. Rec. Shells 2:644 (ref. Seba, vol. 3, fig. A, and Chemnitz, vol. 4, pl. 154, fig. 1448)

1844 *Terebra tigrina* (Gmelin), Hinds in Sowerby, Thes. Conch. 1:155, pl. 42, fig. 30.

1869 *Terebra suffusa* Pease, American Journ. Conch. 5(2):65.

1952 *Terebra felina* (Dillwyn), Tinker, Pacific Sea Shells p. 196, figs. upper row right and left; 1960 Weaver, Hawaiian Mar. Moll. 1(4):14, pl. 4, figs. top row; 1967 Cernohorsky, Mar. Shells Pacific 1:199, pl. 50, fig. 364; 1972 Hinton, Shells New Guinea Cent. Indo-Pacific p. 48, pl. 24, fig. 12; 1979 Kay, Hawaiian Mar. Shells p.

397, fig. 129C; 1982 Abbott & Dance, Compendium of Seashells p. 273; 1984 Trondle, Xenophora 15 p. 10; 1984 Aubry, Terebridae pl. 15; 1984 Salvat & Rives, Shells of Tahiti p. 136, pl. 35, fig. 5;

1961 *Subula (Oxymeris) felina* (Dillwyn), Oyama & Takemura, Moll. Shells, Resources Exploitation Inst. 5: *Subula* fig. 2. 1962 Kira, Shells W. Pacific Color 1:209, pl. 71, fig. 16.

1964 *Terebra (Oxymeris) felina* (Dillwyn), Shikama, Sel. Shells World Color 1:126, pl. 102, fig. 3; 1966 Cernohorsky & Jennings, Veliger 9(1):42, pl. 4, fig. 9.

Description: Shell to 89 mm, quite heavy and broad; color dull creamy white, usually decorated with rows of small orange-brown dots posterior to the suture, at and below the periphery of the body whorl; outline of whorls slightly convex; protoconch of 2½ conical whorls; subsutural band flat, ribbed on early whorls, both ribs and band fading completely on large specimens, band marked by a punctate groove, becoming obsolete, then disappearing; axial ribs well-developed on early whorls, becoming obsolete with only macroscopic axial growth striae on later whorls; spiral sculpture absent; aperture quadrate; columella straight.
Type locality: *T. felina*, none given. Piti, Guam Is., Marianas Is., is here designated as the type locality; *T. suffusa* Pease, "Oahu, Hawaii."
Distribution: From East Africa throughout the tropical Indo-Pacific; intertidal to 20 m.
Type: *T. felina*, there is no type. Figure 1448 on plate 154 from Chemnitz, (1780) is here designated as the illustrated lectotype; *T. suffusa* lectotype (pl. 5, fig. 17b) MCZ no. 248802, 29.3 mm.
Discussion: This species has the same feeding habits as *T. chlorata* (**16**). It is found frequently in Hawaii and occasionally elsewhere without any markings on the creamy white base, a color phase described by Pease as *T. suffusa*. In shape both *T. chlorata* (**16**) and *T. troendlei* (**18**) resemble this species, but *T. chlorata* has orange blotches, axial stripes and markings throughout the entire shell. *T. troendlei* has quite large orange spots limited to the subsutural band.

18. *Terebra troendlei* Bratcher, 1981

(Pl. 6, figs. 18a-b)
(Color pl. E, fig. 10)

1981 *Terebra troendlei* Bratcher, Veliger 23(4):330, figs. 3-4. 1984 Salvat and Rivês, Shells of Tahiti p. 136, pl. 35, fig. 6; 1984 Trondle, Xenophora 15:13, fig. 31.

Description: Shell to 34 mm; color white with low gloss, with orange blotches on subsutural band; outline of whorls weakly convex; protoconch of 3 conical whorls; subsutural band flat, defined by microscopically punctate groove and suture; axial ribs on early whorls fading to macroscopic striae; spiral sculpture absent; aperture quadrate; columella curved, with one weak plication.
Type locality: "Entrance to Hana Hevane Bay, Tahuata Is., Marquesas Is., 10 m." Collected by and named after Jean Tröndle of Tahiti.
Distribution: Endemic to the Marquesas Is.; subtidal to 10 m.
Type: Holotype MNHN Paris; 27.6 × 7.9 mm.
Discussion: No work has been done on the feeding type of this species, but it would be interesting to know if it has the same feeding habits and apparatus as *T. felina* and *T. chlorata*, as all three species have shells of the same shape. Refer to *T. felina* (**17**) for comparison.

19. *Terebra cerithina* Lamarck, 1822

(Pl. 6, figs. 19a-c)
(Color pl. C, fig. 1)

1822 *Terebra cerithina*, Lamarck, Hist. Nat. Anim. S. Vert. 7:288; 1839 Kiener, Spéc Gén. Coq. Viv. 9:33, pl. 11, fig. 25; 1967 Cernohorsky, Mar. Shells Pacific 1:198, pl. 49, fig. 356; 1975 Salvat & Rives, Coq. Polynésie p. 359, fig. 382; 1979 Kay, Hawaiian Mar. Shells p. 395, fig. 129B; 1982 Abbott & Dance, Compendium of Seashells p. 274; 1984 Trondle, Xenophora 15:9, fig. 8; 1984 Salvat & Rives, Shells of Tahiti p. 136, pl. 35, fig. 3; 1984 Aubry, Terebridae pl. 4.

1844 *Terebra pulchra* Hinds, Proc. Zool. Soc. Lond. pt. 11:151.

1921 *Terebra spaldingi* Pilsbry, Proc. Acad. Nat. Sci. Philadelphia 65:308, pl. 12, fig. 13.

1960 *Terebra cerithina spaldingi* Pilsbry, Weaver, Hawaiian Mar. Moll. 1(4):14, figs. bottom row.

1961 *Abretiella cerithina* (Lamarck), Oyama & Takemura, Moll. Shells, Resources Exploitation Inst. 5: *Abretiella* fig. 8; 1967 Habe & Kosuge, Stand. Book Jap. Shells Col. 3:102, pl. 40, fig. 16.

1964 *Terebra (Abretiella) cerithina* Lamarck, Shikama, Shells World Color p. 125, pl. 101, fig. 6; 1966 Cernohorsky & Jennings, Veliger 9(1):44, pl. 5, fig. 24.

Description: Shell to 70 mm; color shiny ivory with yellowish tan, rust, or fawn color between axial ribs and often in spiral bands, sometimes giving a tessellated appearance; outline of whorls moderately convex; protoconch of 2½ conical white whorls; subsutural band convex, ribbed, defined by deep punctations in early whorls, becoming flat, marked by very shallow groove, inconspicuous later; axial ribs heavy, rounded, becoming flatter or obsolete on later whorls; interspaces narrower than ribs, spiral sculpture absent; body whorl moderately elongate columella straight, with parietal callus in mature specimens.

Type locality: *T. cerithina*, "Timor Sea, Indonesia;" *T. pulchra*, "Marquesas Is.;" *T. spaldingi*, "Off Honolulu, Hawaii, 6-8 fms."

Distribution: From the Red Sea to Hawaii and the Marquesas Is.; intertidal to 115 m.

Type: *T. cerithina* lectotype MNH Geneva no. 1102/26/1, 56.4 × 12.8 mm; *T. pulchra* holotype BM(NH) no. 1844.6.7.81, 24.7 mm; *T. spaldingi* (pl. 6, fig. 19b) holotype ANSP no. 177,044, 30.3 × 6.9 mm.

Discussion: This is the type species of the genus *Abretia* H. & A. Adams, 1853, [non Rafinesque, 1815] and the replacement name for that genus, *Abretiella* Bartsch, 1923. Bartsch's diagnosis of the genus was, "whorls longitudinally ribbed; columella simple, straight, produced anteriorly; outer lip not sinuated at the fore part." There are no consistent differences in the Hawaiian form Pilsbry described as *T. spaldingi* and those elsewhere in the Pacific.

20. *Terebra eburnea* Hinds, 1844

(Pl. 6, fig. 20a-c)

1844 *Terebra eburnea* Hinds, Proc. Zool. Soc. Lond. pt. 11:153; 1844 Hinds in Sowerby, Thes. Conch 1:166, pl. 45, fig. 123; 1980 Cernohorsky, Rec. Auckland Inst. Mus. 17:149, figs. 31-33.

1853 *Acus eburnea* (Hinds), H. & A. Adams, Genera Rec. Moll. 1: 224.

Description: Shell to 38 mm; color ivory, protoconch and early whorls frequently orange; outline of early whorls almost straight, later whorls obese, somewhat turreted; protoconch of 3½ to 4 smooth whorls; subsutural band defined by deep punctations; axial ribs straight on early whorls, fading completely on mature whorls; spiral sculpture of 2 to 8 rows of punctations; body whorl with additional spiral lines; aperture quadrate; columella curved.

Type locality: "Seychelles Islands."

Distribution: From the Seychelles to Western Samoa; subtidal.

Type: Type lost. The specimen illustrated in Hinds in Sowerby (1844) on plate 45, fig. 123, designated as the illustrated lectotype of *T. eburnea* (Cernohorsky, 1980).

Discussion: Specimens of this little known species were recently collected in Western Samoa. See *T. cingulifera* (**23**) for comparison.

21. *Terebra circumcincta* Deshayes, 1857

(Pl. 6, fig. 21, Pl. 7, 21b)

1857 *Terebra circumcincta* Deshayes, Journ. Conchyl. 6:77, pl. 3, fig. 9; 1860 Reeve, Conch. Icon. 12: pl. 15, sp. 70; 1971 Powell, Rec. Auckland Inst. Mus. 8:225, fig. 26; 1979 Powell, New Zealand Moll. p. 246, pl. 47, fig. 18; 1980 Cernohorsky, Rec. Auckland Inst. Mus. 16:183, figs. 29, 30 (figured lectotype).

1908 *Terebra (Perirhoe) circumcincta* Deshayes, Dall, Nautilus 21 (11):125.

Description: Shell to 51 mm; color white to ivory, rarely with brown blotches; outline of whorls slightly convex, slightly turreted; protoconch of 1½ very large dome-shaped whorls; subsutural band flat, defined by punctate groove; axial sculpture absent; spiral rows of narrow, punctate grooves, often deeply incised on flat surface, 5 to 6 rows, 10 to 12 on body whorl; aperture quadrate; columella straight.

Type locality: "Red Sea." probably in error.

Distribution: From Northern Australia and New Caledonia to northern New Zealand.

Type: Holotype BM(NH) no. 1978150; 38.0 × 8.0 mm.

Discussion: This is the type species of the section *Perirhoe* Dall, 1908. Dall's diagnosis of the section was "Shell short, small, axial sculpture obsolete, spiral sculpture emphatic." The spiral grooves appear to divide the whorls of this rare species into a series of even bands. Refer to *T. cingulifera* (**23**) Lamarck for comparison.

Plate 6

FIGS. **18a,b,** *Terebra troendlei* Bratcher. Marquesas Is. **18a,** Holotype Museum National Histoire Naturelle, Paris; 27.6 mm. **18b,** Middle whorls of holotype. **19a-c,** *T. cerithina* Lamarck. **19a,** Efate I, Vanuatu; 51.2 mm. **19b,** Holotype of *T. spaldingi* Pilsbry, Academy Natural Sciences, Philadelphia No. 117044; 30.3 mm. **19c,** Middle whorls. **20a-c,** *T. eburnea* Hinds, Samoa Is. **20a,** 34.0 mm. **20b,** 38.0 mm. **20c,** Middle whorls. **21a,** *T. circumcincta* Deshayes, Holotype British Museum (N.H.) No. 1978150; 38.0 mm.

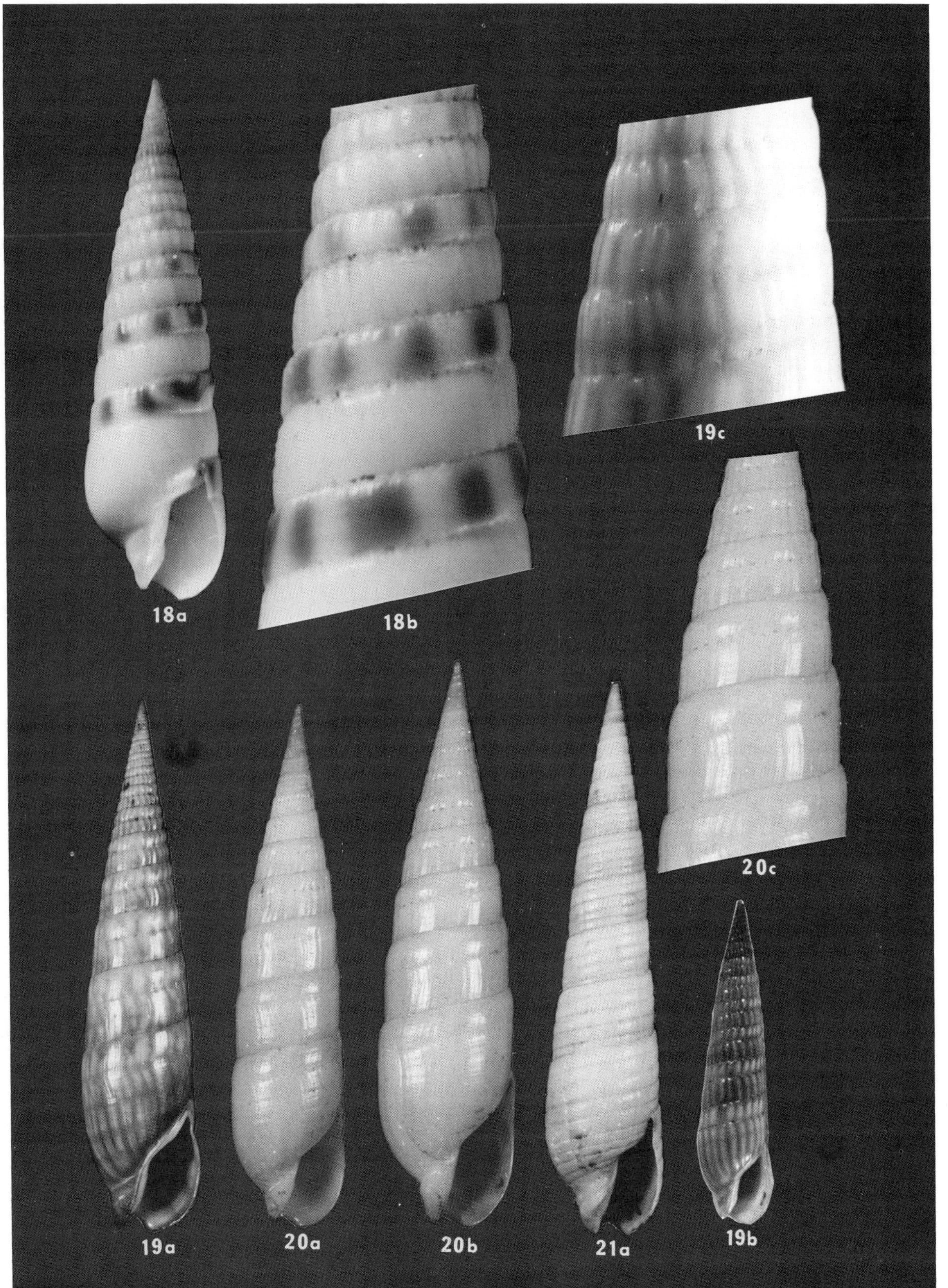

Plate 6. Indo-Pacific *Terebra*.

22. *Terebra montgomeryi* R. D. Burch, 1965

(Pl. 7, fig. 21a-c)
(Color pl. E, fig. 8)

1965 *Terebra (Dimidacus) montgomeryi* R. D. Burch, Veliger 7 (4): 250, pl. 31, fig. 7.

1966 *Terebra (Perirhoe) montgomeryi* R. D. Burch, Cernohorsky & Jennings, Veliger 9 (1): 52, pl. 5, fig. 20.

1967 *Terebra montgomeryi* R. D. Burch, Cernohorsky, Mar. Shells Pacific 1: 204, pl. 51, fig. 378; 1982 Abbott & Dance, Compendium of Seashells p. 275 (holotype illust.)

Description: Shell to 57 mm; color background white, early whorls with flammules of rosy orange which coalesce in later whorls to form large blotches extending into the subsutural band; outline of whorls slightly turreted; subsutural band evenly crenulate, defined by deep punctations; axial sculpture of microscopic growth striae only; spiral sculpture of 4 to 6 rows of punctations; aperture moderately quadrate; columella recurved, with a single weak plication; siphonal fasciole curved, with weak posterior keel.
Type locality: "Piti Bay, Guam, Is., Marianas Is."
Distribution: From the Mariana Is. to Papua New Guinea and the Fiji Is.; intertidal to 20 m.
Type: Holotype MCZ 251240; 57.0 × 10.0 mm.
Discussion: This uncommon species has an unusually beautiful shell with its rosy orange flammules on a rather shiny background. The rows of punctations are well-marked.

23. *Terebra cingulifera* Lamarck, 1822

(Pl. 7, figs. 23a-e)

1822 *Terebra cingulifera* Lamarck, Hist. Nat. Anim. S. Vert. 7:289; 1838 Kiener, Spéc. Gén. Icon. Coq. Viv. 9:39, pl. 13, fig. 30; 1917 Hirase, Terebridae Jap. Emp. p. 20, pl. 3, figs. 21, 22; 1967 Cernohorsky, Mar. Shells Pacific 1:198 pl. 49, fig. 358; 972 Hinton, Shells New Guinea & Central Indo-Pacific p. 46, pl. 23, fig. 30; 1975 Salvat & Rives, Coq. Polynésie p. 361, fig. 389; 198; 1984 Trondle, Xenophora 15:9, fig. 10; 1984 Aubry Terebridae pl. 14; 1984 Salvat & Rives, Shells of Tahiti p. 136, pl. 35, fig. 1.

1825 *Terebra punctulata* Sowerby, Cat. Shells Tankerville App. 24

1859 *Terebra columnaris* Deshayes, Proc. Zool. Soc. Lond. pt. 27: 310.

1859 *Terebra obsoleta* Deshayes, Proc. Zool. Soc. Lond. pt. 27:310.

1923 *Terebra (Terebrina) cingulifera* Lamarck, Bartsch, Nautilus 37 (2):63.

1929 *Perirhoe melamans* Iredale, Austral. Zool. 5(4):341, pl. 38, fig. 7.

1929 *Terebra (Dimidacus) cingulifera* Lamarck, Iredale, Australian Zoologist 5(4):341; 1966 Cernohorsky & Jennings, Veliger 9(1):50, pl. 5, figs. 18, 18a.

1967 *Dimidacus cinguliferus* (Lamarck), Habe & Kosuge, Stand. Book Jap. Shells Color 3:102, pl. 40, fig. 15.

Description: Shell to 101 mm; color ivory to yellow or pinkish orange; outline of whorls straight, turreted on early whorls, less so on later ones; protoconch of 3 elongate whorls set at an angle; subsutural band finely crenulate sometimes, divided by punctate lines; axial sculpture of growth striae only; spiral sculpture of 2 to 7 shallow punctate grooves; aperture elongate-ovate; columella curved.
Type locality: *T. cingulifera*, none given originally. Burias Is., Philippines, is here designated as the type locality; *T. punctulata*, none; *T. columnaris*, none; *T. obsoleta*, none; *P. melamans*, "Sydney Harbour, N.S.W.," Australia.
Distribution: From Mauritius to the Tuamotu Archipelago; intertidal to 20 m.
Type: *T. cingulifera* holotype MHN Geneva no. 1102/27/1, 71.3 × 10.6 mm; *T. punctulata* type lost; *T. columnaris* holotype MNHNP, 46.5 × 8.0 mm; *T. obsoleta* holotype (pl. 7, fig. 23b) MNHNP, 39.9 × 7.0 mm; *T. melamans* holotype (pl. 7, fig. 23c) AMS, 73.5 mm.
Discussion: This is the type species of the subgenus, *Terebrina* Bartsch, 1923, [non Rafinesque, 1815] and its replacement name, *Dimidacus* Iredale, 1929. Bartsch diagnosed the subgenus as "Subsutural cords nodulose on early whorls only; spiral lines punctate." This species is the most common of the several species with which it might be confused. They are *T. laevigata* **(25)** *T. quoygaimardi* **(30)** *T. punctatostriata* **(24)** and *T. tricolor* **(31)** none of which

Plate 7
FIG. **21b,** *Terebra circumcincta* Deshayes. False Entrance, New Caledonia; 34.8 mm. **22a-c,** *T. montgomeryi* R. D. Burch. **22a,** Holotype Museum Comparative Zoology, Harvard No. 251240; 57.0 mm. **22b,** Papua New Guinea; 50.0 mm. **22c,** Middle whorls. **23a-e,** *T. cingulifera* Lamarck. **23a,** Holotype Museum d'Histoire Naturelle, Geneva No. 1102/27/1; 71.3 mm. **23b,** Holotype of *T. obsoleta* Deshayes, Museum National Histoire Naturelle, Paris; 39.9 mm. **23c,** Holotype of *Perirhoe melamans* Iredale, Australian Musuem, Sydney No. C-57863; 73.5 mm. **23d,** Middle whorls. **23e,** Upper whorls. **24a,** *T. punctatostriata* Gray. Holotype British Museum (N.H.) No. 1874.11.10.1.; 70.2 mm.

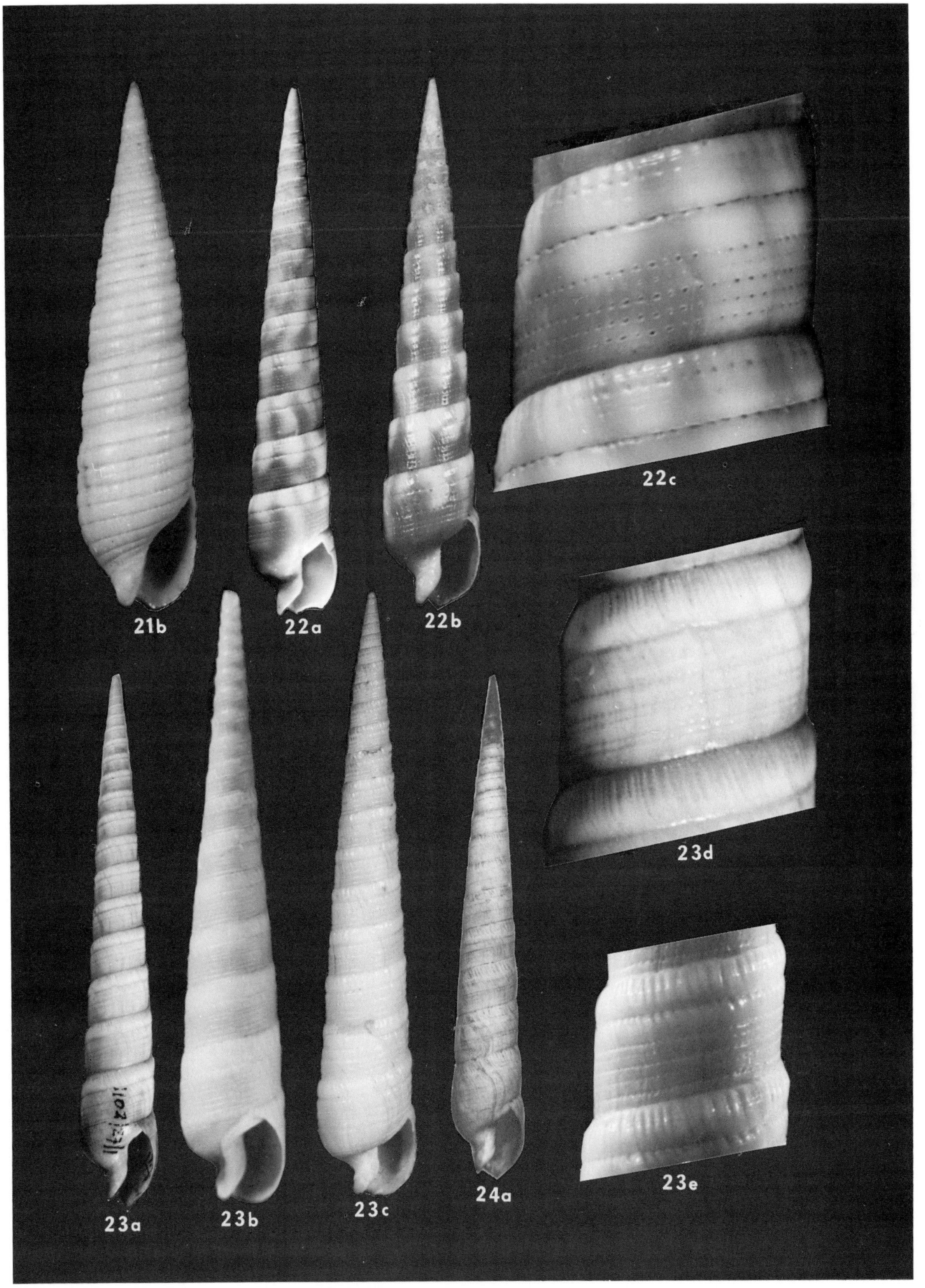

Plate 7. Indo-Pacific *Terebra.*

have punctate spiral sculpture. Additional differences are that *T. laevigata* has a smooth, narrow subsutural band with no crenulations. *T. quoygaimardi* has a shinier shell with white, flattish nodes instead of crenulations on the subsutural band. *T. punctatostriata* has reddish orange coloration, a convex outline, and curved axial growth striae. *T. tricolor* has a shorter shell with two deep, smooth, spiral (often red) grooves per whorl, three on the body whorl, plus the subsutural groove. *T. circumcincta* **(21)** and *T. eburnea* **(20)** both have punctate sculpture, but both have shorter, broader shells. *T. circumcincta* has a protoconch of 1½ large whorls and does not have the slightly crenulate subsutural band of *T. cingulifera*. *T. eburnea* is very similar to *T. circumcincta*, but the early whorls are axially ribbed, with deep punctations between the ribs defining the subsutural band. *T. montgomeryi* **(22)** is shaped like *T. cingulifera* but has a distinctive color pattern of rosy orange blotches and lacks the crenulate subsutural band.

24. *Terebra punctatostriata* Gray, 1834

(Pl. 7, fig. 24a; pl. 8, figs. 24b-d)
(Color pl. F, fig. 3)

1834 *Terebra punctatostriata* Gray, Proc. Zool. Soc. Lond. pt. 2: 61; 1976 Cernohorsky & Bratcher, Rec. Auckland Inst. & Mus. 13:134, figs. 10-13; Trondle, Xenophora 15:12, fig. 27.

1857 *Terebra pallida* Deshayes, Journ. Conchyl. 6:87, pl. 4, fig. 3; (Hab. Marquesas Islands; lectotype BM(NH) no. 197967; 71.8 mm); 1967 Cernohorsky, Mar. Shells Pacific 1:205, pl. 51, pl. 51, fig. 382; 1972 Hinton, Shells New Guinea & Cent. Indo-Pacific, p. 48, pl. 24, fig. 19.

1860 *Terebra cingulifera* Lamarck (*pars*), Reeve, Conch. Icon. 12, pl. 11, fig. 44a only.

1931 *Perirhoe exulta* Iredale, Rec. Australian Mus. 18(4):224, pl. 25 fig. 13 (Hab.: Sydney Harbor, Australia; holotype AM Sydney no. C-57805; 72.7 × 10.0 mm).

1966 *Terebra (Perirhoe) pallida* Deshayes, Cernohorsky & Jennings, Veliger, 9(1):52, pl. 5, fig. 19.

1967 *Dimidacus cinguliferus* (Lamarck), Habe & Kosuge, Stand. Book Jap. Shell Color 3: 102, pl. 40, fig. 15 (non *Terebra cingulifera* Lamarck, 1822).

Description: Shell to 75 mm; color reddish orange to reddish brown; outline of whorls slightly convex; protoconch mamillate, larger than the following whorl; subsutural band noded in early whorls, with weak ribs or axial striae only in later ones, band marked by narrow subsutural groove; axial sculpture of curved striae; spiral sculpture of 4 to 8 narrow groove; aperture quadrate; columella recurved, short, with parietal callus on mature individuals.

Type locality: *T. punctatostriata*, none originally given. Laing Island, Papua New Guinea, is here selected as the type locality; *T. pallida*, "Marquesas Is.;" *T. exulta*, "Sydney Harbour," Australia.

Distribution: Tropical west Pacific; intertidal to 34 m.

Type: *T. punctatostriata* holotype BM(NH) no. 1874.11.10.1, 70.2 mm; *T. pallida* lectotype (pl. 8, fig. 24b) BM(NH) no. 197967, 71.8 mm; *P. exulta* holotype (pl. 8, fig. 24c) AMS no. C-57805, 72.7 × 10.0 mm.

Discussion: The juveniles of this species are often a brilliant reddish orange color. *Terebra pallida* Deshayes is a composite species. Only the lectotype is conspecific with *T. punctatostriata*. The two paralectotypes are specimens of *T. cingulifera* **(23)**. *T. punctatostriata* Gray, in spite of its name, lacks any punctate sculpture and has many curved axial growth striae. The color usually is reddish orange. See *T. cingulifera* **(23)** for comparison.

25. *Terebra laevigata* Gray, 1834

(Pl. 8, fig. 25a-c)

1834 *Terebra laevigata* Gray, Proc. Zool Soc. Lond. pt. 2:61. 1860 Reeve, Conch. Icon. 12: pl. 21, sp. 108; 1967 Cernohorsky, Mar. Shells Pacific 1:202, pl. 51, fig. 373; 1971 Bratcher & Burch in Keen, Seashells Trop. W. America 2nd ed: 680, fig. 1546; Aubry, Terebridae pl. 1.

1844 *Terebra (Myurella) laevigata* Gray, Hinds in Sowerby, Thes. Conch. 1:171, pl. 44, fig. 93.

Plate 8
FIGS. **24b-d,** *Terebra punctatostriata* Gray. **24b,** Lectotype of *T. pallida* Deshayes, British Museum (N.H.) No. 197967; 71.8 mm. **24c,** Holotype of *Perirhoe exulta* Iredale, Australian Museum, Sydney No. C-57805; 72.7 mm. **24d,** Middle whorls. **25a-c,** *T. laevigata* Gray. **25a,** Lectotype British Museum (N.H.) No. 1874.11.10.9.; 28.0 mm. **25b,** Holotype of *T. stylus* Dall, National Museum Natural History, Washington No. USNM-32773; 28.6 mm. **25c,** Fiji Is.; 44.6 mm. **26a-d,** *T. funiculata* Hinds. **26a,** Lectotype British Museum (N.H.) No. 197960; 50.3 mm. **26b,** Holotype of *T. langfordi* Pilsbry, Academy Natural Sciences, Philadelphia No. 117083; 50.4 mm. **26c,** Holotype of *T. angustior* Pilsbry, Academy Natural Sciences, Philadephia No. 47184; 29.0 mm. **26d,** Lower whorls of holotype of *T. archimedis* Deshayes, Museum National Histoire Naturelle, Paris; 31.0 mm.

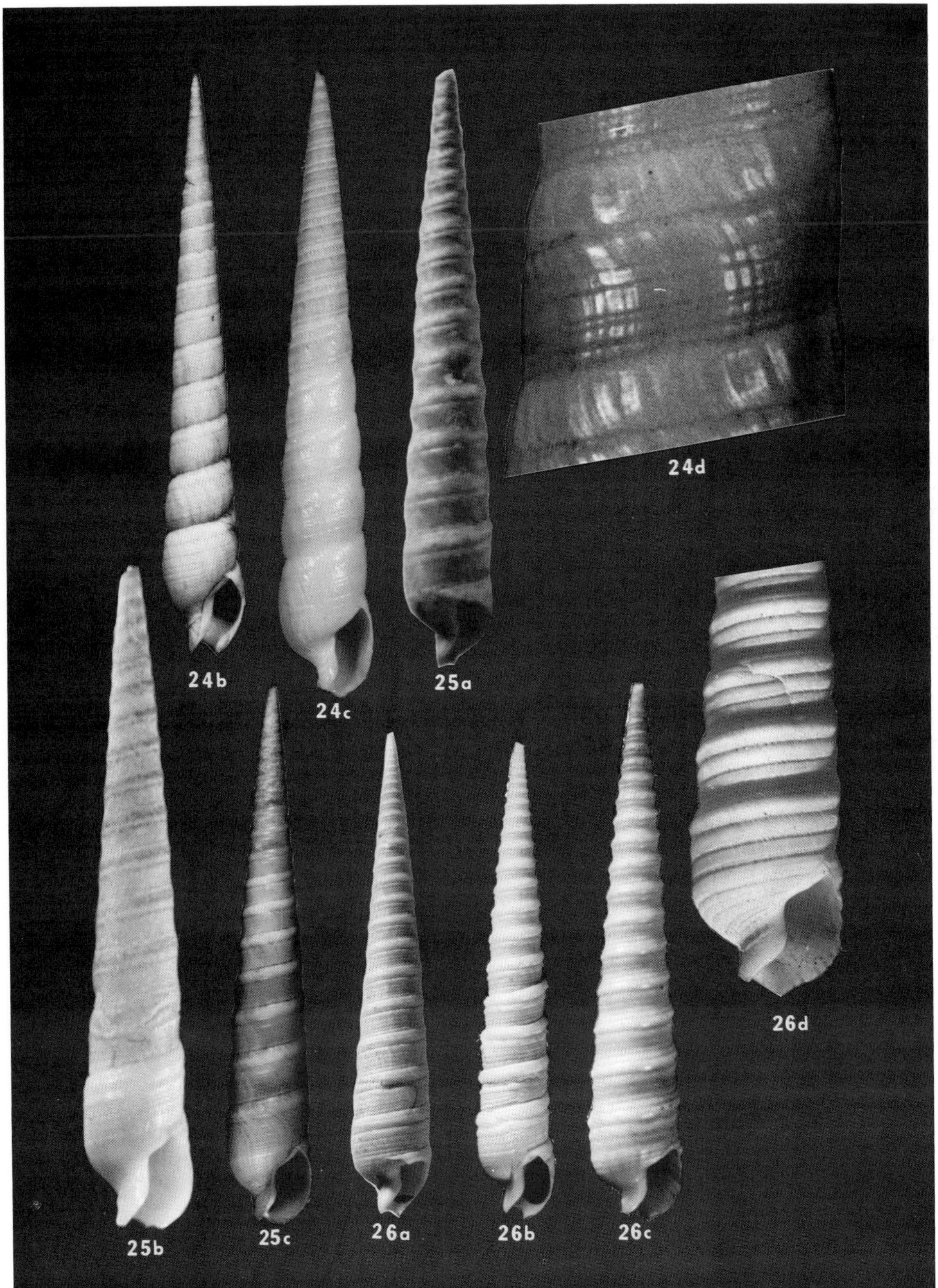

Plate 8. Indo-Pacific *Terebra*.

1908 *Terebra (Perirhoe?) stylus* Dall, Bull. Mus. Comp. Zool. 43 (6):253.
1917 *Terebra cingulifera laevigata* Gray, Hirase, Terebridae Jap. p. 20, pl. 3, figs. 23, 24.
1961 *Dimidacus laevigata* (Gray), Oyama & Takemura, Moll. Shells, Resources Exploitation Inst. 5: *Dimidacus* fig. 3.
1966 Habe & Kosuge, Shells World Color 2:98, pl. 39, fig. 6.
1966 *Terebra (Perirhoe) laevigata* Gray, Cernohorsky & Jennings, Veliger 9(1):51, pl. 5, fig. 21.

Description: Shell to 59 mm; color pure white, cream, or yellowish tan; outline of whorls slightly turreted; protoconch 4 whorls, the last being long; subsutural band smooth, moderately convex, defined by subsutural groove; narrow bands, sometimes obsolete, occurring both posterior and anterior to subsutural band on most specimens; axial sculpture absent except for faint arcuate growth striae; remainder of whorls with 6 to 10 spiral grooves; aperture moderately elongate; columella curved, with light parietal callus.
Type locality: *T. laevigata*, none originally given. Madang, Papua New Guinea is here designated as the type locality; *T. stylus*, "Panama Bay," error?
Distribution: From South Africa to Japan and the Fiji Islands; intertidal to 20 m.
Type: *T. laevigata* lectotype BM(NH) no. 1874.11.10.9, 28.0 × 4.4 mm; *T. stylus* holotype (pl. 8, fig. 25b) USNM no. 3773, 38.6 × 4.8 mm.
Discussion: The type specimen of *T. stylus* Dall, described from Panama, is undoubtedly conspecific with *T. laevigata*, but it is highly doubtful that the species really lives in the Panamic region. *T. funiculata* **(26)** has quite similar sculpture, but it is not as slender and has heavier sculpture and a quadrate aperture and a very short columella. *T. cingulifera* has punctate spiral sculpture.

26. *Terebra funiculata* Hinds, 1844
(Pl. 8, figs. 26a-d)

1844 *Terebra funiculata* Hinds, Proc. Zool. Soc. Lond. pt. 11:153; 1844 Hinds in Sowerby, Thes. Conch. I:168, pl. 43, fig. 63; 1860 Reeve, Conch. Icon. 12, pl. 12, sp. 48; 1952 Tinker, Pacific Sea Shells p. 22, figs. upper row center; 1960 Weaver, Hawaiian Mar. Moll. 1(8):6, pl. 2, figs. lower left; 1967 Cernohorsky, Mar. Shells Pacific I: 200, pl. 50, fig. 369; 1967 Kay, Hawaiian Shell News 15 (7): 1, 2 figs. on right; 1976 Cernohorsky & Bratcher, Rec. Auckland Inst. Mus. 13: 133, figs. 5-9; 1978 Cernohorsky, Trop. Pacific Mar. Shells p. 147, pl. 52, fig. 5; 1978 Salisbury, Hawaiian Shell News 26(8):8, 4 text figs; 1979 Kay, Hawaiian Mar. Shells, p. 398, fig. 129g; 1982 Abbott & Dance, Compendium of Seashells p. 275; Trondle, Xenophora 15:10, fig. 18; 1984 Aubry, Terebridae pl. 8; 1984 Salvat & Rives, Shells of Tahiti p. 136, pl. 35, fig. 4.
1859 *Terebra archimedis* Deshayes, Proc. Zool. Soc. Lond. pt. 27: 314; 1967 Kay, Hawaiian Shell News 15(7):1, figs on left; 1977 Bratcher, Nautilus 91(2):39, fig. 1.
1921 *Terebra langfordi* Pilsbry, Proc. Acad. Nat. Sci. Philadelphia 72:303, pl. 12, fig. 5; 1975 Salvat & Rives, Coq. Polynésie p. 362, fig. 390.
1921 *Terebra langfordi angustior* Pilsbry, Proc. Acad. Nat. Sci. Philadelphia 72:304, pl. 12, fig. 6.
1961 *Dimidacus funiculata* (Hinds), Oyama & Takemura, Moll. Shells Resources Exploitation Inst. *Dimidacus* fig. 2.
1966 *Terebra (Perirhoe) funiculata* Hinds, Cernohorsky & Jennings, Veliger 9 (1):51, pl. 5, fig. 22.
1966 *Terebra (Perirhoe) langfordi* Pilsbry, Cernohorsky & Jennings, Veliger 9(1):51, pl. 5, fig. 23.

Description: Shell to 69 mm; color light to dark beige; outline of whorls concave; protoconch of 4½ whorls; sculpture variable; subsutural band of double or triple convex, almost smooth spiral bands; axial sculpture absent except for fine striae on some individuals; spiral sculpture of 2 or 3 additional convex or flat cords, smooth or roughened by shallow axial striae; in some individuals the subsutural bands are followed by indistinct cords and curved axial growth striae; aperture moderately quadrate; columella recurved; the operculum is transluscent pale yellow.
Type locality: *T. funiculata*, none originally given. Honolulu, Hawaii is here designated as the type locality; *T. archimedis*, none; *T. langfordi*, "off Honolulu, Hawaii, 6-8 fms;" *T. langfordi angustior*, "dredger dump at Honolulu, Hawaii."
Distribution: From Mauritius to Japan, French Polynesia and Hawaii.
Type: *T. funiculata* lectotype BM(NH) no. 197960; 50.3 mm; *T. archimedis* holotype (pl. 8, fig. 26d) MNHNP, 31.0 × 5.8 mm; *T. langfordi* holotype (pl. 8, fig. 26b) ANSP no. 117,083, 50.4 × 8,9 mm; *T. langfordi angustior* holotype (pl. 8, fig. 26c) ANSP no. 47,184, 29.0 × 5.3 mm.
Discussion: Intergrades between all sculptural forms of this extremely variable species have been found. The animal is yellow.

27. *Terebra virgo* Schepman, 1913

(Pl. 9, fig. 27)

1913 *Terebra (Strioterebrum) virgo* Schepman, Siboga Exped. 49T (5):376, pl. 25, fig. 13.

Description: Shell to 31 mm; color shiny white; outline of whorls concave; protoconch of 2 or 3 whorls; subsutural band rather flat with flat rectangular ribs, a second band of smaller nodes which becomes almost square in later whorls, the two bands are separated by a deep groove; axial ribs very flat, broken into flat squares by spiral grooves, appearing as spiral rows of very small squares; aperture semiquadrate; columella moderately short, curved.
Type locality: "Bay of Badjo, W. Coast of Flores, Indonesia, 40 m, sand and mud."
Distribution: From the Indian Ocean to Hawaii.
Type: ZMUA in wet collection; 25.0 × 5.8 mm.
Discussion: This species may be distinguished from the similar *T. funiculata* **(26)** by its spiral cords being broken into small squares and rectangles and by its white color. The shell of *T. funiculata* is beige.

28. *Terebra stearnsii* Pilsbry, 1891

(Pl. 9, figs. 28a-c)

1891 *Terebra stearnsii* Pilsbry, Nautilus 5(8):93; 1892 Pilsbry, Proc. Acad. Nat. Sci. Philadelphia 43: 472, pl. 19, fig. 5; 1917 Hirase, Terebridae Jap. Emp. p. 19, pl. 4, fig. 40, pl. 7, figs. 105, 106.; 1973 Bratcher, Hawaiian Shell News 21 (4):3, text fig. 1978 Hinton, Guide Austral. Shells. pl. 57, fig. 1; 1978 Hinton, Guide Shells Papua New Guinea, pl. 48, fig. 1.

1936 *Terebra (Strioterebrum?) stearnsi* Pilsbry, Hirase, Coll. Jap. Shells ed. 5:89, pl. 117, fig. 11.

1961 *"Myurella" stearnsii* (Pilsbry), Oyama, Venus, Jap. Journ. Malac. 21(2):188.

1961 *Cinguloterebra stearnsii* (Pilsbry), Oyama & Takemura, The Molluscan Shells, Resources Exploitation Inst. 5: *Cinguloterebra* pl, fig. 9; 1962 Kira, Shells W. Pacific Col. 1: 110, pl. 39, fig. 7.

1963 *Terebra (Triplostephanus) stearnsii* Pilsbry, Kira, Sel. Shells W. Pacific Color p. 126, pl., 102, fig. 11.

Description: Shell to 137 mm; color yellowish cream with orange brown irregular spots on subsutural band and orange-brown flammules placed irregularly on remainder of whorls; outline of whorls concave; subsutural band noded, convex, defined by a narrow groove; axial sculpture absent; spiral cords, 4 or 5, heavy, the posterior and anterior ones being larger and somewhat noded; body whorl short; aperture quadrate; columella sharply recurved.
Type locality: "Japan."
Distribution: Japan; subtidal to 100 fms.
Type: Presumed lost. Original dimensions 105 × 13.5 mm.
Discussion: Although the type is presumed lost, Pilsbry's description and figure leave no doubt as to the identity of this species. It was named for Fredrick Stearns of Michigan.

29. *Terebra sowerbyana* Deshayes, 1857

(Pl. 9, figs. 29a-b)

1857 *Terebra sowerbyana* Deshayes, Journ. Conchyl. 6:93, pl. 3, fig. 8; 1859 Deshayes, Proc. Zool. Soc. Lond. pt. 27:277; 1960 Reeve, Conch. Icon. 12: pl. 4, sp. 63.

Description: Shell to 56 mm; color ivory with a slight rosy tint; outline of whorls moderately convex; subsutural band quite flat, ribbed; axial ribs wide-spaced, not contiguous with those on band; spiral grooves almost obsolete, about 3 per whorl; aperture quadrate; columella extremely recurved, with 2 heavy plications.
Type locality: "Gambia." Evidently an error. Bouchet (pers. comm.) states that this species has not been found in West Africa as yet.
Distribution: Unknown.
Type: Holotype BM(NH) no. 197973; 56.3 mm.
Discussion: This species is known only from the holotype, and the locality is in doubt.

30. *Terebra quoygaimardi* Cernohorsky & Bratcher, 1976

(Pl. 9, figs. 30a-b)
(Color pl. E, fig. 6)

1833 *Terebra monile* Quoy & Gaimard, Voy. L'Astrolabe, Zool. 2: 467, pl. 36, figs. 21, 22; 1838 Küster Syst. Conch. Cab. ed. 2, 5(2):29, pl. 6, fig. 10 [non *Buccinum monile* Linnaeus, 1771 = *Terebra*]

1844 *Terebra monilis* Quoy & Gaimard, Hinds, Proc. Zool. Soc. Lond. pt. 11:163; 1844 Hinds in Sowerby, Thes. Conch. 1:168, pl. 43, figs. 65, 66; 1860 Reeve, Conch Icon. 12: pl. 11, fig. 42a; 1967 Cernohorsky, Marine Shells Pacific 1: 204, pl. 51, fig. 377.

1964 *Terebra (Cinguloterebra) monilis* Quoy & Gaimard, Shikama, Select. Shells World Color 2: p. 69, fig. 19.

1966 *Cinguloterebra monilis* (Quoy & Gaimard), Habe & Kosuge, Shells World Col. 2:99, pl. 39, fig. 14.

1967 *Dimidacus monilis* (Quoy & Gaimard), Habe & Kosuge, Stand. Book Jap. Shells Color 3:102, pl. 40, fig. 17.

1976 *Terebra quoygaimardi* Cernohorsky & Bratcher, Rec. Auckland Inst. Mus. 13:135, figs. 14, 15. 1978 Cernohorsky, Trop. Pacific Mar. Shells p. 146, pl. 52, fig. 3;

(replacement name for *T. monile* Quoy & Gaimard, 1833 [non Linnaeus, 1771]; 1984 Aubry, Terebridae pl. 5.

Description: Shell to 50 mm; color shiny yellow to yellowish brown or orange-brown with white subsutural nodes; outline of whorls straight; protoconch of 2 whorls, the first short, the second elongate; subsutural band noded, defined by well marked suture and narrow, moderately deep, subsutural groove, with occasionally a row of obsolete nodes anterior to band; sculpture on remainder of whorl variable, smooth in some individuals or with 1 to 3 spiral grooves; axial striae present in some specimens; aperture quadrate; columella extremely recurved, with heavy parietal callus on mature specimens.
Type locality: "Marianas and Caroline Is."
Distribution: Andaman Is. to Japan, N. Australia, and Micronesia; intertidal to 30 m.
Type: Type of *Terebra monile* [= *T. quoygaimardi*] MNHN Paris; 43.8 × 6.9 mm.
Discussion: Dodge (1956), who treated Linnaeus' *Buccinum* species in detail, and several authors before him, concluded that *B. monile* Linnaeus, 1771, is a *Terebra* of unknown identity. A specimen in the Linnean collection of the Linnean Society, London, contains a *Terebra* species 36.7 × 7.2 mm, which appears to be a very faded *T. undulata* Gray, 1834. This particular specimen is undocumented and was segregated by Hanley from an unmarked box. Since the origin of this probable type specimen is unknown, it is better to consider *T. monile* (Linnaeus) a *nomen dubium*. *T. monile* Quoy & Gaimard is a secondary homonym. This species was named for Quoy & Gaimard, who first recognized it as a separate species.

31. *Terebra tricolor* Sowerby, 1825

(Pl. 9, figs. 31a-b)
(Color pl. E, fig. 7)

1825 *Terebra tricolor* Sowerby, Cat. Shells Tankerville App. xxiv; 1844 Hinds in Sowerby, Thes. Conch. 1:168, pl. 44, fig. 82; 1978 Cernohorsky, Trop. Pacific Mar. Shells p. 146, pl. 52. fig. 2; 1982 Abbott & Dance, Compendium of Seashells p. 275.
1833 *Terebra taeniolata* Quoy & Gaimard, Voy. Astrolabe p. 466. pl. 36, figs. 25, 26.
1966 *Cinguloterebra tricolor* (Sowerby), Habe, Shells W. Pacific Col. 127, pl. 40, fig. 11.

Description: Shell to 57 mm; color pink with cream subsutural band and 1 to 3 very narrow red grooves per whorl, occasionally pale peach color with only slightly lighter subsutural band and uncolored grooves; outline of whorls straight; protoconch mamillate, larger than following whorls; subsutural band slightly convex, without nodes, defined by shallow groove; 1 or 2 additional shallow grooves per whorl, usually colored red; axial sculpture absent; body whorl with 3 to six red grooves, rounded periphery; aperture quadrate; columella recurved, with light to moderate parietal callus.
Type locality: *T. tricolor*, "St. Thomas," error. Limu Is., Tonga Is., is here designated as the type locality; *T. taeniolata*, "Tonga."
Distribution: Japan and Indonesia to the Tonga Is.; subtidal.
Type: *T. tricolor* type lost; *T. taeinolata* holotype (pl. 9, fig. 31a) MNHNP, 56.7 mm.
Discussion: There is little chance of this beautiful and colorful *Terebra* being mistaken for any other species. Although the type has been lost, Sowerby's description and Hinds' subsequent figure leave little doubt as to its identity. The holotype of *T. taeniolata* is rather faded, but the sculpture matches perfectly with Hinds' figure of *T. tricolor.*

32. *Terebra babylonia* Lamarck, 1822

(Pl. 10, figs. 32a-c)
(Color pl. E, fig. 25)

1822 *Terebra babylonia* Lamarck, Hist. Nat. Anim. S. Vert. 7:187; 1839 Kiener, Spéc. Gén. Icon. Coq. Viv. 9:39, pl. 14, figs. 35, 35a; 1917, Hirase, Terebridae Jap. Emp. pl. 21, pl. 4, fig. 41; p. 21, pl. 4, fig. 41; 1960 Weaver, Hawaiian Mar. Moll. 1(2):6, pl. 2, fig. top left; 1967 Cernohorsky, Mar. Shells Pacific p. 198, pl. 49, fig. 355; 1969 Rudman, Veliger 12(1):59, anatomy; 1975 Salvat & Rives, Coq. Polynésie p. 361, fig. 388; 1979 Kay, Hawaiian Mar. Shells p. 393, fig. 130h; 1982 Abbott & Dance, Compendium of Seashells p. 275;

Plate 9
FIG. 27, *Terebra virgo* Schepman. Holotype Zoological Museum, University of Amsterdam; 25.0 mm. **28a-c,** *T. stearnsii* Pilsbry. **28a,** Type-figure from Pilsbry, 1892, pl.19, Fig. 5; 105.0 mm. **28b,** Tanabe, Kii, Japan, 95 m; 130.9 mm. **28c,** Lower whorls. **29a,b,** *T. sowerbyana* Deshayes. **29a,** Holotype British Museum (N.H.), No. 197973; 56.3 mm. **29b,** Middle whorls. **30a,b,** *T. quoygaimardi* Cernohorsky & Bratcher. **30a,** Type-specimen Museum National d,Histoire Naturelle, Paris; 43.8 mm. **30b,** Middle whorls. **31a,b,** *T. tricolor* Sowerby. **31a,** Holotype of *T. taeniolata* Quoy & Gaimard, Museum National d'Histoire Naturelle, Paris, 56.7 mm. **31b,** Tonga Is.; 38.8 mm.

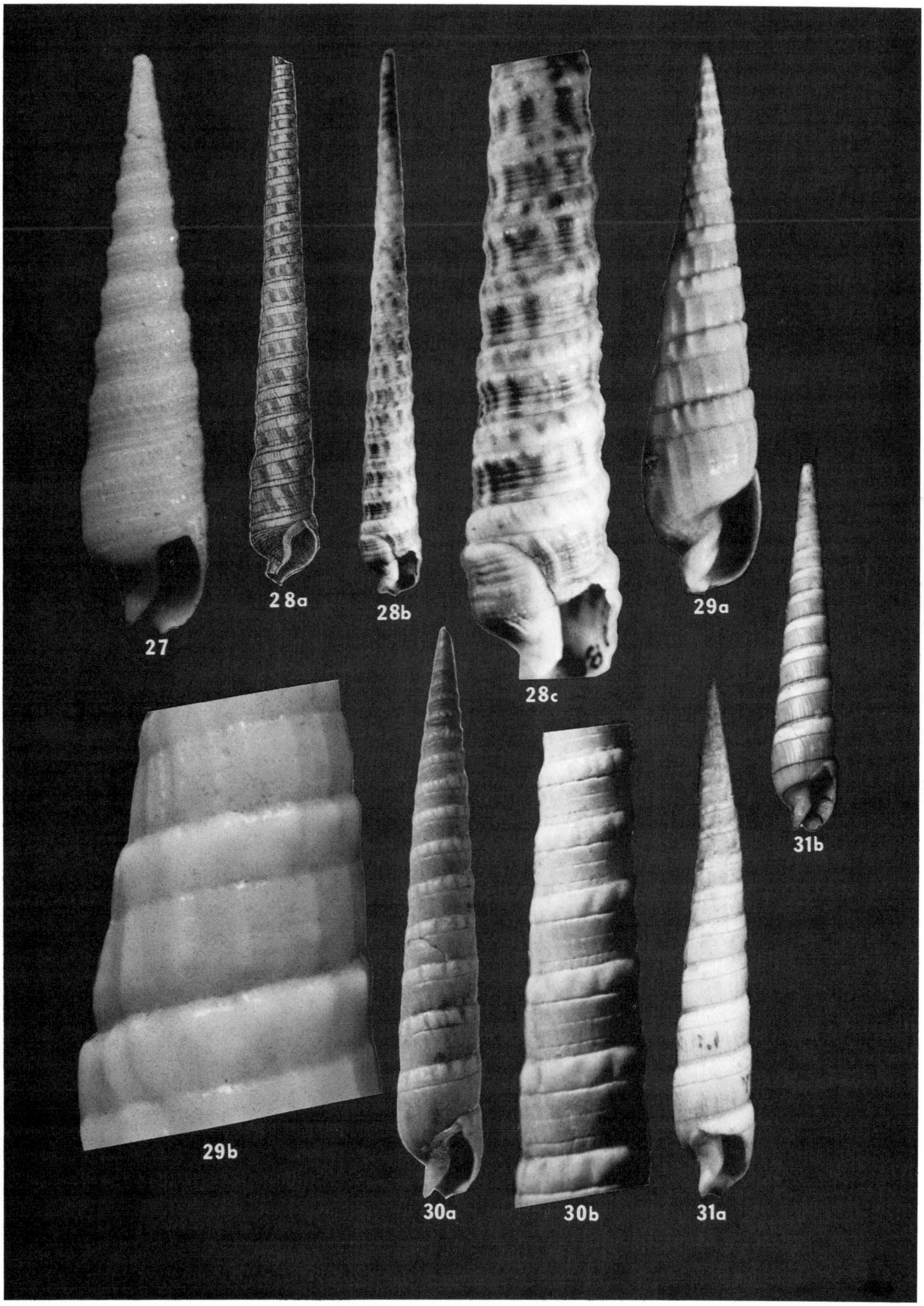

Plate 9. Indo-Pacific *Terebra*.

1984 Trondle, Xenophora 15:9, fig. 7; 1984 Aubry, Terebridae pl. 4.

1834 *Terebra striata* Gray, Proc. Zool. Soc. Lond. pt. 2:60. [non Basterot, 1825; nec. Quoy & Gaimard, 1833].

1961 *Dimidacus babylonia* (Lamarck), Oyama, Venus 21(2):187; 1962 Kira, Shells W. Pacific Color 1:110, pl. 39, fig. 10.

1964 *Terebra (Dimidacus) babylonia* Lamarck, Shikama, Sel. Shells World Color 1:125, pl. 101, fig. 14.

1966 *Terebra (Perirhoe) babylonia* Lamarck, Cernohorsky & Jennings, Veliger 9(1):50, pl. 5, fig. 25.

Description: Shell to 76 mm; color ivory to dark flesh-pink with ivory subsutural band and pinkish brown grooves and striae; outline of whorls slightly convex; protoconch 3 to 3½ conical whorls; subsutural band defined by deep groove; axial grooves unevenly spaced, flexuous, from suture to suture; spiral grooves, usually 2, crossing axial grooves, entire shell appearing broken into flat squares or rectangles by dark axial and spiral grooves; aperture semi-elongate, columella recurved, with somewhat heavy parietal callus on mature specimens.
Type locality: *T. babylonia*, none originally given. Samar, Philippines, is here designated as the type locality; *T. striata*, none.
Distribution: From the Red Sea to French Polynesia and the Hawaiian Islands; intertidal to 200 m.
Type: *T. babylonia* lectotype MHNG no. 1102/2/1; 67.1 × 12.4 mm; *T. striata* holotype BM(NH) no. 1874.10.11.3, 35.0 mm.
Discussion: The eyes of this species are on short stalks, and it has a poison gland and harpoon-shaped radular teeth (Rudman, 1969). *T. knudseni* **(34)** has the same color range and somewhat similar sculpture but differs by having a mamillate protoconch and a swollen subsutural band occupying two thirds of the whorl. Refer to *T. deshayesi* for comparison with that species.

33. ***Terebra deshayesii*** Reeve, 1860

(Pl. 10, figs. 33a-c)

1860 *Terebra deshayesii* Reeve, Conch. Icon. 12: pl. 11, sp. 45. 1984 Aubry, Terebridae pl. 8.

Description: Shell to 55 mm; color orange-red with pinkish white areas marked into squares by darker orange-red lines; subsutural band broad, flat, occupying about half of whorl, divided in center and bordered on anterior by deeply punctate spiral grooves; axial sculpture absent; spiral sculpture of an additional row of punctations dividing remainder of whorl; body whorl with 3 additional rows of punctations below periphery; aperture orange red, elongate; columella curved.
Type locality: None originally given. Comoro Is., off East Africa, is here designated as the type locality.
Distribution: East Africa.
Type: Type lost.
Discussion: At casual glance this species could be mistaken for a particularly colorful and beautiful *T. babylonia* **(32)**, both having somewhat the same or pink coloring marked into squares or rectangles with dark lines. Closer examination shows the lines of *T. deshayeii* to be in color only, not grooves. In addition, *T. deshayesi* has three spiral rows of punctations. Although the type has been lost, Reeve's description is excellent, and with his figure there is no doubt as to the identity of this beautiful species, which is one of the most colorful of terebrids.

34. ***Terebra knudseni*** Bratcher, 1983

(Pl. 10, figs. 34a, b)

1983 *Terebra knudseni* Bratcher, Steenstrupia, Zool. Mus. Univ. Copenhagen 8(15):301, fig. 1.

Description: Shell to 45 mm; color pinkish beige; outline of whorls concave in early whorls, becoming flat; protoconch of 1½ mamillate whorls; subsutural band with double row of flat, squarish nodes, the nodes often fading and band becoming a swollen area scarcely marked by a subsutural groove; axial grooves numerous, flexuous, unevenly-spaced, extending from suture to suture, about 48 to 84 on penultimate whorl; spiral grooves, about 3 to 5, cutting through axial grooves to form small flat rectangles; body whorl with axial grooves ending at periphery; aperture quadrate, columella curved.

Plate 10
FIGS. **32a-c,** *Terebra babylonia* Lamarck. **32a,** Lectotype Museum d'Histoire Naturelle, Geneva, No. 1102/2/1; 67.1 mm. **32b,** Holotype of *T. striata* Gray, British Museum (N.H.) No. 1874.10.11.3.; 35.0 mm. **32c,** Middle whorls. **33a-c,** *T. deshayesii* Reeve. **33a,** Type-figure from Reeve, 1844, pl.11, FIG. **45. 33b,** S.E. of M'Zamboro Pass, Mayotte I., Comoro Is., 16 m-18m; 54.6 mm. **33c,** Middle whorls. **34a,b,** *T. knudseni* Bratcher. **34a,** Holotype University Zoological Museum, Copenhagen; 41.4 mm. **34b,** Middle whorls. **35a,** *T. amanda* Hinds. Holotype of *T. albomarginata* Deshayes, British Museum (N.H.) No. 197979; 44.5 mm.

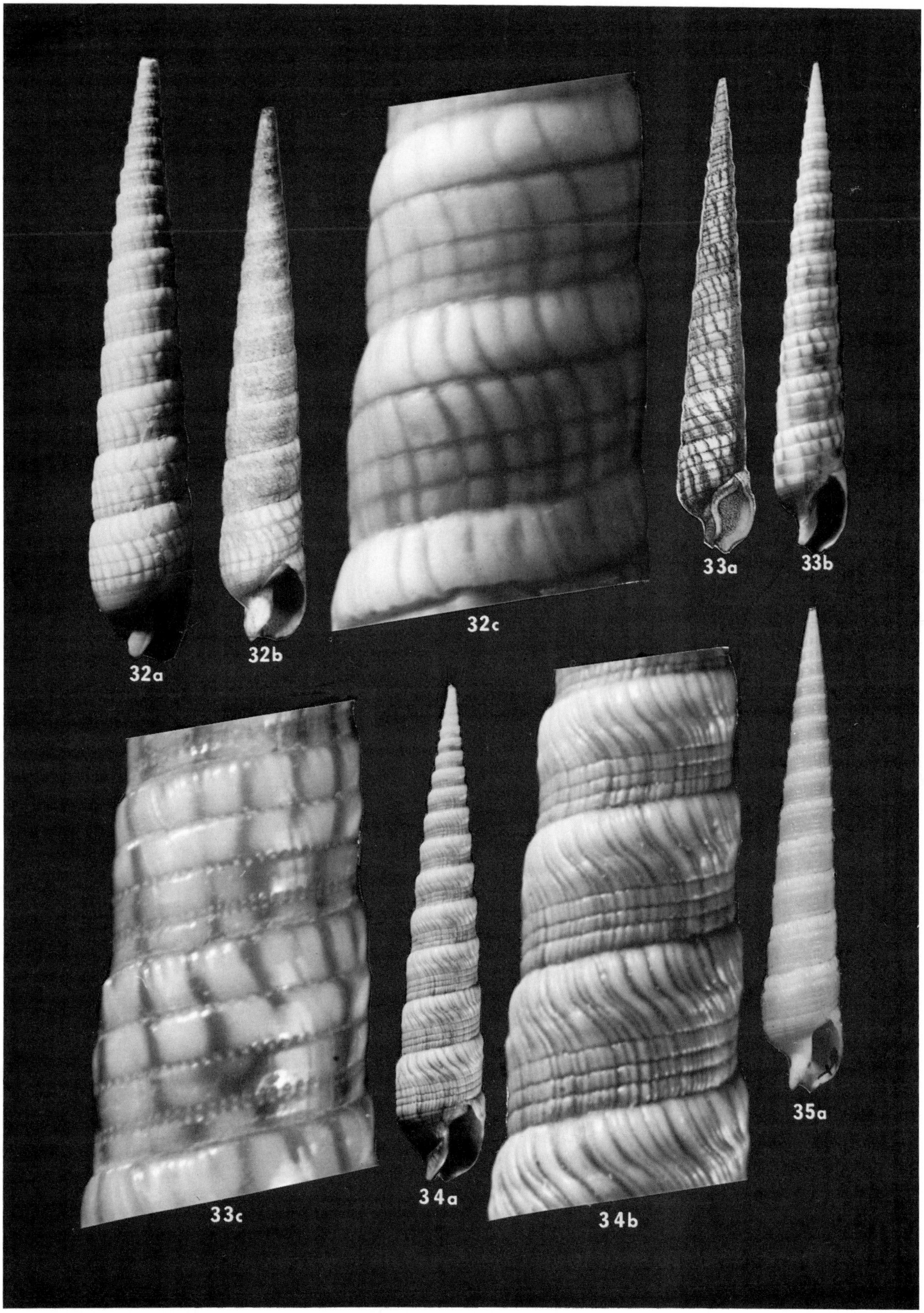

Plate 10. Indo-Pacific *Terebra*.

Type locality: "Malawai Is., Kudat District, North Borneo."
Distribution: From Indonesia to the Philippines; subtidal.
Type: Holotype UZMC; 41.4 × 8.1 mm.
Discussion: Some individuals retain the early sculpture with the subsutural band of flat squares into maturity, although the second band appears to become almost obsolete, while others develop a rather narrow convex band. For comparison, see *T. babylonia* **(32)**.

35. *Terebra amanda* Hinds, 1844

(Pl. 10, fig. 35a,
pl. 11, figs. 35b-d)

1844 *Terebra amanda* Hinds, Proc. Zool Soc. Lond. pt. 11:154; 1844 Hinds in Sowerby, Thes. Conch. 1:166, pl. 45, fig. 100; 1967 Cernohorsky, Mar. Shells Pacific 1:196. pl. 49, fig. 351; 1972 1972 Hinton, Shells New Guinea & Cent. Indo-Pacific p. 46, pl. 23, fig. 21; 1976 Cernohorsky & Bratcher, Rec. Auckland Inst. & Mus. 13: 131, figs. 1-4 (fig'd. holotype of *T. albomarginata* Deshayes); 1978 Cernohorsky, Trop. Pacific Mar. Shells p. 147, pl. 52, fig. 4; 1979 Kay, Hawaiian Mar. Shells p. 392, fig. 130i; 1984 Trondle, Xenophora 15:8, fig. 3.

1859 *Terebra albomarginata* Deshayes, Proc. Zool. Soc. Lond. pt. 27:314.

1908 *Terebra unicolor* Preston, Rec. Indian Mus. 1(2):190, pl. 17, holotype Zool. Survey India no. M3999/1; 43.5 × 7.5 mm).

1960 *Terebra straminea* Gray, Weaver, Hawaiian Mar. Moll. 1(2):6, pl. 7, fig. top right [non Gray, 1834].

1966 *Terebra (Perirhoe) amanda* Hinds, Cernohorsky & Jennings, Veliger 9(1):49, pl. 5, fig. 17.

1967 *Terebra (Dimidiacus) amanda* Hinds, Bratcher & Burch, Veliger 10(1):9.

1967 *Dimidiacus stamineus (sic)* (Gray), Habe & Kosuge, Standard Book Jap. Shells Color. 3:102, pl. 40, fig. 19 (non *straminea* Gray, 1834).

Description: Shell to 61 mm; color yellow or orange to brown, with or without a white subsutural band; outline of whorls concave; protoconch 1½ conical whorls; subsutural band of double row of nodes, weak or well-developed; axial sculpture absent; spiral rows of punctations, 3 to 5; aperture quadrate, columella recurved.
Type locality: *T. amanda*, "Straits of Macassar, 11 fms." Indonesia; *T. albomarginata*, "Australia;" *T. unicolor*, "off Port Blair, Andaman Ids, 7-10 fms."
Distribution: From the Red Sea through the tropical Indo-Pacific; 10 to 23 m.
Type: *T. amanda* type specimens are lost. Cernohorsky & Bratcher (1976) have designated Hinds in Sowerby's well-executed figure (1844, pl. 45, fig. 100) as the representative of the lectotype; *T. albomarginata* holotype (pl. 10, fig 35a) BM(NH) no. 197979, 44.5 × 8 mm; *T. unicolor* holotype (pl. 11, fig. 35b) ZSI no. M3999/1; 43.5 × 7.5 mm.
Discussion: The type of *T. amanda* was reported lost as early as 1860, and several unsuccessful searches for it have been made in the British Museum (Nat. Hist.). Hinds' original description and well-executed illustration of the missing type show all the salient points of the species. *T. albomarginata* Deshayes is without doubt conspecific with *T. amanda*. *T. mariesi* **(36)** may be distinguished from this species by its lack of spiral rows of punctations. *T. straminea* Gray is conspecific with *T. anilis* **(39)** not with *T. amanda*. Refer to *T. anilis* for comparison with that species.

36. *Terebra mariesi* E. A. Smith, 1880

(Pl. 11, figs. 36a-b)

1880 *Terebra mariesi* E. A. Smith, Proc. Zool. Soc. Lond.: 480, pl. 48, fig. 5; 1917 Hirase, Terebridae Jap. Emp. p. 9, pl. 6, fig. 96.

1961 *Cinguloterebra jefreysii* (sic) (Smith), Oyama & Takemura, Moll. Shells, Resources Exploitation Inst. 5: *Cinguloterebra* fig. 6. [non jeffreysii E. A. Smith, 1879]

Description: Shell to 45 mm; color beige with streaks or splotches of brown; outline of early whorls flat, later whorls flat with convex subsutural band; protoconch of 1½ mamillate whorls; subsutural band with double row of nodes, separated by a groove; axial ribs, usually on last 3 whorls only, reaching from second row of nodes on band to suture; spiral cords fine, 6 to 8; aperture quadrate; columella curved, with faint parietal callus.

Plate 11
FIGS. **35b-d,** *Terebra amanda* Hinds. **35b,** Holotype of *T. unicolor* Preston, Zoological Survey of India, Calcutta, No. M-3999/1; 43.5 mm. **35c,** Red Sea; 42.6 mm. **35d,** Middle whorls. **36a,b,** *T. mariesi* E. A. Smith. **36a,** Holotype British Museum (N.H.) No. 1880.3.1.94.; 43.0 mm. **36b,** Middle whorls. **37,** *T. russetae* (Garrard). Holotype Australian Museum, Sydney, No. C-102587; 29.0 mm. **38a-d,** *T. adamsii* E. A. Smith. **38a,** Lectotype British Museum (N.H.) No. 1979132; 35.5 mm. **38b,** Lectotype of *T. jeffreysii* E. A. Smith, British Museum (N.H.) No. 1878.10.16.4.; 25.0 mm. **38c,** Holotype of *T. hedleyi* Pilsbry and *T. hedleyana* Pilsbry, Academy Natural Sciences, Philadelphia, No. 85946; 33.6 mm. **38d,** Middle whorls.

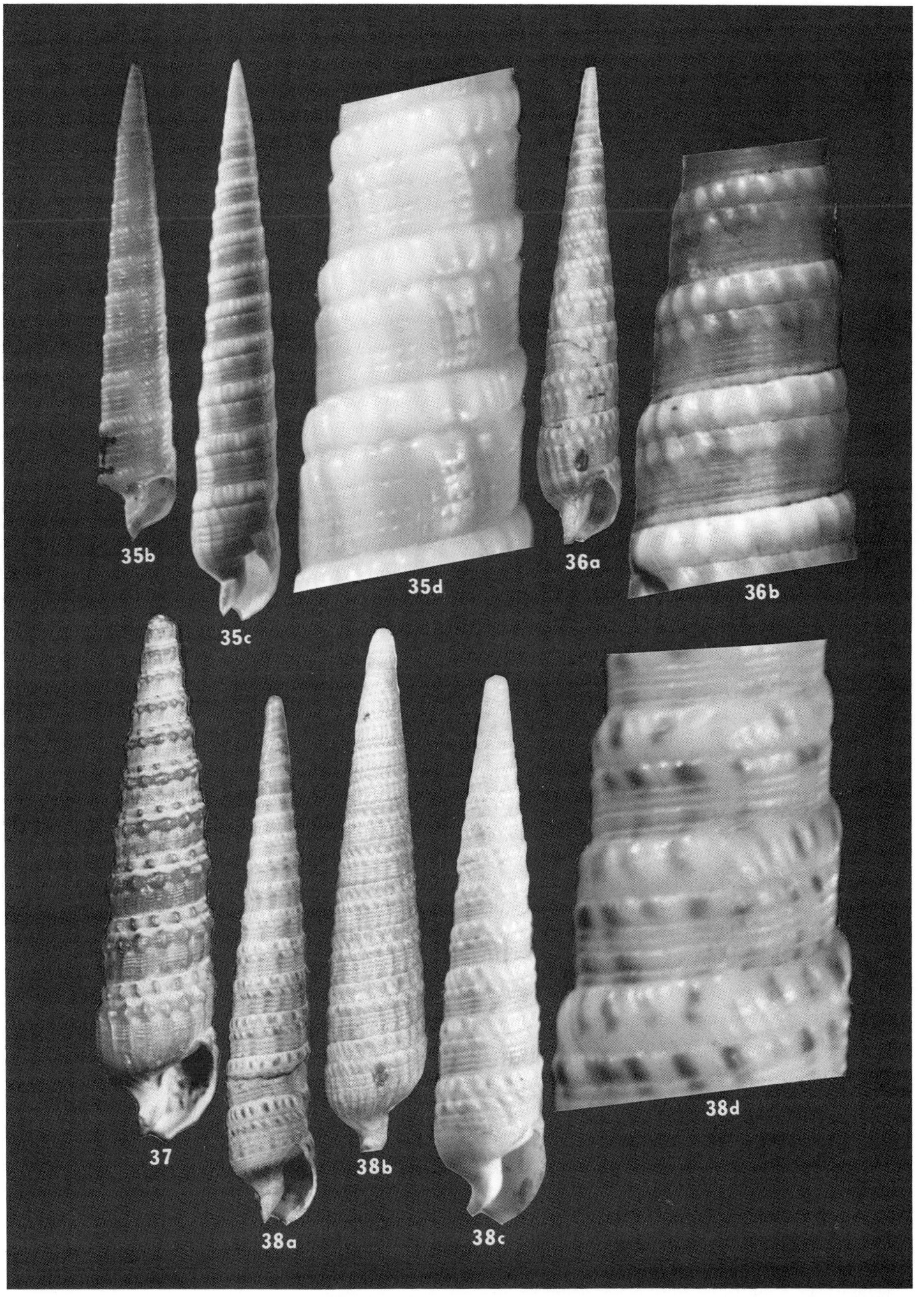

Plate 11. Indo-Pacific *Terebra*.

Type locality: "Japan."
Distribution: From Japan to Australia.
Type: Holotype BM(NH) no. 1880.3.1.94; 43.0 × 6.8 mm.
Discussion: This species has been confused with *T. adamsii* **(38)** a species which lacks the double row of subsutural nodes. This species is more similar to *T. russetae* (37) in both color and sculpture, although it is more slender, has shorter whorls, and the posterior row of nodes is smaller.

37. *Terebra russetae* (Garrard, 1976)

(Pl. 11, fig. 37)

1976 *Pervicacia russetae* Garrard, Journ. Malac. Soc. Australia 3(4):184.

Description: Shell to 29 mm; color light reddish fawn, darker on base of the body whorl; outline of whorls flat with convex subsutural nodes; subsutural band of small, widely-spaced nodes followed by a row of round larger nodes; axial riblets obsolete on early whorls becoming slightly strengthened later; spiral threads fine, about 10 on penultimate whorl, crossing nodes and obsolete riblets; aperture moderately quadrate; columella curved, with light parietal callus.
Type locality: "8 km. E. of Port Jackson, N.S.W." Australia.
Distribution: Known from the type locality only.
Type: Holotype AMS no. C 102587; 29.0 × 7.3 mm.
Discussion: The holotype of this species is a dead and worn specimen. The sculpture is similar to that of *T. mariesi* **(36)** from Japan except that the posterior row of subsutural nodes in that species is larger. Additional material may prove the two to be conspecific.

38. *Terebra adamsii* E. A. Smith, 1873

(Pl. 11, figs. 38a-d)
(Color pl. E, fig. 16)

1873 *Terebra adamsii* E. A. Smith, Ann. & Mag. Nat. Hist. (ser. 4) 11: 264.
1879 *Terebra jeffreysii* E. A. Smith, Proc. Zool. Soc. Lond. p. 184, pl. 19, fig. 2.
1904 *Terebra hedleyi* Pilsbry, Proc. Acad. Nat. Sci. Philadelphia 56. 3, pl. 1, figs. 1, 1a. [non Tate, 1901].
1905 *Terebra hedleyana* Pilsbry, Proc. Acad. Nat. Sci. Philadelphia 57: 101. [nom. nov. pro. *Terebra hedleyi* Pilsbry, 1904]

Description: Shell to 47 mm; color beige to off-white, with spiral rows of small fulvous dots or fulvous streaks on some individuals; outline of whorls straight, somewhat turreted; protoconch of 1½ inflated, glassy whorls; subsutural band of inconspicuous oblique nodes followed by a cord of smaller nodes; early whorls of teleoconch quite smooth, with faint spiral lines beginning to develop about 3rd whorl; remainder of whorl evenly divided by 3 to 5 spiral cords; aperture quadrate; columella extremely recurved.
Type locality: *T. adamsii*, "Japan;" *T. jeffreysii*, "E. of Kii, Japan, 56 fms;" *T. hedleyana*, "Japan."
Distribution: From Japan to the Philippine Is.
Type: *T. adamsii* lectotype BM(NH) no. 1979132; 35.5 mm; *T. jeffreysii* holotype (pl. 11, fig. 38b) BM(NH) no. 1878.10.16.4, 25.0 × 5.0 mm; *T. hedleyana* holotype (pl. 11, fig. 38c) ANSP no. 85946, 33.6 mm.
Discussion: It is surprising that E. A. Smith could have described *T. jeffreysi* as a new species six years after he originally described it as *T. adamsii*, especially as the type lots of the two show no variation, except that the latter specimens are more worn. In describing *T. jeffreysi* he said, "This species is remarkable on account of the unusually large size of the nuclear whorls." The nuclear whorls on his types of *T. adamsii* are equally large. The holotype of *T. hedleyana* Pilsbry also matches well in both color and sculpture. *T. hedleyana* [=*T. adamsii*] is the type species of the genus *Cinguloterebra* Oyama, 1961.

39. *Terebra anilis* (Röding, 1798)

(Pl. 12, figs. 39a-g)

1798 *Epitonium anile* Röding, Mus. Bolten p. 95 (ref. Chemnitz, 1780, vol. 4:pl. 155, fig. 1456).
1798 *Epitonium cancellatum* Röding, Mus. Bolten, p. 95, no. 1225. (ref. Chemnitz, fig. 1456).
1834 *Terebra straminea* Gray, Proc. Zool. Soc. Lond. pt. 2:62. 1844 Hinds in Sowerby, Thes. Conch. 1:169, pl. 43, figs. 22, 23; 1860 Reeve, Conch. Icon. 12: pl. 12, sp. 47.

Plate 12
FIGS. **39a-g**, *Terebra anilis* (Röding). **39a**, Type-figure from Chemnitz, 1780, vol. 4, pl.155, FIG.**1456. 39b**, Holotype of *T. circinata* Deshayes, British Museum (N.H.) No. 1979150; 41.3 mm. **39c**, Probable holotype of *T. straminea* Gray, British Museum (N.H.) No. 1979158; 63.0 mm. **39d**, Holotype of *T. rubrobrunnea* Preston, Zoological Survey of India, Calcutta, No. 3998/1; 46.8 mm. **39e**, Middle whorls. **39f**, Middle whorls of smooth form. **39g**, Upper whorls. **40a-c**, *T. commaculata* (Gmelin). **40a**, Type-figure from Chemnitz, 1780, vol. 4, pl.154, FIG. 1452. **1452. 40b**, Holotype of *T. myros* Lamarck, Museum d'Histoire Naturelle, Geneva, No. 1102/28/1; 74.4 mm. **40c**, Middle whorls.

Plate 12. Indo-Pacific *Terebra*.

1850 *Terebra serotina* Adams & Reeve, Zool. Voy. *H.M.S. Samarang* p. 30, pl. 10, fig. 20; 1860 Reeve, Conch. Icon. 12: pl. 15, sp. 66; 1917 Hirase, Terebridae Jap. Emp. p. 11, pl. 5, fig. 97.

1857 *Terebra circinata* Deshayes, Journ. Conchyl. 6:99, pl. 4, figs. 6, 7.

1857 *Terebra acuta* Deshayes, Journ. Conchyl. 6:100, pl. 4, figs. 4, 5.

1908 *Terebra rubrobrunnea* Preston, Rec. Indian Mus. 2:189, pl. 2, fig. 64.

1961 *Cinguloterebra serotina* (Adams & Reeve), Oyama, Venus, 21 (2): 183; 1962 Kira, Shells W. Pacific Color 1:208, pl. 71, fig. 4.

1964 *Terebra anilis* (Röding), R. D. Burch, Veliger 6(1):146; 1967 Cernohorsky, Mar. Shells Pacific 1: 196, pl. 49, fig. 352; 1972 Hinton, Shells New Guinea & Cent. Indo-Pacific p. 48, pl. 24, figs. 20, 21; 1975 Coleman, "What Shell is That" p. 236, fig. 658; 1984 Aubry, Terebridae pl. 11.

1964 *Dimidacus stramineus* (Gray), Habe, Shells W. Pacific Col. 2: 127, pl. 40, fig. 10; 1967 Habe & Kosuge, Stand. Book Jap. Shells Col. 3:102, pl. 40, fig. 19.

Description: Shell to 88 mm; color cream, tan, or brown; outline of whorls concave in early whorls, almost flat in later ones; protoconch of $2^1/_2$ conical whorls; subsutural band convex, with oblique axial ribs followed by a row of small nodes, the two being separated by a groove, band becoming progressively flatter; axial sculpture often limited to subsutural band; occasionally, weak ribs persist on remainder of whorl; spiral cords strong, 3 to 6; columella recurved, with parietal callus on mature specimens.
Type locality: *T. anilis*, none originally given. Bileau Island, Madang Harbour, Papua New Guinea is here selected as the type locality; *T. straminea*, none; *T. serotina*, "Nagasaki Bay, Japan;" *T. circinata*, China Seas;" *T. acuta*, "China Seas;" *T. rubrobrunnea*, "off Port Blair, Andaman Is."
Distribution: From Borneo to Samoa.
Type: The type of *T. anilis* is lost, and the specimen illustrated by Chemnitz, 1780, is here designated as the illustrated lectotype; *T. straminea* lectotype (pl. 12 fig. 39c) BM(NH) no. 1979158, 63.0 mm; *T. serotina* type lost; *T. circinata* holotype (pl. 12, fig. 39b) BM(NH) no. 979150, 41.3 × 6.0 mm; *T. acuta* type not found; *T. rubrobrunnea* holotype pl. 12, fig. 39d) ZSI no. M3998/1, 46.8 × 6.4 mm.
Discussion: The cream-colored specimens tend to be somewhat larger than the darker ones, and there is a resemblance to *T. cumingii* (**44**) but the latter retains its concave outline into maturity and also retains two rows of well-developed, rounded subsutural nodes. *T. amanda* (**35**) has spiral rows of punctations below the subsutural band which this species does not.

40 *Terebra commaculata* (Gmelin, 1791)

(Pl. 12, fig. 40a-c)
(Color pl. B, fig. 6)

1791 *Buccinum commaculatum* Gmelin, Syst. Nat. ed. 13:3502 (ref. Chemnitz, vol. 4, pl. 154, fig. 1452).

1798 *Epitonium acutissimum* var. Röding, Mus. Bolten. 5(2):94 (ref. Chemnitz, vol. 4, pl. 154, fig. 1452).

1822 *Terebra myuros* Lamarck, Hist. Nat. Anim. S. Vert. 7:289; 1839 Kiener, Spéc. Gén. Icon. Coq. Viv. 9: 40, pl. 14, figs. 34, 34a; 1860 Reeve, Conch. Icon. 12, pl. 8, sp. 31; 1972 Hinton, Shells New Guinea & Cent. Indo-Pacific, p. 46, pl. 23, fig. 29; 1978 Hinton, Guide Austr. Shells, pl. 57, fig. 9.

1844 *Terebra commaculata* (Gmelin), Hinds in Sowerby, Thes. Conch. 1:170, pl. 42, fig. 37; 1973 Bratcher, Hawaiian Shell News 21(4):1, text fig.; 1982 Abbott and Dance, Compendium of Seashells, p. 274, fig.

1961 *Myurella myuros* (Lamarck), Oyama & Takemura, Moll. Shells, Resources Exploitation Inst. 5, *Myurella* sp. 4, 10; 1966 Habe & Kosuge, Shells World Col. 2: 100, pl. 39, fig. 17; 1972 Hinton, Shells New Guinea & Cent. Indo-Pacific p. 46, pl. 23, fig. 29.

Description: Shell to 154 mm, slender; color beige with brown rectangular markings, sometimes broken into 2 axially oriented squares; outline of whorls flat; protoconch of $1^1/_2$ mamillate whorls, larger than following whorls; double subsutural band, very weakly noded and blotched with brown, coalescing into a single broad band without nodes on later whorls; sculpture of axial cords crossed by spiral cords of about equal strength; aperture moderately quadrate; columella short, extremely recurved.
Type locality: *T. commaculata*, none originally given. Chemnitz specimen originated in the "East Indies." Moluccas, Indonesia, is here designated as the type locality; *E. acutissimum*, none; *T. myuros*, "Indian Ocean and Moluccas."
Distribution: From Zanzibar and Natal to the Solomon Islands; 30 to 100 m.
Type: *T. commaculata* type is lost, and the immature specimen depicted by Chemnitz, 1780, in the figure 1452 on plate 154, is here designated as the illustrated lectotype; *E. acutissimum*, same type; *T. myuros* holotype (pl. 12, fig. 40b) MHNG no. 1102/28/1, 74.4 × 10.1 mm.
Discussion: Although they are very similar, *T. vicdani* (**41**) may be separated from this species by its coarser sculpture, lack of a double subsutural band on the early whorls, by having a

broader shell, and square spots that seldom coalesce. *T. lima* **(42)** is equally flat in outline, but it has a broader shell, lacks the rectangular maculations, and has much coarser sculpture. *T. pretiosa* **(43)** has strong curved axial ribs which *T. commaculata* lacks.

41. *Terebra vicdani* Kosuge, 1981

(Pl. 13, figs. 41a-b)
(Color pl. B, fig. 15)

1981 *Terebra (Myurella) vicdani* Kosuge, Bull. Inst. Malac. Tokyo 1(6):93, pl. 31, fig. 8; 1986 Springsteen & Leobrera, Shells of the Philippines p. 330, pl. 94, fig. 7 (not 6).

Description: Shell to 73 mm; color ivory with 2 rows of brown square axially aligned spots which sometimes coalesce, body whorl with 2 rows of spots above and one below the periphery; outline of whorls flat; subsutural band of numerous fine axial ribs which in later whorls are crossed by about 3 fine spiral cords; band defined by shallow groove ornamented with fine axial threads; sculpture on remainder of whorl consisting of about 31 fine spiral threads, crossed by about 7 fine spiral threads of equal value, forming small shallow square pits between the intersections; body whorl squared off with weak keel at periphery; aperture quadrate; columella curved.
Type locality: "Bohol Island, Philippines."
Distribution: Known only from type locality.
Type: Holotype Institute of Malacology of Tokyo no. IMT-81-9; 72.6 × 10.8 mm.
Discussion: Although this rare species is known only from the type locality, a number of specimens have been netted there from deep water. It is similar to *T. commaculata* **(40)**, but the differences appear to be constant. Refer to that species for comparison.

(42) *Terebra lima* Deshayes, 1857

(Pl. 13, figs. 42a-c)
(Color pl. F, fig. 7)

1857 *Terebra lima* Deshayes, Journ. Conchyl. 6:69, pl. 4, fig. 2; 1860 Reeve, Conch. Icon. 12, pl. 12., sp. 50; 1917 Hirase, Terebridae Jap. Emp. p. 14, pl. 3, figs. 19, 20; 1973 Bratcher, Hawaiian Shell News 21(4):3, text fig.

1877 *Terebra loebbeckeana* Dunker, Malakozool. Blätter 24:74 1882 Dunker, Index Moll. Mar. Japonici p. 72, pl. 5, figs. 17, 18; 1917 Hirase, Terebridae Jap. Emp. p. 13, pl. 6, figs. 87, 90.

1961 *Triplostephanus lima* (Deshayes), Oyama & Takemura, Moll. Shells, Resources Exploitation Inst. 5, *Triplostephanus* figs. 6, 7, 10, 11; 1962 Kira, Shells W. Pacific Col. 1: 209, pl. 71, fig. 12.

Description: Shell to 94 mm; color, brown, or beige, sometimes with brownish blotches; outline of whorls flat; protoconch mamillate, larger than the following whorl of teleoconch; subsutural band with double row of elongate nodes, separated by a spiral groove, the posterior row being larger, nodes connected by 2 to 4 rows of fine spiral cords; remainder of whorl finely cancellate, with arcuate axial cords or striae crossed by 3 or 4 coarse spiral cords, and sometimes additional finer ones; aperture small, quadrate; columella recurved, with slight parietal callus.
Type locality: *T. lima*, "China Seas;" *T. loebbeckeana*, "Japan Seas".
Distribution: Indian Ocean to Japan and Papua New Guinea; 35 to 350 m.
Type: *T. lima* holotype BM(NH) no. 197965, 76.5 × 10.5 mm; *T. loebbeckeana* holotype (pl. 13, fig. 42b) MNHUB, 76.5 × 10.0 mm.
Discussion: The extremely flat profile of this shell and its lack of maculations make it easy to identify. Refer to *T. commaculata* **(40)** for comparison.

43. *Terebra pretiosa* Reeve, 1842

(Pl. 13, figs. 43a-c)
(Color pl. B, fig. 7)

1842 *Terebra pretiosa* Reeve, Conch. Syst. 2: 245, pl. 274, fig. 2; 1843 Reeve, Proc. Zool. Soc. Lond. pt. 10: 200; 1846 Reeve, Elem. Conch. 1 (5): 55, pl. 3, fig. 15; 1860 Reeve, Conch. Icon. 12, pl. 8, sp. 30a, b; Abbott and Dance, Compendium of Seashells, p. 274, fig.

1952 *Terebra fujitai* Kuroda & Habe, Check List & Bibl. Rec. Moll. Japan p. 89 (refers to *T. pretiosa* Reeve, 1860, pl. 8, fig. 30b); 1973 Bratcher, Hawaiian Shell News 21 (4): 3, text fig.

1961 *Myurella fujitai* (Kuroda & Habe) Oyama & Takemura, Moll. Shells, Resources Exploitation Inst. 5: *Myurella* pl., figs. 7, 13.

1964 *Terebra (Myurella) fujitai* Shikama, Sel. Shells World Color 1:126, pl. 102, fig. 9.

1984 *Terebra stearnsi* Pilsbry, Aubry, Terebridae pl. 1 [non *Terebra stearnsi* Pilsbry, 1892].

Description: Shell to 142 mm; color orange-tan with a broken band of rich brown between subsutural band and following suture; outline of whorls flat, slightly turreted; protoconch of $2^{1}/_{2}$ conical whorls; subsutural band on early whorls with a single row of elongate nodes, defined by a broad shallow groove, band later becoming a row of square pits separated by narrow slanting ribs, defined by a concave row of square pits; axial

ribs numerous, narrow, well-developed, sinuous, slanting to the right on the band, curving to left on remainder of whorl, about 29 on penultimate whorl; spiral threads or cords 6 to 9, unevenly spaced, some very fine, others heavy, crossing ribs; aperture quadrate; columella recurved, with parietal callus.
Type locality: None given originally by Reeve, who later gave the locality as China in 1860.
Distribution: Japan and China to the Philippines, principally in Japan.
Type: Holotype BM(NH) no 1979116; 140.0 × 17.4 mm.
Discussion: Although Reeve figured two specimens of *T. pretiosa* in his 1860 monograph, he had only one specimen at the time he first described the species in 1842, stating in the original description that it was unique. His illustration of the holotype is a perfect example of the shell that was subsequently described as *T. fujitai* Kuroda & Habe, 1952. The holotype in the British Museum matches the figure well. Refer to *T. commaculata* (**40**).

44. *Terebra cumingii* Deshayes, 1857

(Pl. 13, figs. 44a-b)
(Color pl. C, fig. 4)

1857 *Terebra cumingii* Deshayes, Journ. Conchyl 6:66, pl. 3, fig. 1; 1860 Reeve, Conch. Icon. 12: pl. 8, sp. 29; 1885 Tryon, Man. Conch. 7(5):28, pl. 8, fig. 49; 1967 Bratcher & Burch, Veliger 10(1): pl. 2, fig. 4; 1984 Aubry Terebridae pl. 11.

Description: Shell to 95 mm; color white to yellowish tan; outline of whorls concave in early whorls, almost flat later; subsutural band with double row of round bead-like nodes, the posterior being slightly longer; axial striae curved; spiral cords, 5 to 9, broken by axial striae, sometimes forming spiral rows of small nodes, the anterior row being somewhat enlarged on some individuals; aperture elongate-quadrate; columella recurved, with parietal callus on mature specimens.
Type locality: "China Seas."
Distribution: From Japan to Papua New Guinea.
Type: Lectotype BM(NH) no. 1979571 93.0 × 11.8 mm.
Discussion: As with all the species he described in 1857, Deshayes had a beautiful illustration of this species. It is unfortunate that he illustrated none of the many species he described in 1859. Refer to *T. triseriata* (**45**) for comparison with other long slender species.

45. *Terebra triseriata* Gray, 1834

(Pl. 14, figs. 45a-c)
(Color pl. C, fig. 5)

1834 *Terebra triseriata* Gray, Proc. Zool. Soc. Lond. pt. 2:62; 1844 Hinds in Sowerby, Thes. Conch. 1:171, pl. 45, fig. 119; 1967 Cenohorsky, Mar. Shells Pacific 1:206, pl. 52, fig. 389; 1972 Hinton, Shells New Guinea Cent. Pacific p. 46, pl. 23, fig. 34; 1974 Dance, Encycl. Shells, p. 219, upper fig., right column; 1978 Hinton, Guide Austral. Shells, pl. 56, fig. 14; 1979 Kay, Hawaiian Mar. Shells p. 403; 1982 Abbott & Dance, Compendium of Seashells p. 274; 1984 Aubry, Terebridae pl. 1.

1859 *Terebra praelonga* Deshayes, Proc. Zool. Soc. Lond., pt. 27: 315.

1963 *Terebra (Triplostephanus) triseriata* Gray, Shikama, Sel. Shells World Col. 1:126, pl. 102, fig. 10; 1966 Cernohorsky Jennings, Veliger 9(1):57, pl. 4, fig. 14.

1966 *Triplostephanus triseriata* Gray, Habe & Kosuge, Shells World Color 2:100, pl. 39, fig. 21.

1967 *Triplostephanus triseriatus* (Gray), Habe & Kosuge, Stand. Book Jap. Shells Color 3:101, pl. 40, fig. 11.

Description: Shell to 103 mm; color cream to pale orange or light brown; outline of whorls concave; protoconch of 4 conical whorls; subsutural band of 2 rows of round evenly spaced nodes, the posterior one larger, a row of smaller nodes forming immediately posterior to the subsutural band; axial sculpture of fine threads or striae only; spiral cords, 3 or 4, broken by axial threads, sometimes forming a file-like or bead-like surface; body whorl short, with a row of nodes forming a keel at the periphery; aperture quadrate, columella sharply curved.
Type locality: *T. triseriata*, none originally given. Ticao Is., Philippines, is here designated as the type locality; *T. praelonga*, "Port Curtis, Qld., Australia."
Distribution: The tropical Pacific; subtidal to 150 m.

Plate 13
FIGS. **41a,b**, *Terebra vicdani* Kosuge. **41a**, Cebu I., Philippines; 64.0 mm. **41b**, Middle whorls. **42a-c**, *T. lima* Deshayes. **42a**, Holotype British Museum (N.H.) No. 197965; 76.5 mm. **42b**, Holotype of *T. loebbekeana* Dunker, Zoological Museum, Humboldt University Berlin; 75.0 mm. **42c**, Middle whorls. **43a-c**, *T. pretiosa* Reeve. **43a**, Holotype British Museum (N.H.) No. 1979116; 140.0 mm. **43b**, Type-figure of *T. fujitai* Kuroda & Habe, from Reeve, 1860, pl.8, Fig. **30b.** **43c**, Middle whorls. **44a,b**, *T. cumingii* Deshayes. **44a**, Lectotype British Museum (N.H.) No. 1979571; 93.0 mm. **44b**, Lower whorls.

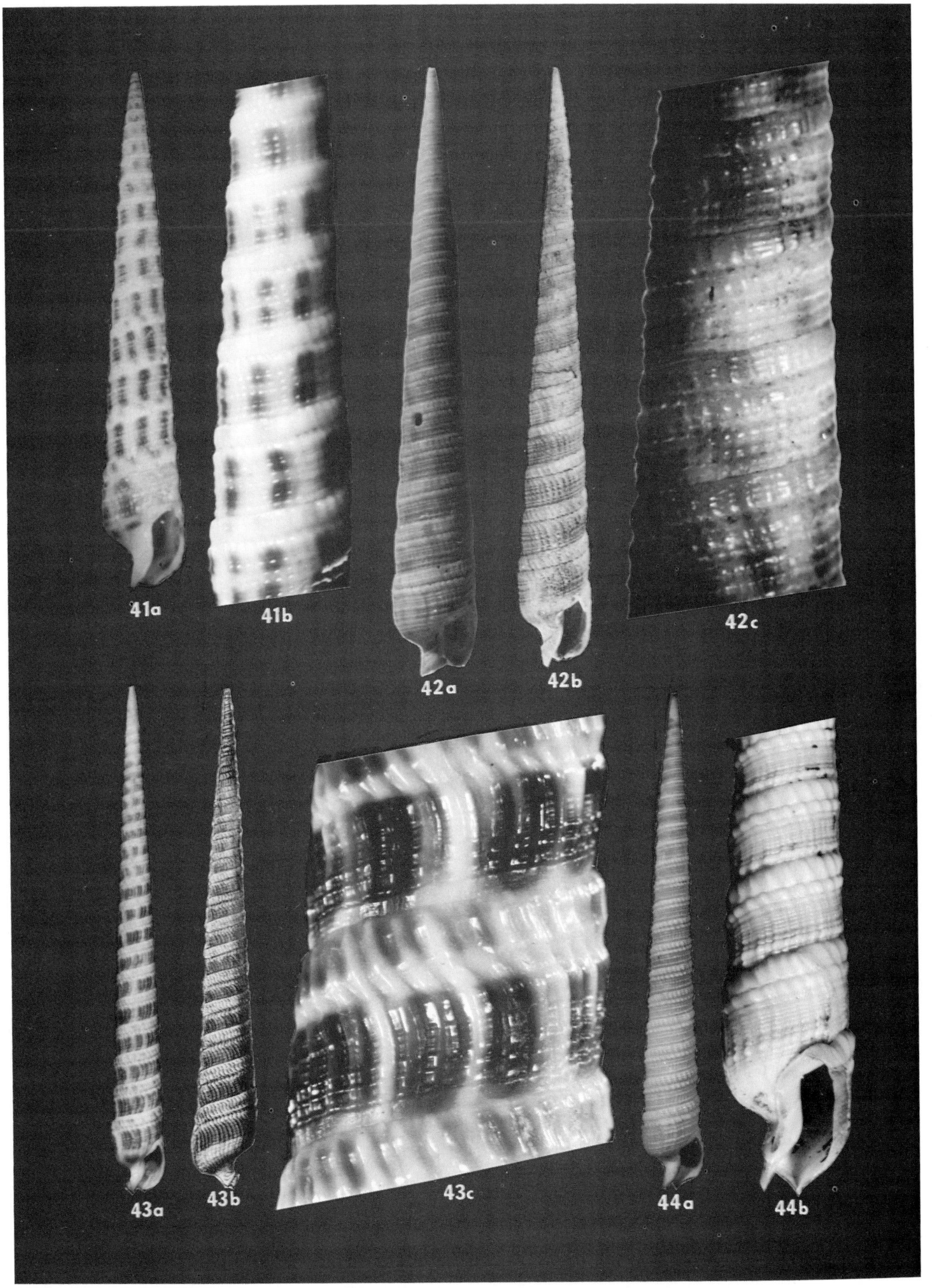

Plate 13. Indo-Pacific *Terebra*.

Type: *T. triseriata* lectotype BM(NH) no. 1874.9.9.1, 46.7 mm; *T. praelonga* lectotype (pl. 14, fig. 45b) BM(NH) no. 1979110, 92.4 mm.
Discussion: This is the type species of the subgenus of *Terebra*, *Triplostephanus* Dall, 1908, which he diagnosed as "Shell elongate, whorls medially constricted, whorls nodulous at both margins." This is the most slender and graceful of all the long terebrids. It is extremely similar in sculpture and color to *Terebra cumingii* **(44)** but it is much more slender, and the whorls are almost half as long. It also is more slender and has shorter whorls than those of *T. jenningsi* **(49)** or *T. caddeyi* **(50)**.

46. *Terebra boucheti* Bratcher, 1981

(Pl. 14, figs. 46a-c)
(Color pl. F, fig. 10)

1981 *Terebra boucheti* Bratcher, Veliger 23(4):329, figs. 1, 2; 1984 Aubry, Terebridae pl. 14.

Description: Shell to 62 mm; color shiny white or white blotched with orange-brown; outline of whorls concave with convex double subsutural band and projecting row of nodes; protoconch of 3½ extremely elongate whorls, the last whorl being twice the length of the preceding one; the first 2 whorls of the teleoconch being extremely angulate because of sharp nodes projecting from the center of whorl, with a row of smaller, less conspicuous nodes both anterior and posterior to projecting row, the 2 posterior rows of nodes becoming equal in size after the 4th whorl, forming the subsutural band; sculpture below double subsutural band consisting of 3 or 4 rows of nodes, connected both axially and spirally by shiny raised cords; nodes on band becoming elongate on later whorls; aperture quadrate; columella extremely recurved.
Type locality: "Philippine Is., 14 °16′N,120 ° 31′E.; 67 to 70 m.
Distribution: From the Philippine Is. to the Solomon Is.; 30 to 70 m.
Type: Holotype MNHN Paris; 47.5 × 7.7 mm.
Discussion: The most outstanding features of this species are the elongate protoconch whorls followed by extremely angulate early whorls caused by a keel of small nodes in the center of the first whorls of the teleoconch. This species was named for Philippe Bouchet of the MNHN Paris. For comparison refer to *T. insalli* **(47)**

47. *Terebra insalli* Bratcher & Burch, 1976

(Pl. 14, figs. 47a-b)

1976 *Terebra (Triplostephanus) insalli* Bratcher & Burch, Veliger 10(1):7, pl. 2, figs. 1-3; 1971 Mienis, Argamon 2(1-2):11; 1984 Aubry, Terebridae pl. 5.

Description: Shell to 83 mm; color pale beige; outline of whorls slightly convex; protoconch of 2½ whorls; subsutural band of double row of pearly nodes, the posterior being slightly larger; remainder of whorl with fine cancellate sculpture of somewhat unevenly spaced axial and spiral cords; body whorl short, with enlarged cord at periphery; aperture moderately elongate; columella recurved.
Type locality: "Gulf of Aqaba, Red Sea, 29° 32′20″N, 34°57′00″E."
Distribution: Eilat to southern tip of the Sinai, Red Sea; 3 to 6 m.
Type: Holotype CAS no. 12946; 59.9 × 8.0 mm.
Discussion: This shell was named for the collector D. C. Insall. The shell of *T. boucheti* **(46)** bears some resemblance to that of this species but has a protoconch of 3½ extremely long whorls and a sculpture of nodes anterior to the double rows of nodes below the suture. The shell of *T. elliscrossi* **(48)** is larger, white rather than beige, and has paired small brown dots.

48. *Terebra elliscrossi* Bratcher, 1979

(Pl. 14, fig. 48a-b)

1979 *Terebra elliscrossi* Bratcher, Veliger 22(1):65, fig. 1; 1984 Aubry, Terebridae pl. 14.

Description: Shell to 83 mm; color white with small round fulvous dots, usually in pairs, scattered at random; outline of whorls slightly concave; protoconch of 3½ conical whorls; subsutural band with double row of nodes, the

---

Plate 14
FIGS. **45a-c,** *Terebra triseriata* Gray. **45a,** Lectotype British Museum (N.H.) No. 1874.9.9.1.; 46.7 mm. **45b,** Lectotype of *T. praelonga* Deshayes, British Museum (N.H.) No. 1979110; 92.4 mm. **45c,** Lower whorls. **46a-c,** *T. boucheti* Bratcher. **46a,** Holotype Museum National d'Histoire Naturelle, Paris; 47.5 mm. **46b,** Middle whorls of holotype. **46c,** Middle whorls of paratype. **47a,b,** *T. insalli* Bratcher & Burch. **47a,** Holotype California Academy of Sciences, No. 12946; 59.9 mm. **47b,** Lower whorls. **48a,b,** *T. elliscrossi* Bratcher. **48a,** Holotype Los Angeles County Museum No. 1257; 42.0 mm. **48b,** Lower whorls.

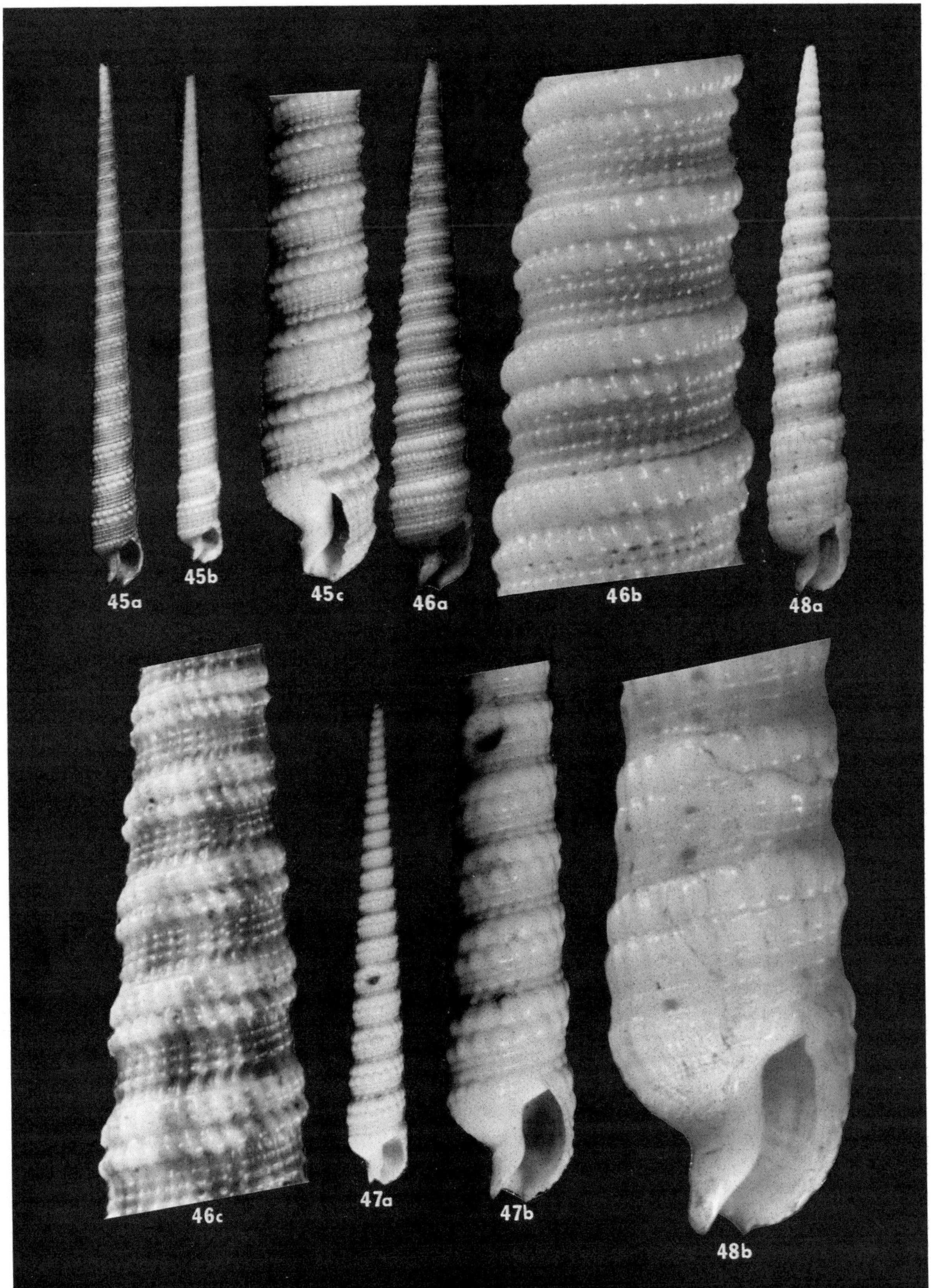

Plate 14. Indo-Pacific *Terebra*.

posterior one being smaller, band occupying about half of whorl; sculpture cancellate with axial and spiral cords of about equal strength forming small nodes at intersections on some individuals, others with finer sculpture less likely to form nodes; cancellate sculpture less obvious on some individuals with spiral sculpture predominating; body whorl with a row of slightly enlarged nodes at periphery; aperture quadrate, columella recurved.
Type locality: "Honolulu side of Pearl Harbor entrance, Honolulu, Hawaii, 21°17′N, 157°56′W, 300 m., sand and rubble bottom."
Distribution: From Madagascar to the Hawaiian Islands.
Type: Holotype LACM no. 1257; 42.0 × 7.2 mm.
Discussion: Some specimens of this species have many tiny fulvous dots, usually in vertical pairs; others have a few scarcely noticeable ones. Larger specimens tend to become less coarsely sculptured on later whorls. The species was named for Ellis Cross, former editor of the Hawaiian Shell News, who dredged the holotype. For comparison refer to *T. insalli* **(47).**

49. *Terebra jenningsi* R. D. Burch, 1965

(Pl. 15, figs. 49a-b)
(Color pl. C, fig. 6)

1965 *Terebra (Triplostephanus) jenningsi* Burch, Veliger 7(4):248, pl. 31, fig. 9; 1966 Cernohorsky & Jennings, Veliger 9(1):56, pl. 4, fig. 13, 13a.
1967 *Terebra jenningsi* Burch, Cernohorsky, Mar. Shells Pacific 1:202, pl. 50, fig. 371; 1972 Hinton, Shells New Guinea & Cent. Indo-Pacific p. 46, pl. 23, fig. 33; 1978 Hinton, Guide Australian Shells pl. 56, figs. 10, 10a; 1984 Aubry, Terebridae pl. 11.

Description: Shell to 80 mm, very slender; color ivory, beige, or pale yellow, often with a white subsutural band; outline of early whorls concave, flat to concave later; protoconch of 2½ conical whorls; subsutural band narrow, usually smooth, rarely with obsolete elongate nodes, a much narrower band occurring posterior to the weak suture, and occasionally anterior to weakly punctate subsutural groove; axial sculpture of arcuate growth striae, sometimes heavy; spiral grooves, 3 to 6, usually weak grooves, body whorl with a raised keel created by the continuation of the narrow convex band posterior to the suture; aperture short, quadrate; columella recurved; operculum yellow.
Type locality: "1 mile N.W. of Namotu Is., Fiji Is. 17°50′S, 177° 25′E; 6 to 8 m."
Distribution: From Madagascar to the Fiji Is; intertidal to 15 m.
Type: Holotype MCZ no. 251248; 76.2 × 8.5 mm.
Discussion: This species, named for Mr. A. Jennings of Fiji, is sometimes almost as slender as *T. triseriata* **(45)**, but it lacks the noded sculpture of that species. *T. caddeyi* **(50)** may be separated from this species by its higher shine, somewhat broader shell, fewer and deeper spiral grooves, and heavy parietal callus.

50. *Terebra caddeyi* Bratcher & Cernohorsky, 1982

(Pl. 15, figs. 50a-b)

1982 *Terebra caddeyi* Bratcher & Cernohorsky, Nautilus 96(2): figs. 4, 12.

Description: Shell to 53 mm; color shiny tan; outline of whorls straight; sculpture of 3 spiral bands, each defined by a spiral groove, posterior band narrow, without nodes, center band of obsolete nodes, anterior band flatter and scarcely noded; axial sculpture absent except for striae; body whorl with deep groove posterior to periphery, followed by 3 weak grooves; aperture small, quadrate; columella curved, with moderately heavy parietal callus; siphonal notch broad.
Type locality: "Korere Village, Rabaul, Papua New Guinea, 3 m."
Distribution: Known only from the type locality.
Type: Holotype LACM no. 1967; 52.7 × 6.9 mm.
Discussion: This shiny caramel-colored shell, named for its original collector, P. T. Caddey, is monochromatic from the protoconch to the parietal callus. For comparison refer to *T. triseriata* **(45).**

Plate 15
FIGS. **49a,b,** *Terebra jenningsi* Burch. **49a,** Holotype Museum of Comparative Zoology, Harvard, No. 251248; 76.2 mm. **49b,** Lower whorls of paratype. **50a,b,** *T. caddeyi* Bratcher & Cernohorsky. **50a,** Holotype Los Angeles County Museum No. 1967; 52.7 mm. **50b,** Middle whorls. **51a-c,** *T. marrowae* Bratcher & Cernohorsky. **51a,** Holotype Los Angeles Museum No. 1969; 26.1 mm. **51b,** Upper whorls and protoconch. **51c,** Middle whorls. **52a,b,** *T. fenestrata* Hinds. **52a,** Lectotype British Museum (N.H.) No.1968233; 32.7 mm. **52b,** Middle whorls.

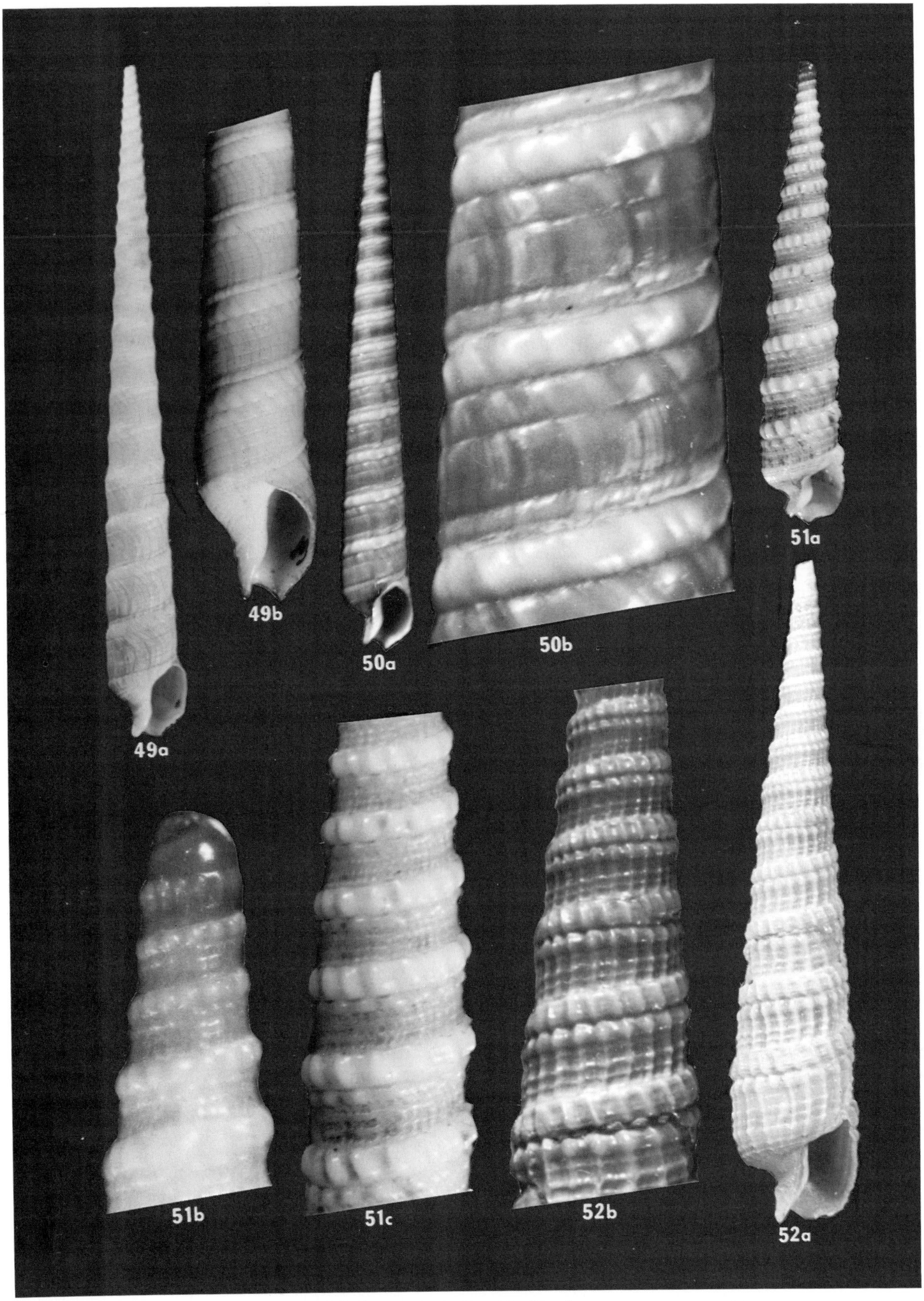

Plate 15. Indo-Pacific *Terebra*.

51. *Terebra marrowae*
Bratcher & Cernohorsky, 1982
(Pl. 15, figs. 51a-c)

1982 *Terebra marrowae* Bratcher & Cernohorsky, Nautilus 96(2): figs. 2, 9.

Description: Shell to 32 mm; color cream with a few small scattered orange-brown dots, protoconch and first whorls of teleoconch brown; outline of whorls concave on early whorls, flat on later ones; protoconch of 1½ very large, mamillate, blackish brown whorls; subsutural band of large pearl-like nodes, occasionally spotted between by orange brown, defined by narrow deep subsutural groove; sculpture on remainder of whorl consisting of another narrow, almost smooth, inconspicuous band which becomes slightly nodulous on last 2 whorls, followed by finely cancellate sculpture of 2 or 3 spiral cords broken by axial striae leaving tiny pinpoint pits; aperture short, quadrate; columella recurved.
Type locality: "Cleaverville, N.W. Australia; intertidal."
Distribution: Northwest Australia; intertidal.
Type: Holotype LACM no. 1969; 26.1 × 5.0 mm.
Discussion: The most outstanding feature of this species is the extremely large, blackish brown or purplish brown, mamillate protoconch with a lighter brown or purplish area extending through two to four whorls of the teleoconch, becoming progressively lighter. The sculpture of some specimens is finer than that of others, with smaller nodes on the subsutural band. The species was named for its original collector, Lorna Marrow, of Australia.

52. *Terebra fenestrata* Hinds, 1844
(Pl. 15, figs. 52a-b; pl. 16, figs. 52c-e)
(Color pl. E, fig. 15)

1844 *Terebra fenestrata* Hinds, Proc. Zool. Soc. Lond. pt. 11:153; 1844 Hinds in Sowerby, Thes. Conch. I:176, pl. 44, fig. 86; 1860 Reeve, Conch. Icon. 12, pl. 22, sp. 114; 1967 Cernohorsky, Mar. Shells Pacific I:200, pl. 50, fig. 365; 1972 Hinton, Shells New Guinea & Cent. Indo-Pacific p. 46, pl. 23, fig. 31; 1978 Hinton, Guide Australian Shells pl. 57, fig. 8; 1984 Aubry, Terebridae pl. 7.

1850 *Terebra caelata* Adams & Reeve, Zool. Voy. H.M.S. "Samarang" p. 30, pl. 10, fig. 22; 41.0 mm; 1860 Reeve, Conch. Icon. 12: pl. 15, sp. 64.

1850 *Terebra torquata* Adams & Reeve, Zool. Voy. H. M. S. Samarang p. 30, pl. 10, fig. 13; type BM(NH); 45.5 mm); 1860 Reeve, Conch. Icon. 12: pl. 15, sp. 69; 1917 Hirase, Terebridae Jap. Emp. p. 11, pl. 3, figs. 30-32.

1961 *Cinguloterebra? torquata* (Adams & Reeve), Oyama, Venus, Jap. Journ. Malac. 21(2):183.

1964 *Cinguloterebra torquata* (Adams & Reeve), Habe, Shells W. Pacific Color 2:128, pl. 40, fig. 18.

1966 *Terebra (Strioterebrum) fenestrata* Hinds, Cernohorsky & Jennings, Veliger 9(1):53, pl. 6, fig. 37.

1967 *Myurella torquata* (Adams & Reeve), Habe & Kosuge, Standard Book Jap. Shells Color 3: 101, pl. 40, fig. 5.

1975 *Myurella pretiosa* Okutani, Moll. Japan p. 155, fig. 2 [non *pretiosa* Reeve].

Description: Shell to 79 mm; color white, beige, brown, or mottled with light brown; outline concave; protoconch mamillate, larger than first whorls of teleoconch; subsutural band of double row of round or elongate nodes, the posterior usually larger, 3 to 5 spiral cords may or may not ornament one or both rows of nodes; sculpture on remainder of whorl extremely cancellate with numerous axial and spiral cords crossing and forming small nodes at intersections, either may vary in strength or in uniformity; aperture quadrate; columella recurved.
Type locality: *T. fenestrata*, "San Nicholas, Island of Cebu, Philippines"; *T. caelata*, "Philippine Is."; *T. torquata*, "China Sea."
Distribution: Japan to the Philippine Is. and Fiji Is.; subtidal to 60 m.
Type: *T. fenestrata* lectotype BM(NH) no. 1968233, 32.7 mm; *T. caelata* holotype BM(NH) no. 979130, 41.0 mm; *T. torquata* questionable type BM(BH), 45.5 mm.
Discussion: This is an extremely variable species. Specimens having round nodes and no spiral sculpture on the subsutural band gradually integrating with those having elongate nodes and numerous spiral grooves on the band, or individuals may have the former sculpture on

---

Plate 16
FIGS. **52c-e,** *Terebra fenestrata* Hinds. **52c,** Paralectotype British Museum (N.H.) No. 1968233; 31.1 mm. **52d,** Type-figure of *T. torquata* Adams & Reeve from Adams & Reeve, 1848-50, pl.10,fig.13. **52e,** Middle whorls of questionable holotype of *T. caelata* Adams & Reeve. **53a,b,** *T. nodularis* Deshayes. **53a,** Middle whorls of lectotope. **53b,** Ohau, Hawaiian Is.; 32.5 mm. **54a,b,** *T. waikikiensis* Pilsbry. **54a,** Ohau, Hawaiian Is.; 35.0 mm. **54b,** Middle whorls of holotype. **55a-c,** *T. mamillata* Watson. **55a,** Lectotype British Museum (N.H.) No. 87.2.9.1159.; 33.7 mm. **55b,** Paralectotype British Museum (N.H.) No. 82.2.9.1159.; 29.6 mm. **55c,** Holotype of *T. lillianae* Whitney, Los Angeles County Museum No. 1724; 42.2 mm.

Plate 16. Indo-Pacific *Terebra*.

early whorls and the latter on mature ones. *T. boucheti* **(46)** when mottled with reddish brown may look quite similar to some forms of *T. fenestrata*, but *T. boucheti* has a protoconch of 3½ extremely elongate whorls, and a shiny shell. *T. fenestrata* does not. *T. fenestrata* is a species that emerges directly from the eggs, while *T. boucheti* has planktonic development.

53. *Terebra nodularis* Deshayes, 1859

(Pl. 16, figs. 53a-b)

1859 *Terebra nodularis* Deshayes, Proc. Zool. Soc. Lond. pt. 27: 295; 1860 Reeve, Conch. Icon. 12: pl. 22, fig. 105; 1978 Cernohorsky, Trop. Pacific Mar. Shells p. 147, pl. 52, fig. 7; 1979 Kay, Hawaiian Mar. Shells p. 401, fig. 129F; 1984 Aubry, Terebridae pl. 9.

Description: Shell to 40 mm; color white to tan; outline of whorls slightly concave; protoconch of 1½ whorls; subsutural band of double row of nodes, the posterior being larger, divided by a deep narrow groove; axial ribs weak, contiguous with nodes on band; spiral grooves fine, 3 to 5, lightly, crossing ribs; aperture quadrate; columella short, recurved, with parietal callus.
Type locality: "Hawaiian Is."
Distribution: Endemic to the Hawaiian Is.; 2 to 130 m.
Type: Lectotype BM(NH) no. 1979106; 34.0 × 6.0 mm.
Discussion: This species, found only in the Hawaiian Islands, somewhat resembles *T. mamillata* **(55)** which has a broader shell, flat outline, very large mamillate protoconch, and color pattern of yellow and white stripes.

54. *Terebra waikikiensis* Pilsbry, 1921

(Pl. 16, figs. 54a-b)
(Color pl. D, fig. 14)

1921 *Terebra waikikiensis* Pilsbry, Proc. Acad. Nat. Sci. Philadelphia 69:305, pl. 12, fig. 12; 1960 Weaver, Hawaiian Mar. Moll.1(3):10, pl. 3, fig. lower right; 1974 Quirk & Wolfe, Sea shells Hawaii, pl. 14, fig. 10; 1978 Cernohorsky, Trop. Mar. Shells p. 147, pl. 52, fig. 8; 1979 Kay, Hawaiian Mar. Shells p. 403, fig. 130j; 1981 Kosuge, Bull. Inst. Malac. Tokyo 1(6):93, pl. 31, fig. 7.

Description: Shell to 60 mm; color shiny white with vertical pairs of orange-brown, regularly spaced dots between subsutural band and following suture; outline of whorls flat, turreted; protoconch of 2 conical whorls; noded subsutural band followed by a second row of nodes; remainder of whorl with axial ribs, 22 to 26 on penultimate whorl, crossed by 2 or 3 coarse spiral cords, forming rectangular grooves or pits between intersections, aperture moderately quadrate; columella with fold and moderate parietal callus in mature specimens.
Type locality: "Off Waikiki, Oahu, Hawaii, 25-50 fms."
Distribution: From the Philippines to the Hawaiian Is. In the Hawaiian Is. the species occurs at depths of 50 to 200 m, often in beds of *Pinna* shells.
Type: Holotype ANSP no. 117030; 29.7 × 6.8 mm.
Discussion: This uncommon shell usually is dredged from deep water. *T. elliscrossi* **(48)** also has paired brown dots, but they are smaller and scattered at random, not regularly spaced. It is slightly longer and is more slender, reaching a length of 82 mm. With holotypes of approximately the same width, the length of *T. elliscrossi* is 42 mm while that of *T. waikikiensis* is 29.7 mm.

55. *Terebra mamillata* Watson, 1886

(Pl. 16, figs. 55a-c)

1886 *Terebra (Myurella) mamillata* Watson, Rept. Sci. Res. Voy. H.M.S. *Challenger*, Zool. 15: 381, pl. 16, fig. 1.
1976 *Terebra lillianae* Whitney, Veliger 18(4):381, figs. 1-4.

Description: Shell to 46 mm; color cream with a broad brownish yellow band on the anterior third of whorls; outline of whorls flat; protoconch of 1½ mamillate whorls; subsutural band, occupying about two thirds of a whorl, consisting of double row of nodes, occasionally with grooves in interspaces; remainder of whorl sometimes having 2 additional rows of nodes, sometimes with about 20 to 23 axial ribs crossed by 2 to 5 spiral cords, forming small nodes at intersections; aperture quadrate; columella short, extremely recurved; interior columella with two heavy plications, scarcely visible exteriorly.
Type locality: *T. mamillata*, "Philippine Is., 12° 43′S,122°10′E; 100 to 115 fms, green mud;" *T. lillianae*, "off Mossuril, Mozambique," error.
Distribution: Apparently endemic to the Philippine Is.; 177 to 350 m.
Type: *T. mamillata* lectotype BM(NH) no.

87.2.9.1159, 33.7 mm; *T. lillianae* holotype (pl. 16, fig. 55c) LACM no. 1724, 42.2 × 9.1 mm.
Discussion: The large mamillate protoconch and the distinctive color pattern would make it difficult to confuse with any other terebrid species. Faded beach shells with the protoconch missing might possibly be mistaken for *T. fenestrata* (**52**) which has shorter whorls, a much more concave outline in the early whorls, and does not have the broad, mamillate protoconch. The type locality of the junior synonym, *T. lillianae* is probably erroneous.

56. *Terebra swobodai* Bratcher, 1981

(Pl. 17, figs. 561-b)

1981 *Terebra swobodai* Bratcher, Veliger 23(4):330, figs. 5, 6.

Description: Shell to 27 mm, slender; color beige; outline of whorls angulate; subsutural band noded; sculpture coarsley cancellate, with 2 to 4 spiral rows of heavy cords bisecting axial cords of equal strength, forming bead-like nodes at intersections and square pits between; body whorl with 2 rows of spiral cords crossing ribs to form strong nodes and 3 rows of square pits ending at the periphery; aperture narrow and elongate; columella slightly curved.
Type locality: "Zamboanga, Mindanao, Philippines."
Distribution: Philippine Is.
Type: Holotype LACM no. 1185; 24.3 × 4.3 mm.
Discussion: The most outstanding feature of this species is the large square pits formed by the crossing of heavy spiral and axial cords. In all specimens examined, the axial and spiral cords are of equal strength. All specimens of the type lot were removed from specimens of *Xenophora pallidula* (Reeve) caught in net traps. Subsequent to the original description, live taken specimens in the collections of the USNM and the MNHNP have been examined. This species was named in honor of the Californian shell collector, Edward Swoboda.

57. *Terebra bathyrhaphe* E. A. Smith, 1875

(Pl. 17, figs. 57a-e)

1875 *Terebra (Myurella) bathyrhaphe* E. A. Smith, Ann. & Mag. Nat. Hist. ser. 4, 15:415.
1877 *Terebra bathyrhaphe* E. A. Smith, Ann. & Mag. Nat. Hist ser. 4, 19:226; 1917 Hirase, Terebridae Jap. Emp. p. 10, pl. 3, fig. 27, 29.
1877 *Terebra persica* E. A. Smith, Ann. & Mag. Nat. Hist. ser. 4, 19: 225.
1886 *Terebra bathyrhaphe* (sic) E. A. Smith, Watson, Rept. Sci. Res. Voy. H.M.S. *Challenger* 15:377, pl. 14, fig. 9.
1925 *Terebra constricta* Thiele, Wiss. Ergeb. Deut. Tief. Exped. "Valdiva" 17:347, pl. 30, fig. 5.
1961 *Granuliterebra bathyrhaphe* (E. A. Smith), Oyama, Moll. Shells, Resources Exploitation Inst. 5: *Granuliterebra* figs. 1, 2. 1964 Habe, Shells W. Pacific. Color 2:127, pl. 40, fig. 16; 1967 Habe & Kosuge, Stand. Book Jap. Shells Color 3:101, pl. 40, fig. 12; 1975 Okutani, Moll. Japan p. 154, fig. 5.
1968 *T. (Strioterebrum) tricincta* Smith, Ray, Indian Mus. Bull. Calcutta 3(1/2):27, text fig. 25. (non E. A. Smith, 1877)

Description: Shell to 36 mm; color yellowish beige to gray or beige-brown, often with a slight chalky deposit; outline of whorls concave; protoconch of 4 convex whorls; sculpture extremely variable; subsutural groove absent on most individuals, large nodes at posterior of ribs give the effect of a subsutural band; axial ribs heavily noded at both ends, usually becoming somewhat obsolete in center, on some individuals fading completely, on others scarcely at all, occasionally developing a third row of nodes; spiral sculpture may be absent or there may be numerous spiral cords on the subsutural band and remainder of the whorl, to 16 on penultimate whorl; periphery of the body whorl often with row of nodes; aperture quadrate; columella recurved.
Type locality: *T. bathyrhaphe*, "Gulf of Yedo, Japan; 6 to 25 fms." *T. persica*, "Persian Gulf;" *T. constricta*, "Padang," Sumatra.
Distribution: From the Persian Gulf and Arabian Sea to Japan and Northern Australia; 12 to 100 m.
Type: *T. bathyrhaphe* lectotype BM(NH) no. 1873.8.6.10, 24.8 × 5.0 mm; *T. persica* lectotype (pl. 17, fig. 57b) BM(NH) no. 873.7.5.4, 11.4 × 3.0 mm; *T. constricta* lectotype, MNHUB, 7.4 × 1.8 mm.
Discussion: The texture and dull color of this species, which often has a "worn" look, tend to make even fresh specimens look like fossils, although an occasional specimen may be seen in good condition. This is the type species of the genus *Granuliterebra* Oyama, 1961. For comparison refer to *T. tricincta* (**58**).

58. *Terebra tricincta* E. A. Smith, 1877

(Pl. 17, figs. 58a-b)

1877 *Terebra tricincta* E. A. Smith, Ann & Mag. Nat. Hist. ser. 4, 15:225; 1913 Schepman, Siboga Exped. 49 I (5): 376, pl. 25, fig. 11.

1961 *Granuliterebra tokunagai* Oyama & Takemura, Moll. Shells Resources Exploitation Inst. 5: *Granuliterebra* fig. 3.

Description: Shell to 11 mm; color dirty white; outline of whorls concave; subsutural band of large nodes; axial cords very thin, almost obsolete, widely spaced; two thin spiral cords, one immediately anterior to the subsutural band, the other immediately posterior to the suture; bands noded where the axial cords intersect; body whorl with a noded band at the periperiphery, followed by a second row of nodes; aperture quadrate, columella curved.
Type locality: "Persian Gulf."
Distribution: Persian Gulf to Japan and northern Australia.
Type: Lectotype BM(NH) no. 1873.7.5.6; 10.9 × 2.8 mm.
Discussion: This species is very similar to *T. bathyrhaphe* (**57**) and further research could prove them to be conspecific. *T. bathyrhaphe* lacks the band anterior to the subsutural band, and the other two bands are connected by well-developed ribs rather than by thin cords.

59. *Terebra puella* Thiele, 1925

(Pl. 17, fig. 59)

1925 *Terebra puella* Thiele, Wiss. Ergeb. Tief. Exped. Valdivia 17:347, pl. 30, fig. 4.

Description: Shell to 18 mm; color tan with a brown nebulous stripe running through subsutural area and white or light band at periphery of body whorl followed by brown at the anterior of shell; outline of whorls straight, slightly turreted; protoconch of 1½ small dome shaped whorls; subsutural band absent; axial ribs, about 19 on penultimate whorl; spiral sculpture absent; outer lip thin; aperture elongate; columella recurved.
Type locality: "Padang," Sumatra, Indonesia.
Distribution: Indonesia.
Type: Holotype MNHUB, 18.4 × 3.0 mm.
Discussion: We have seen no specimens other than the unique holotype.

60. *Terebra irregularis* Thiele, 1925

(Pl. 17, fig. 60)

1925 *Terebra irregularis* Thiele, Wiss. Ergeb. Deut. Tief. Exped. "Valdivia" 17:346, pl. 30, fig. 3.

Description: Shell to 20 mm; color fulvous brown with white subsutural band; outline of whorls faintly convex; protoconch of 3½ whorl; subsutural band with a "squared-off" appearance, decorated with slender inconspicuous ribs, defined by groove, becoming obsolete on body whorl; axial ribs obsolete, slanting; spiral sculpture absent; sutures deep; outer lip and columella broken on holotype.
Type locality: Padang, Sumatra, Indonesia.
Distribution: Known only from the type locality.
Type: Holotype MNHUB, 19.5 mm.
Discussion: This species is known only from the holotype and has not been reported since its original description. It may prove to be a pathological specimen of another species.

61. *Terebra bifrons* Hinds, 1844

(Pl. 18, figs. 61a-b)
(Color pl. E, fig. 21)

1844 *Terebra bifrons* Hinds, Proc. Zool. Soc. Lond. pt. 11:155; 1844 Hinds in Sowerby, Thes. Conch. 1:174, pl. 43, fig. 57; 1917 Hirase, Terebridae Jap. Emp. p. 18, pl. 4, figs. 45, 46.

1928 *Terebra bellanodosa* Grabau & King, Peking Soc. Nat. Hist. Hand-book no;. 2:119, pl. 214, pl. 9, fig. F80.

1961 *Noditerebra (Pristiterebra) bifrons* (Hinds), Oyama, Molluscan Shells, Resources Exploitation Inst. 5: *Noditerebra* (2) figs. 2,10; 1962 Kira, Shells W. Pacific Color 1:110, pl. 39, fig. 11.

Description: Shell to 64 mm, sturdy; color brownish; outline of whorls almost straight for

---

Plate 17
FIGS. **56a,b,** *Terebra swobodai* Bratcher. Philippines. **56a,** Holotype Los Angeles County Museum No. 1185; 24.3 mm. **56b,** Middle whorls. **57a-e,** *T. bathyraphe* E. A. Smith. **57a,** Lectotype British Museum (N.H.) No. 1873.8.6.10.; 24.8 mm. **57b,** Lectotype of *T. persica* E. A. Smith, British Museum (N.H.) No. 1873.7.5.4.; 11.4 mm. **57c,** Middle whorls. **57d,** Juvenile lectotype of *T. constricta* Thiele, Zoological Museum Humboldt University, Berlin; 7.4 mm. **57e,** Keppel Bay, Queensland, Australia; 18.6 mm. **58a,b,** *T. tricincta* E. A. Smith. **58a,** Lectotype British Museum (N.H.) No. 1873.7.5.6.; 10.9 mm. **58b,** Paralectotype, same data; 10.8 mm. **59,** *T. puella* Thiele. Holotype Zoological Museum Humboldt University, Berlin; 18.4 mm. **60,** T. irregularis Thiele. Holotype Zoological Museum, Humboldt University, Berlin; 19.5 mm.

Plate 17. Indo-Pacific *Terebra*.

first 9 or 10 whorls, then becoming moderately convex; protoconch multispiral; subsutural band of axial ribs, defined by a space rather than a groove; axial ribs on early whorls, becoming axially and spirally aligned rows of pustules, sometimes coalescing again into weak, often interrupted ribs; spiral sculpture usually absent, occasionally showing as narrow weak cords bisecting ribs, forming pustules; body whorl elongate, with 6 to 10 spiral rows of pustules below the subsutural band; aperture quadrate; columella almost straight.
Type locality: "Japan, sandy mud."
Distribution: Japan; 9 to 20 m.
Type: Holotype BM(NH) no. 1968237; 51.0 mm.
Discussion: *T. bellanodosa* Grabau & King is questionably included as a synonym. We have not seen the type, but the figure and description seem to fit this species. *T. pustulosa* **(62)** is another Japanese species with a brownish shell ornamented with pustules, but the shell is more delicate, almost translucent and has a keel-like cord at the periphery of the body whorl.

62. *Terebra pustulosa* (E. A. Smith, 1879)

(Pl. 18, figs. 62a-b)
(Color pl. E, fig. 20)

1873 *Myurella granulosa* E. A. Smith, Ann. & Mag. Nat. Hist. Ser. 4, 11:268 [non *T. granulosa* Lamarck, 1822].
1879 *Terebra pustulosa* E. A. Smith, Proc. Zool. Soc, Lond. p. 186 [nom nov. pro *Myurella granulosa* E. A. Smith, 1873].

Description: Shell to 38 mm, thin and delicate; color yellowish brown; outline of early whorls straight, later ones convex; subsutural band flat, extremely narrow, defined by a row of spirally elongate pustules instead of a groove; remainder of whorl with 2 or 3 spiral rows of pustules, axially aligned, in some individuals forming obsolete ribs; additional spiral sculpture of a heavy cord immediately posterior to the suture and at the periphery of the body whorl, also a few fine spiral threads occurring on some specimens; aperture moderately quadrate; outer lip thin; columella almost straight.
Type locality: "Japan."
Distribution: Japan; to 37 m.
Type: Lectotype BM(NH) no. 1979183; 24.6 mm.
Discussion: This species has a delicate shell with even the larger specimens looking slightly translucent. See *Terebra bifrons* **(61)** for comparison.

63. *Terebra miranda* (E. A. Smith, 1873)

(Pl. 18, figs. 63a-b)
(Color pl. D, fig. 13)

1873 *Myurella miranda* E. A. Smith, Ann. & Mag. Nat. Hist. ser. 4 11:267.

Description: Shell to 35 mm; color brownish tan; outline of first few whorls angulate, remainder convex; subsutural band narrow and noded on early whorls, flat, without nodes, inconspicuous later; sculpture completely cancellate throughout with numerous, unevenly spaced axial cords crossed by about 6 spiral cords forming little bead-like nodes at intersections; sutures deep; body whorl elongate; aperture elongate; columella straight; siphonal fasciole with sharp posterior keel and sometimes a rounded keel toward center.
Type locality: "Malacca, Indonesia."
Distribution: Indonesia to Thailand; 6 to 10 m.
Type: Lectotype BM(NH) no. 1979136; 24.7 × 5.8 mm.
Discussion: Both specimens in the type lot are white but have a faded look and probably were tan when fresh like other specimens in the UZMC. They were collected in the Gulf of Thailand. There is no other species with which this could be mistaken.

64. *Terebra albida* Gray, 1834

(Pl. 18, figs. 64a-b)

1834 *Terebra albida* Gray, Proc. Zool. Soc. Lond. pt. 2:63; 1844 Hinds in Sowerby, Thes. Conch. 1:158; pl. 43, fig. 56; Reeve, Conch. Icon. 12: pl. 17, sp. 79; 1932 Cotton & Godfrey, S. Aust. Nat. 13(2):64, pl. 3, fig. 7.
1853 *Acus albida* (Gray), H. & A. Adams, Gen. Rec. Moll. 1:224.

Plate 18
FIGS. **61a,b,** *Terebra bifrons* Hinds. **61a,** Holotype British Museum (N.H.) No. 1968237; 51.0 mm. **61b,** Middle whorls. **62a,b,** *T. pustulosa* E. A. Smith. **62a,** Lectotype British Museum (N.H.) No. 1979183; 24.6 mm. **62b,** Paralectotype, same data; 21.8 mm. **63a,b,** *T. miranda* (E. A. Smith). **63a,** Paralectotype British Museum (N.H.) No. 1979136; 24.7 mm. **63b,** Specimen from unknown locality; 23.4 mm. **64a,b,** *T. albida* Gray. **64a,** Holotype British Museum (N.H.) No. 74.11.10.4.; 34.3 mm. **64b,** Holotype of *Nototerebra flindersi* Cotton, South Australian Museum, Adelaide, No. D-14435; 23.8 mm. **65a,b,** *T. affinis* Gray. **65a,** Lectotype British Museum (N.H.) No. 1874.11.10.17.; 42.5 mm. **65b,** Middle whorls of lectotype.

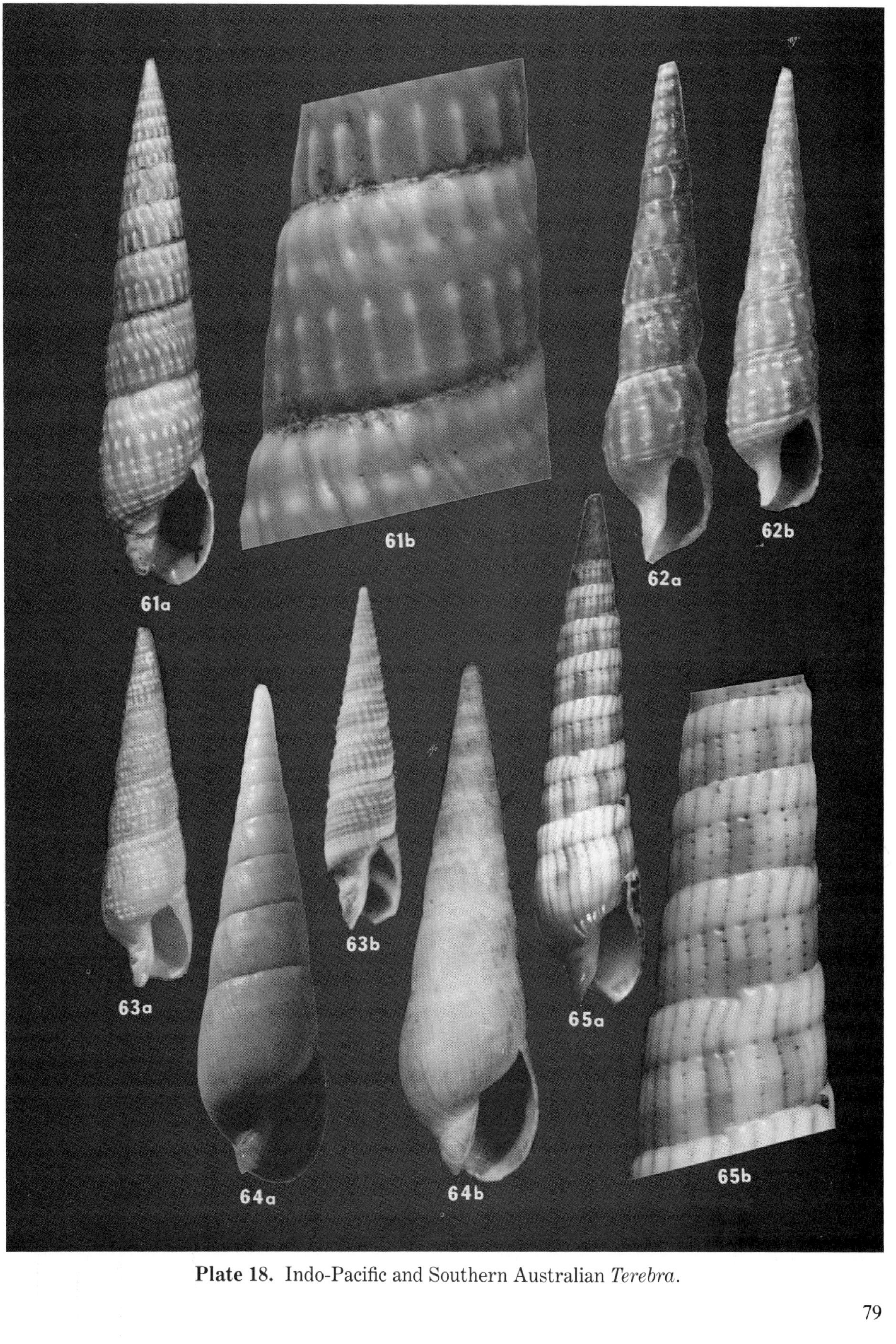

Plate 18. Indo-Pacific and Southern Australian *Terebra*.

1952 *Nototerebra albida* (Gray), Cotton, Trans. Royal Soc. S. Aust. 75:40.

1952 *Nototerebra flindersi* Cotton, Trans. Royal Soc. S. Aust. 75:41.

Description: Shell to 46 mm; color ivory-white to beige, sometimes with rusty brown subsutural spots or splotches; outline of whorls convex; protoconch 1½ dome shaped whorls; subsutural band weakly indicated, in some individuals band is depressed and quite distinct; teleoconch sculptured only with fine axial striae, with occasionally heavy axial growth striae gathered into groups forming flat ribs most noticeable anterior to the suture; body whorl slightly inflated; aperture quadrate; columella curved; siphonal fasciole coarsely striate, with a sharp keel.
Type locality: *T. albida*, none originally given. Portsea, Victoria, Australia is here designated as the type locality; *T. flindersi*, "Beachport, S. Australia, 150 fms."
Distribution: Victoria to Tasmania and Esperance, West Australia; to 275 mm.
Type: *T. albida* holotype BM(NH) no. 74.11.10.4, 34.3 mm; *T. flindersi* holotype (pl. 18, fig. 64b) SAM no. D-14435, 23.9 mm.
Discussion: This species has been confused with *T. buccinulum* Deshayes which is not a *Terebra* but is a *Bullia*. The lack of a terebrid siphonal fasciole makes the *Bullia* easy to distinguish. *T. albida* is one of the few terebrid species that appears to have no sculpture. *T. albida* is the type species of two genera, *Acus* H. and A. Adams and *Nototerebra* Cotton, 1947.

65. ***Terebra affinis*** Gray, 1834

(Pl. 18, figs. 65a-b; pl. 19, figs. c-d)

1833 *Terebra striata* Quoy & Gaimard, Voy. l'Astrolabe Zool. 2: 468, pl. 36, figs. 23, 24. [non Basterot, 1825].

1834 *Terebra affinis* Gray, Proc. Zool. Soc. Lond. pt. 2:60; 1917 Hirase, Terebridae Jap. Emp. p. 4, pl. 2, fig. 5; 1960 Weaver, Hawaiian Mar. Moll. 1(5):18, pl. 5, fig. top left; 1966 Cernohorsky & Jennings, Veliger 9(1):45, pl. 5, fig. 26; 1967 Cernohorsky, Mar. Shells Pacific 1:195, pl. 49, fig. 394; 1969 Rudman, Veliger, 12(1):59 (anatomy); 1971 Bratcher & Burch in Keen, Seashells Trop. W. America, 2nd ed: 670, fig. 1519; 1972 Hinton, Shells New Guinea & Cent. Indo-Pacific p. 46, pl. 23, fig. 24; 1975 Salvat & Rives, Coq. Polynésie p. 360, fig. 383; 1979 Kay, Hawaiian Mar. Shells p. 391, fig. 130e; 1982 Abbott & Dance, Compendium of Seashells p. 275; 1984 Trondle, Xenophora 15:7, figs. 7,2; 1984 Aubry, Terebridae pl. 10.

1834 *Terebra flava* Gray, Proc. Zool. Soc. Lond. pt. 2:60

1839 *Terebra pertusa* Sowerby, (pars) Kiener, Spéc. Gén. Icon. Coq. Viv. 8:34, pl. 11, fig. 24, 24b only. [non Born, 1778].

1859 *T. puncticulata* Deshayes, Proc. Zool. Soc. Lond. pt. 27:280.

1859 *Terebra peasii* Deshayes, Proc. Zool. Soc. Lond. pt. 27:302.

1960 *Terebra flavescens* Deshayes, Weaver, Hawaiian Mar. Moll. 1(7):26, pl. 7, fig. bottom right. [non Deshayes, 1859].

1964 *Terebra (Decorihastula) affinis* Gray, Shikama, Sel. Shells World Color pl. 69, fig. 18.

1975 *Decorihastula affinis* Okutani, Gakken illust. Nat. Encyclo. Moll. Japan, p. 154, fig. 15.

1984 *Terebra affinis affinis* Salvat & Rives, Shells of Tahiti p. 138, pl. 36, fig. 2.

1984 *Terebra affinis puncticulata* Salvat & Rives, Shells of Tahiti p. 138, pl. 36, fig. 1.

Description: Shell to 58 mm; color creamy white ornamented with blotches of brown or yellowish orange, interspaces between ribs often brownish, body whorl usually with a light tan band; outline of whorls convex; protoconch of 3½ conical whorls; subsutural band marked with deep punctations between ribs; axial ribs, usually broad, with only a punctate line between, occasionally narrow with broader interspaces filled with punctations; aperture semi-elongate; columella straight; operculum reddish brown, transluscent.
Type locality: *T. affinis*, none originally given. Ponape, Caroline Islands, Micronesia, is here designated as the type locality; *T. striata* Q. & G., "Caroline Is.;" *T. flava*, none; *T. puncticulata*, none; *T. peasii*, "Hawaiian Is."
Distribution: From the Red Sea throughout the tropical Indo-Pacific to the Gulf of California, Mexico; from the intertidal zone to 100 meters.
Type: T. affinis lectotype BM(NH) no.

Plate 19

FIGS. **65c,d**, *Terebra affinis* Gray. **65c,** Holotype of *T. peasii* Deshayes, British Museum (N.H.) No. 1979207; 45.6 mm. **65d,** Lower whorls of lectotype of *T. puncticulata* Deshayes. **66a,b**, *T. columellaris* Hinds. **66a,** Lectotype British Museum (N.H.) No. 1968228; 43.0 mm. **66b,** Holotype of *T. carnicolor* Preston, Zoological Survey India, Calcutta, No. M3997/1; 22.0 mm. **67a-c**, *T. kilburni* Burch. **67a,** Holotype Museum Comparative Zoology, Harvard, No. 251247; 32.0 mm. **67b,** Lectotype of *T. areolata* Adams & Reeve, British Museum (N.H.) No. 1979129; 24.0 mm. **67c,** Middle whorls. **68**, *T. nebulosa* Sowerby, Fiji Is.; 75.0 mm. **69a,b**, *T. undulata* Gray. **69a,** Lectotype British Museum (N.H.) No. 1874.11.10.5.; 33.6 mm. **69b,** Juvenile holotype of *T. undulata* var. *laevior* Schepman, Zoological Museum, University Amsterdam; 17.0 mm.

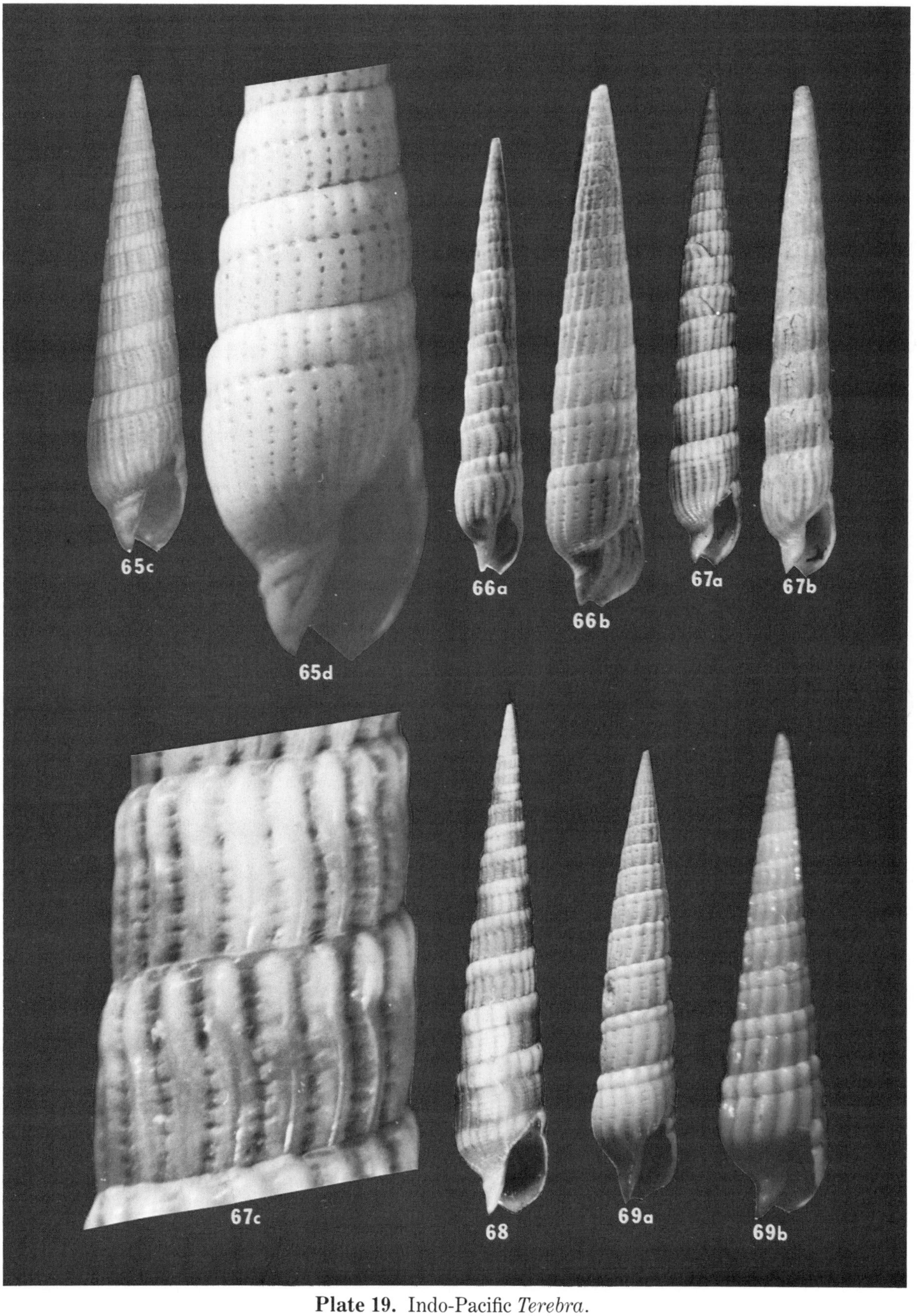

Plate 19. Indo-Pacific *Terebra*.

1874.11.10.17, 42.5 mm; *T. flava* holotype BM(NH no. 1979122, 23.4 mm; *T. puncticulata* lectotype (p. 19, fig. 65d) MNHNP, 24.6 mm; *T. peasii* holotype (pl. 19, fig. 65c) BM(NH) no. 1979207, 45.6 mm.

Discussion: This common species, with neither radula nor poison gland, is variable in apical angle, sculpture and markings. While axial ribs are separated by obviously punctate lines, they may be separated by lines which look merely narrow and impressed except under a microscope. Light brown markings may be mottled or may appear as narrow axial lines or as faint spiral lines. A single population may contain both slender and obese forms. The form Deshayes described as *T. puncticulata* is an extreme one found rarely in Hawaii with deep, heavily marked punctations. It intergrades into the slightly less extreme form Deshayes called *T. peasii*, found more commonly in Hawaii, both of which intergrade into the more typical form, also found in Hawaii but more rarely. The extreme forms are occasionally found elsewhere in the Indo-Pacific. *T. affinis* is the type species of two genera, *Myurella* Hinds, 1844 and *Decorihastula* Oyama, 1961. For comparison refer to *T. nebulosa* **(68).**

66. ***Terebra columellaris*** Hinds, 1844

(Pl. 19, figs. 66a-b)
(Color pl. D, fig. 19)

1844 *Terebra columellaris* Hinds, Proc. Zool. Soc. Lond. pt. 11: 151; 1844 Hinds in Sowerby, Thes. Conch. 1:172, pl. 44, fig. 77; 1860 Reeve, Conch. Icon. 12: pl. 22, sp. 113; 1960 Weaver, Hawaiian Mar. Moll. 1(1):2, pl. 1, fig. lower right; 1967 Cernohorsky, Mar. Shells Pacific 1:198, pl. 49, fig. 359; 1979 Kay, Hawaiian Mar. Shells p. 395, fig. 130b; 1982 Abbott & Dance, Compendium of Seashells p. 275; 1984 Trondle, Xenophora 15:9, fig. 11; 1984 Aubry, Terebridae, pl. 7; 1984 Salvat & Rives, Shells of Tahiti, p. 136, fig. 8.

1908 *Terebra carnicolor* Preston, Rec. Indian Mus. 2(2):189, pl. 17, fig. 63.

1961 *Decorihastula columellaris* (Hinds), Oyama & Takemura, Moll. Shells, Resources Exploitation Inst. 5: *Decorihastula* figs. 4, 11.

1966 *Terebra (Decorihastula) columellaris* Hinds, Cernohorsky & Jennings, Veliger 9(1):45, pl. 5, fig. 28.

Description: Shell to 44 mm, slender; color shiny white, mottled with tan or rust, white band at periphery of body whorl, most subsutural nodes white, ribs sometimes lighter than background; outline of whorls straight protoconch of 3½ conical whorls; subsutural band defined by deep punctations between ribs; axial ribs broad, curved, noded at posterior end; interspaces about equal to ribs in width; spiral grooves 6 to 13; aperture elongate; columella curved; operculum pale orange-brown.

Type locality: *T. columellaris*, none originally given. Suva, Fiji Is. is here designated as the type locality; *T. carnicolor*, off Port Blair, Andaman Is., 7-10 fms.

Distribution: From the Red Sea throughout the tropical Indo-Pacific to Polynesia and Hawaii.

Type: *T. columellaris* lectotype BM(NH) no. 1968228, 43.0 mm; *T. carnicolor* holotype (pl. 19, fig. 66b) ZSI no. M3997/1, 22 × 4.2 mm.

Discussion: This not uncommon species has a shell with a higher sheen than the others in this group. For comparison refer to *T. nebulosa* **(68).**

67. ***Terebra kilburni*** R. D. Burch, 1965

(Pl. 19, figs. 67a-c)
(Color pl. D, fig. 20)

1850 *Terebra areolata* Adams & Reeve, Zool. H.M.S. *Samarang* p. 30 pl. 10, fig. 23. [non *Vertagus areolatus* Link, 1807 = *Terebra*]

1965 *Terebra (Decorihastula) kilburni* R. D. Burch, Veliger 7(4): 249, pl. 31, fig. 8; 1966 Cernohorky & Jennings, Veliger 9(1):46, pl. 6, figs. 32, 32a.

1967 *Terebra kilburni* R. D. Burch, Cernohorsky, Mar. Shells Pacific 1:202, pl. 51, fig. 372; 1975 Salvat & Rives, Coq. Polynésie p. 360, fig. 384; 1979 Kay, Hawaiian Mar. Shells p. 399, fig. 130a; 1984 Trondle, Xenophora 15:11, fig. 20; 1984 Aubry, Terebridae pl. 12; 1984 Salvat & Rives, Shells of Tahiti, p. 136, pl. 35, fig. 9.

Description: Shell to 43 mm, slender; color ivory mottled with fulvous-brown or lavender-brown; outline of whorls flat-convex, turreted; protoconch 2½ conical whorls; subsutural band of elongate nodes, band defined by deep punctations between ribs; axial ribs close set, from suture to suture, beginning at nodes on band; spiral grooves, 4 to 7, confined to interspaces below subsutural band, rarely with grooves on band; aperture elongate; columella recurved.

Type locality: *T. kilburni*, "Off Wading Is. (Wadigi Is.), Fiji Is., 2 to 3 feet;" *T. areolata* A. & R., "China."

Distribution: South Africa to French Polynesia and Hawaii; 1 to 110 m.

Type: *T. kilburni* holotype MCZ no. 251247; 32.0

× 5.0 mm; *T. areolata* A. & R. lectotype (pl. 19, fig. 67b) BM(NH) no. 1979129, 24 mm.
Discussion: This species, the least common shell in this group, has extremely fine sculpture. It was named for the South African malacologist, Richard N. Kilburn. Refer to *T. nebulosa* **(68)** for comparison.

68. ***Terebra nebulosa*** Sowerby, 1825

(Pl. 19, fig. 68)
(Color pl. D, fig. 18)

1825 *Terebra nebulosa* Sowerby, Cat. Shells Tankerville App p.xxv; 1844 Hinds in Sowerby, Thes. Conch. 1:162, p. 43, fig. 51; 1860 Reeve, Conch. Icon. 12, pl. 1, sp. 1; 1917 Hirase, Terebridae Jap. Emp. p. 4, pl. 2, fig. 10, pl. 4, fig. 55; 1967 Cernohorsky, Mar. Shells Pacific 1:204, pl. 51, fig. 380; 1972 Hinton, Shells New Guinea & Cent. Indo-Pacific p. 48, pl. 24, fig. 16; 1975 Salvat & Rives, Coq. Polynésie p. 360, fig. 385; 1979 Kay, Hawaiian Mar. Shells p. 400, fig. 130c; 1982 Abbott & Dance, Compendium of Seashells p. 175; 1984 Aubry, Terebridae pl. 9.

1961 *Decorihastula nebulosa* (Sowerby), Oyama & Takemura, Moll. Shells, Resources Exploitation Inst. 5, Decorihastula figs. 2, 9.

1963 *Terebra (Decorihastula) nebulosa* Sowerby, Shikama, Sel. Shells World Col. 1:125, pl. 101, fig. 17.

1966 *Terebra (Decorihastula) nebulosa* Sowerby, Cernohorsky & Jennings, Veliger 9 (1):47, pl. 5, fig. 27.

Description: Shell to 84 mm; color dull pinkish tan to reddish brown mottled with ivory, sometimes with ivory subsutural band; outline of whorls slightly convex; protoconch of 3½ conical whorls; sculpture extremely variable; subsutural band may be ornamented with elongate nodes or ribs or (rarely) may be smooth in later whorls, band marked by punctations between ribs which may coalesce into a punctate groove in later whorls; axial ribs heavy, curved, rounded, 15 to 28 on penultimate whorl, sometimes becoming obsolete on later whorls; spiral grooves, about 5 to 8, in interspaces, usually not overriding the summit of ribs except when ribs become obsolete; rare individuals have unbroken spiral lines on later whorls when the axial ribs become completely obsolete; aperture moderately elongate; columella recurved.
Type locality: None originally given. Honiara, Solomon Islands is here designated as the type locality.
Distribution: Throughout the tropical Indo-Pacific; subtidal to 165 m.
Type: Lost. The identity of the species is elucidated through the illustrations of Hinds in Sowerby (1844) and Reeve (1860).
Discussion: There are several Indo-Pacific terebrids that might be confused with this species: *T. affinis* **(65)** *T. columellaris* **(66)** and *T. kilburni* **(67)** all have beige blotches on an ivory background. *T. affinis* may be separated from the others by the narrow punctate interspaces between the axial ribs. *T. columellaris* has a shinier shell and a pattern of ivory rectangles at the periphery of the body whorl. *T. kilburni* is the most slender of the group and has extremely fine sculpture, the blotches usually having a somewhat lavender tone but may be beige. The shell of *T. nebulosa* is solid and has a dull finish and very coarse sculpture with heavy curved ribs.

T. undulata **(69)** has straight or sinuous ribs beginning at a white elongate node on the subsutural band; the interspaces are darkly stained, and the shell has no blotches. *T. parkinsoni* **(70)** also has sinuous ribs beginning at an elongate node (not white) on the subsutural band; some, but not all, of the interspaces on each specimen are brown stained; but it is much more slender than *T. undulata* and has much finer sculpture. *T. flavofasciata* **(72)** is a rather slender shell with numerous spiral cords in the interspaces, which are not stained brown; the color pattern is yellowish beige with two ivory bands on the body whorl. *T. paucistriata* **(71)** is most easily distinguished by what appears to be two rows of white nodes below the suture, one row being an illusion created by white continuing on the ribs a short distance anterior to the band; it has a smaller, more slender shell than that of *T. undulata* and has white subsutural nodes which *T. parkinsoni* lacks.

69. ***Terebra undulata*** Gray, 1834

(Pl. 19, figs. 69, a, b)
(Color pl. D, fig. 16)

1834 *Terebra undulata* Gray, Proc. Zool. Soc. Lond. pt. 2:60; 1960 Weaver, Hawaiian Mar. Moll. 1(1):2, pl. 1, figs. lower left; 1967 Cernohorsky, Mar. Shells Pacific 1:208, pl. 52, fig. 391; 1975 Salvat & Rives, Coq. Polynésie p. 361, fig. 87; 1979 Kay, Hawaiian Mar. Shells p. 402, fig. 130D; 1984 Trondle, Xenophora 15:14, fig. 32; 1984 Aubry, Terebra pl. 2; 1984 Salvat & Rives, Shells of Tahiti p. 136, pl. 35, fig. 7.

1913 *Terebra undulata* var. *laevior* Schepman, Siboga Exped. 49: 373, pl. 25, fig. 9.

1964 *Terebra (Decorihastula) undulata* Gray, Shikma, Sel. Shells World Color 1:125, fig. 2; 1966 Cernohorsky & Jennings, Veliger 9(1):48, pl. 5, fig. 29.

Description: Shell to 45 mm; color cream to tan with white nodes anterior to suture and brown-stained interspaces; outline of whorls slightly concave, weakly turreted; protoconch of 3½ conical whorls; subsutural band indicated only by color and punctations between nodes; axial ribs, 13 to 20, heavy, rounded, straight or sinuous, with white nodes at posterior ends; spiral grooves, 4 to 14, confined to interspaces; body whorl with ribs extending slightly below periphery; aperture semi-elongate; columella slightly recurved; operculum orange-yellow.
Type locality: *T. undulata*, none originally given. Suva Reef, Fiji Is., is here designated as the type locality; *T. undulata* var. *laevior*, "Banda, Indonesia."
Distribution: From the Persian Gulf through the tropical Indo-Pacific to Hawaii; 15 to 24 m.
Type: *T. undulata* lectotype BM(NH) no. 1874.11.10.5, 33.6 mm; *T. undulata* var. *laevoir* holotype (pl. 19, fig. 69b) UZMC, 17.0 mm.
Discussion: The nodes on the subsutural band of this shell are always white or lighter than the remainder of the shell. The type of Schepman's *laevior* is a juvenile specimen showing characteristics of both *T. undulata* and *T. paucistriata* **(71)** making placement in the synonymy of *T. undulata* tentative. For comparison refer to *T. nebulosa* **(68)**.

70. *Terebra parkinsoni* Cernohorsky & Bratcher, 1976

(Pl. 20, figs. 70a-c)
(Color pl. D, fig. 17)

1976 *Terebra parkinsoni* Cernohorsky & Bratcher, Auckland Inst. & Mus. 13:137, figs. 17-21; 1984 Aubry, Terebridae pl. 10.

Description: Shell to 41 mm, slender; color, shiny white, cream or pale beige, ornamented with dark orange-brown between most, but not all of the pale axial ribs of each specimen; outline of whorls slightly convex; protoconch of 3½ smooth whorls; subsutural band weakly noded, defined by punctations between ribs; axial ribs fine, sharp, slightly curved on early whorls, later becoming broader and more arcuate, 13 to 16 on penultimate whorl; spiral grooves in interspaces, 6 to 10 on penultimate whorl; aperture elongate; columella recurved.
Type locality: "Nordup, East New Britain, Papua New Guinea, 24 m."
Distribution: From the Fiji Is. to the Red Sea; 15 to 24 m.
Type: Holotype Auckland Inst. & Mus. no. TM-1374; 29.4 × 5.2 mm.
Discussion: This shell is very similar to *T. undulata* (69). Before it was described we referred to it as "the slender undulata-like" shell, but the differences were constant: It is much more slender; has finer sculpture; and some, but not all, of the interspaces between the ribs are stained brown. Refer to *T. nebulosa* (69) for comparison.

71. *Terebra paucistriata* (E. A. Smith, 1873)

(Pl. 20, figs. 71a-b)
(Color pl. D, fig. 21)

1873 *Myurella paucistriata* E. A. Smith, Ann. & Mag. Nat. Hist. ser. 4, 11:269.

1966 *Terebra (Decorihastula) paucistriata* (E. A. Smith) Cernohorsky & Jennings, Veliger 9(1):47, pl. 5, figs. 31, 31a.

1967 *T. paucistriata* (E. A. Smith), Cernohorsky, Mar. Shells Pacific 1:205, pl. 51, fig. 383, 383a; 1978 Hinton, Guide to Shells of New Guinea pl. 49, figs. 10, 10a; 1984 Trondle, Xenophora 15: 12, fig. 24; 1984 Aubry, Terebridae pl. 12.

Description: Shell to 38 mm; color white to light brown, all with white nodes on subsutural band and with white continuing on ribs a short distance anterior to band, creating the optical illusion of a second row of nodes, interspaces between ribs being stained darker sometimes; outline of whorls slightly convex; protoconch of 3½ conical whorls; subsutural band marked by deeper grooves or punctations between ribs; axial ribs sinuous, 10 to 22; spiral grooves, 1 to 6, confined to interspaces; aperture moderately elongate with sculpture pattern showing through thin outer lip; columella straight; operculum yellowish brown.

Plate 20
FIGS. **69c,d,** *Terebra undulata* Gray. **69c,** Calapan, Mindoro, Philippines; 22.9 mm. **69d,** Middle whorls. **70a-c,** *T. parkinsoni* Cernohorsky & Bratcher. **70a,** Middle whorls. **70b,** Holotype Auckland Institute and Museum No. TM-1374; 29.4 mm. **70c,** Fiji Is.; 24.0 mm. **71a,b,** *T. paucistriata* (E. A. Smith). **71a,** Holotype British Museum (N.H.) No. 1856.11.3.17.; 18.6 mm. **71b,** Luzon Bay, Philippines; 29.5 mm. **72a-c,** *T. flavofasciata* Pilsbry. **72a,** Holotype Academy Natural Sciences, Philadelphia No. 4609; 25.1 mm. **72b,** Middle whorls of holotype. **72c,** Fiji Is.; 35.0 mm.

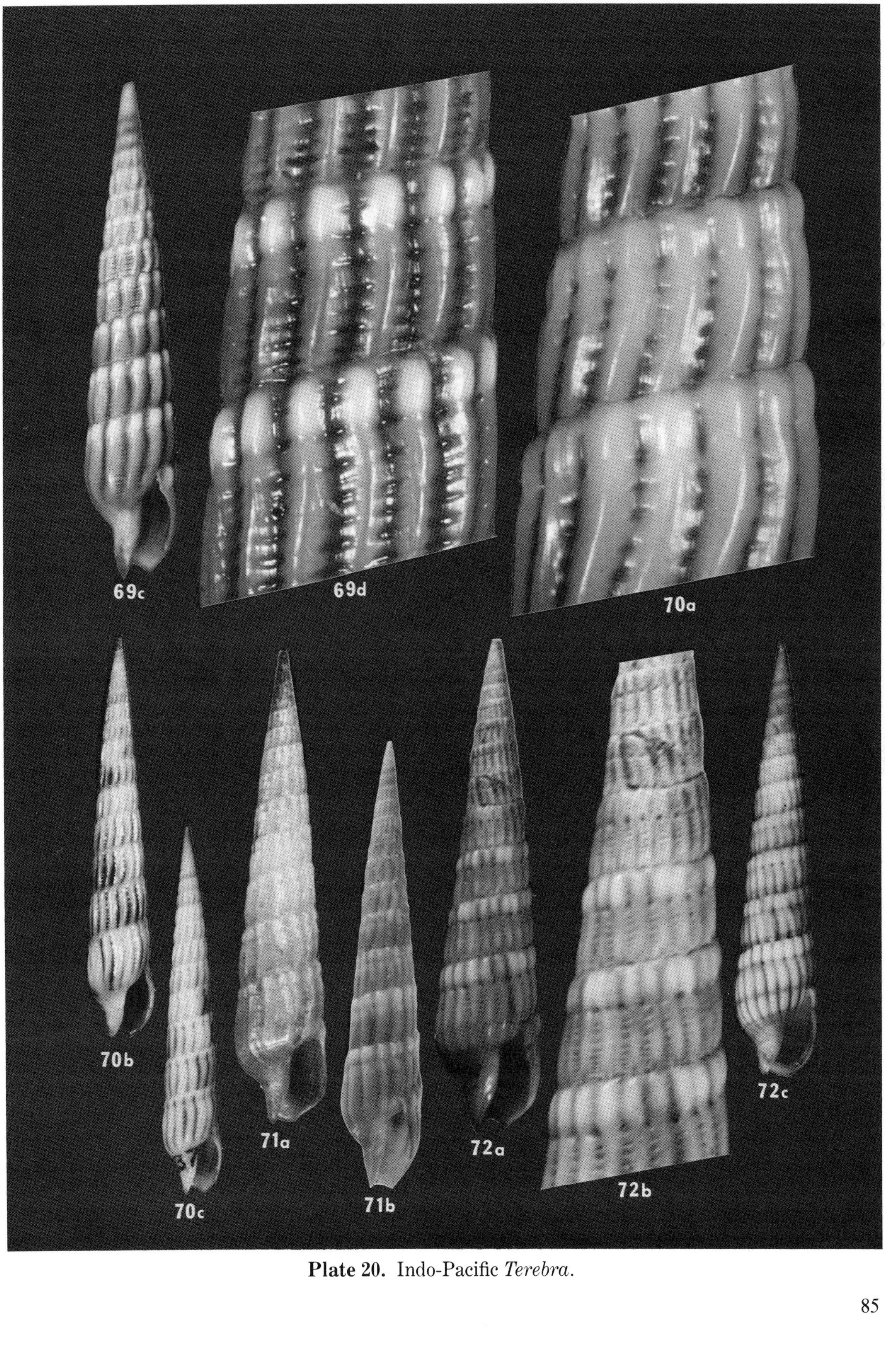

Plate 20. Indo-Pacific *Terebra*.

Type locality: "Ovalau, Fiji Is., 5 fms."
Distribution: From the Philippines to French Polynesia; intertidal to 14 m.
Type: Holotype BM(NH) no. 1856.11.3.17; 18.6 × 4.6 mm.
Discussion: When the shell is in good condition, and the color is good, it is easy to identify because of the white color at the top of the rib just below the subsutural band. It resembles a second row of nodes although that part of the rib is not enlarged into a node. See *T. nebulosa* **(68)** for comparison.

72. *Terebra flavofasciata* Pilsbry, 1921

(Pl. 20, figs. 72a-c)

1921 *Terebra flavofasciata* Pilsbry, Proc. Acad. Nat. Sci. Philadelphia 72:306, pl. 12, fig. 3; 1960 Weaver, Hawaiian Mar. Moll. 1(5):18, pl. 5, fig. top row right; 1967 Cernohorsky, Mar. Shells Pacific I:200, pl. 50, fig. 368; 1984 Aubry, Terebridae, pl. 3.
1967 *Terebra (Decorihastula) flavofasciata* Pilsbry, Cernohorsky & Jennings, Veliger 9(1):46, pl. 6, fig. 35.
1979 *Terebra undulata* Gray, Kay, Hawaiian Marine Shells:402, fig. 130D [non Gray, 1834].

Description: Shell to 35 mm; color cream with white subsutural nodes, usually with 2 white bands on body whorl, sometimes with scattered fulvous blotches; outline of whorls slightly concave; protoconch of 3½ whorls; subsutural band of elongate nodes with up to 6 spiral cords in interspaces, band defined by deep punctations between nodes; axial ribs continuing from nodes on subsutural band; spiral cords, 4 to 9, riding up sides of ribs, not reaching summit, frequently being more crowded immediately below subsutural band; aperture quadrate; columella recurved; operculum reddish brown.
Type locality: "Launiupoko Camp, West Maui, Hawaii."
Distribution: From Mozambique to Hawaii.
Type: Holotype ANSP no. 4609; 25.1 × 5.9 mm.
Discussion: This species was erroneously synonymized with *T. undulata* (69) by Kay (1979). There is no brown staining between the ribs on this species as in *T. undulata*, and the ribs are more closely spaced in this species. It is more common in Hawaii than in any other part of its range. See *Terebra nebulosa* **(68)** for comparison.

73. *Terebra conspersa* Hinds, 1844

(Pl. 21, fig. 72a-d)

1844 *Terebra conspersa* Hinds, Proc. Zool. Soc. Lond. pl. 11:153; 1844 Hinds in Sowerby, Thes. Conch. 1:163, pl. 44, fig. 74; 1967 Cernohorsky, Mar. Shells Pacific 1:199, pl. 49, fig. 360; 1972 Hinton, Shells of New Guinea & Cent. Indo-Pacific p. 46, pl. 23, fig. 27; 1984 Aubry, Terebridae pl. 9.
1857 *Terebra hindsi* Deshayes, Journ. Conchyl. 6:81, pl. 5, fig. 5 [non Carpenter, 1857].
1859 *Terebra bruguieri* Deshayes, Proc. Zool. Soc. Lond. pt. 27: 297 [nom. nov. pro *T. hindsi* Deshayes, 1857].
1966 *Terebra (Decorihastula) conspersa* Deshayes, Cernohorsky & Jennings, Veliger 9(1):46, 4, fig. 12.

Description: Shell to 53 mm; color cream to fawn with dark brown spots on subsutural band, often blending to lighter brown streaks immediately below on whorl, brown anterior to periphery of body whorl; outline of whorls moderately convex; protoconch of 3½ conical whorls; subsutural band marked by punctations between ribs; axial ribs sinuous, 23 to 55 sinuous; spiral grooves, 4 to 10, sometimes faintly crossing ribs; aperture moderately elongate; columella recurved, usually brown.
Type locality: *T. conspersa*, "Catbalonga, Samar Is., Philippines, 8 fms."; *T. hindsi* Deshayes, "China Seas."
Distribution: From China and the Philippines to Tonga; intertidal to 20 m.
Type: *T. conspersa* lectotype BM(NH) no. 1968231; 23.5 mm; *T. hindsi* Deshayes lectotype (pl. 21, fig. 73b) BM(NH), 27.6 mm.
Discussion: Because of the great variability in the number of axial ribs, individuals of this species may be quite dissimilar in appearance. It has been given three names. Refer to *T. pertusa* **(77)** for comparison.

Plate 21
FIGS. **73a-d,** *Terebra conspersa* Hinds. **73a,** Lectotype British Museum (N.H.) No. 1968231; 23.5 mm. (few-ribbed form). **73b,** Lectotype of *T. hindsi* Deshayes, British Museum (N.H.); 27.6 mm (multiribbed form). **73c,** Type specimen of *T. bruguieri* Deshayes, British Museum (N.H.); 39.8 mm. **73d,** Middle whorls. **74a,b,** *T. burchi* Bratcher & Cernohorsky. **74a,** Paratype Bratcher coll.; 20.2 mm. **74b,** Rabaul, Papua New Guinea; 19.2 mm. **75a,b,** *T. loisae* E. A. Smith. **75a,** Holotype British Museum (N.H.) No. 1901.9.22.81.; 28.4 mm. **75b,** Middle whorls of paralectotype. **76a,b,** *T. amoena* Deshayes. **76a,** Holotype British Museum (N.H.) No. 197980; 24.2 mm. **76b,** Middle whorls.

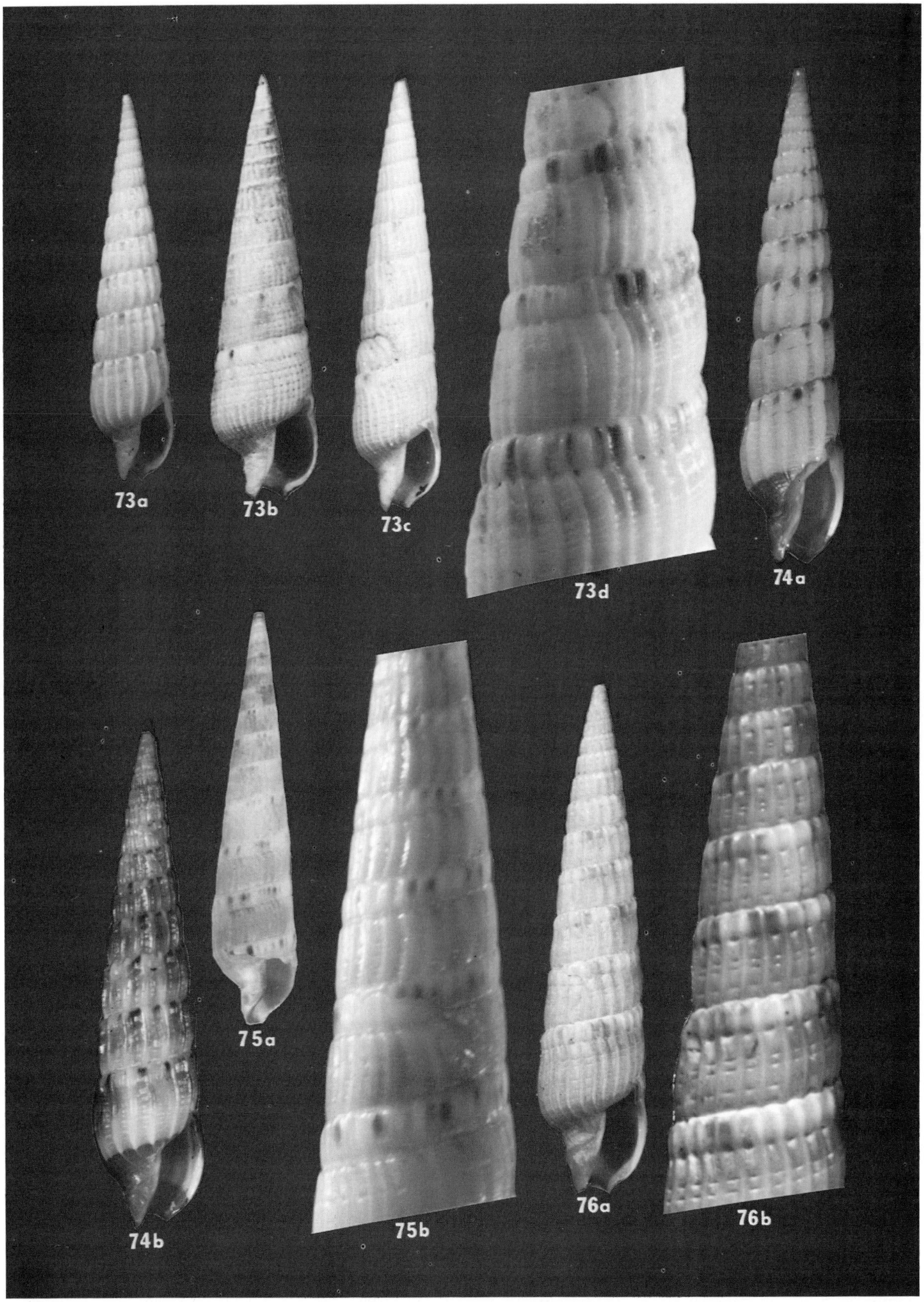

Plate 21. Indo-Pacific *Terebra*.

74. ***Terebra burchi*** Bratcher & Cernohorsky, 1982

(pl. 21, figs. 74a-b)

1982 *Terebra burchi* Bratcher & Cernohorsky, Nautilus 96(2):63, figs, 6, 7.

Description: Shell to 22 mm; color pure white with small yellowish brown dots scattered at random immediately anterior to the suture and with the same color on base of body whorl; outline of whorls convex; protoconch 3½ amber whorls; subsutural grooves confined to interspaces; axial ribs slightly arcuate, well developed, moderately sharp, from suture to suture; spiral grooves evenly spaced, confined to interspaces; aperture quadrate, with basal plication; columella recurved.
Type locality: "Northern Mozambique Channel, 12° 45′S, 45°18′E; to 20 m, Benthedi Exped. Sta. 32."
Distribution: From Mozambique to Papua New Guinea; 2 to 20
Type: Holotype MNHN Paris; 17.9 × 4.3 mm.
Discussion: There are several other Indo-Pacific terebrid shells with the base of the body whorl dark and with spots below the suture, but none of them have a pure white background or are as convex in outline or as small. For comparison refer to *T. pertusa* **(77)**

75. ***Terebra loisae*** E. A. Smith, 1903

(Pl. 21, figs. 75a-b)

1903 *Terebra loisae* E. A. Smith, Proc. Malac. Soc. Lond. 5(6): 360, pl. 15, fig. 1; 1973 Kensley, Sea-shells South Africa, p. 46, fig. 879; 1984 Aubry, Terebridae pl. 4.

Description: Shell to 28 mm; color yellowish tan with light brownish botches and scattered brown dots on subsutural band; outline of whorls straight; subsutural band almost flat, ribbed, defined by narrow groove; axial ribs curved, weak, slender; spiral grooves almost obsolete; aperture broken on type; columella curved.
Type locality: "Umkomaas, Natal, South Africa."
Distribution: Natal, South Africa.
Type: Holotype BM(NH) no. 1901.9.22.81; 28.4 mm.
Discussion: The holotype of *T. loisae* is poorly preserved. This type and subsequent illustrations strongly suggest that *T. loisae* is synonymous with *T. conspersa* **(73)**. This shell was named for Lois Trotter, a Durban shell collector.

76. ***Terebra amoena*** Deshayes, 1859

(Pl. 21, figs. 76a-b, pl. 22, fig. 76c)

1859 *Terebra amoena* Deshayes, Proc. Zool. Soc. Lond. pt. 27:297; 1860 Reeve, Conch. Icon. 12: pl. 18, sp. 80; 1967 Cernohorsky, Marine Shells Pacific 1:196, pl. 49, fig. 350; 1978 Salisbury, Hawaiian Shell News 26(8):7, center fig. on right; Aubry, Terebridae pl. 10.

1860 *Terebra specillata* Hinds, Reeve, Conch. Icon. 12: pl. 17, sp. 75 [non Hinds, 1844].

1898 *Terebra andamanica* Melvill & Sykes, Proc. Malac. Soc. Lond. 3(1):41, pl. 3, fig. 3.

Description: Shell to 48 mm; color mottled shiny beige and brown with darker brown on base of body whorl, bordered on posterior by white peripheral band, subsutural band ornamented with white and brown; outline of whorls slightly convex; protoconch 3½ conical whorls; subsutural band noded, defined by punctations between ribs; axial ribs curved, sharp, attached to nodes on band; spiral grooves, 5 or 6, confined to interspaces; aperture semi-elongate; columella recurved.
Type locality: *T. amoena*, "China Seas;" *T. andamanica*, "Andaman Is."
Distribution: From the Andaman Islands, Indian Ocean to Hawaii; 2 to 233 m.
Type: *T. amoena* holotype BM(NH) no. 197980, 24.2 mm; *T. andamanica* lectotype (pl. 22, fig. 76c) BM(NH) no. 1895.4.30.2, 47.9 mm.
Discussion: The operculum is a uniform rufous-brown and shining. (Melvill, 1898). The animal is transparent red with a few cloudy white markings (Salisbury, 1978). Refer to *T. pertusa* **(77)** and *T. alveolata* **(78)** for comparisons.

77. ***Terebra pertusa*** (Born, 1778)

(Pl. 22, figs. 77a-b)
(Color pl. E, fig. 13)

1778 *Buccinum pertusum* Born, 1778, Index rer. Nat. Mus. Caes. Vindob. p. 263; 1780 Born. Test. Mus. Caes. Vindob. p. 267, pl. 10, fig. 13.

1860 *Terebra pertusa* (Born), Reeve, Conch. Icon. 12, pl. 6, sp. 20; 1917 Hirase, Terebridae Jap. Emp. p. 5, pl. 2, fig. 9; 1958 Tinker, Pacific Sea Shells ed. 2:194, figs. middle row; 1960 Weaver, Hawaiian Mar. Moll. 1(3):10, figs. top left; 1967 Cernohorsky, Mar. Shells Pacific 1:205, pl. 52, fig. 384; 1972 Hinton, Shells of New Guinea & Cent. Indo-Pacific p. 48, pl. 24, fig. 18; 1975 Salvat & Rives, Coq. Polynésie p. 360, fig. 386; 1978 Salisbury, Hawaiian Shell News 26(8):9, text. fig. 1979 Kay, Hawaiian Mar. Shells p. 401, fig. 130G; 19; 1982 Abbott & Dance, Compendium of Seashells p. 275; Trondle, Xenophora 15:12, fig. 25; Aubry, Terebridae pl. 8; 1984 Salvat & Rives, Shells of Tahiti p. 134, fig. lower.

1961 *Decorihastula pertusa* (Born), Oyama & Takemura, Moll. Shells, Resources Exploitation Inst. 5: *Decorihastula* figs. 3, 10; 1964 Habe, Shells W. Pacific Col. 2:128, pl. 40, fig. 23.

1964 *T. (Decolihastula) pertusa* (Born), Shikama, Shells World Col., 1:125, pl. 101, fig. 15.

1966 *T. (Decorihastula) pertusa* (Born), Cernohorsky & Jennings, Veliger 9(1):48, pl. 4, figs. 11, 11a.

Description: Shell to 98 mm; color yellow to yellowish tan with black or brown stains intermittently between white ribs on subsutural band; outline of whorls flat; protoconch of 3½ whorls; subsutural band quite flat, axially ribbed, defined by grooves between ribs and indentation of ribs; axial ribs narrow, sharp arcuate, 25 to 35, narrower than interspaces; spiral grooves, 2 to 12 rows in interspaces; aperture moderately elongate; columella straight.
Type locality: None given by Born. Madang, Papua New Guinea, is here designated as the type locality.
Distribution: Throughout the tropical Indo-Pacific; subtidal to 160 m.
Type: Holotype NHMV 64.9 × 11.2 mm.
Discussion: *Terebra pseudopertusa* **(265)** may be separated from this species by its extra row of brown maculations below the subsutural band and at the periphery of the body whorl and by its mamillate protoconch of 1½ whorls. *T. pseudopertusa* was originally thought to be a color form of this species until a specimen was examined with its mamillate protoconch intact.

This species could be confused also with *T. alveolata*, **(78)** although *T. pertusa* has a flat subsutural band, more widely-spaced, sharper and narrower ribs, and an elongate aperture, all of which the other lacks. *Terebra burchi* Bratcher & Cernohorsky, *T. conspersa* **(73)** *T. amoena* **(76)** Deshayes, and *T. marmorata* **(79)** all are stained brown anterior to the periphery of the body whorl, which *T. pertusa* is not. The whorls of *T. burchi* are more convex in outline, and it has a background color of pure white with occasional tiny brown spots anterior to the suture. *T. amoena*, *T. conspersa*, and *T. marmorata* have brown blotches on the whorls as well as on the subsutural bands. The ribs of *T. amoena* are more wide-spaced and fewer and the outline of the whorls straighter than those of *T. conspersa*. *T. marmorata* may be separated by its purplish early whorls and its turreted outline.

78. *Terebra alveolata* Hinds, 1844

(Pl. 22, fig. 28)
(Color pl. F, fig. 4, 5)

1844 *Terebra alveolata* Hinds, Proc. Zool. Soc. Lond. pt. 11:151. 1844 Hinds in Sowerby, Thes. Conch. 1:162, pl. 45, fig. 120; 1917 Hirase, Terebridae Jap. Emp. p. 48, pl. 7, fig. 110; Aubry, Terebridae pl. 11.

1961 *Hastulopsis alveolata* (Hinds), Oyama, Venus, Jap. Journ. Malac. 21(2):181; 1964 Habe, Shells W. Pacific Col. 2: 128, pl. 40, fig. 19.

Description: Shell to 61 mm; color cream to tan with alternating squares of brown and cream on the subsutural band and at the periphery of body whorl, or of monochromatic brown with first 4 or 5 whorls of teleoconch white; Ribbed subsutural band defined by brown shallow groove, often with punctations between ribs; axial ribs curved, about equal to interspaces; spiral grooves, 2 to 6, not crossing ribs; body whorl with ribs fading at periphery; aperture quadrate; columella recurved.
Type locality: "Straits of Malacca, Indonesia, 31 m."
Distribution: From Japan throughout the Philippine Islands to Indonesia.
Type: Type lost. The type figure from Hinds in Sowerby is here designated as the illustrated lectotype.
Discussion: Although the type is lost, Hinds' excellent figure leaves no doubt as to the identity of his species. *T. amoena* **(76)** may be separated from this species by its smaller size, its color pattern of beige with light brown blotches and brown on the base of the body whorl, and by its lack of a broad shallow groove marking the subsutural band. For another comparison refer to *T. pertusa* **(77)**.

79. *Terebra marmorata* Deshayes, 1859

(Pl. 22, figs. 79a-b)

1859 *Terebra marmorata* Deshayes, Proc. Zool. Soc. Lond. pt. 27: 279; 1860 Reeve, Conch. Icon. 12, pl. 19, sp. 91.

Description: Shell to 59 mm; color of early whorls of teleoconch purplish, remainder of shell cream, blotched with light reddish brown; outline of whorls almost straight, slightly turreted; protoconch of 3½ whorls subsutural band flat, defined by deep punctations between ribs; axial ribs nearly straight, sometimes becoming obsolete on later whorls; spiral grooves, 3 or 4,

confined to interspaces; aperture moderately elongate; columella curved.
Type locality: "Moreton Bay," Queensland, Australia.
Distribution: From Broome, Western Australia to Queensland.
Type: Holotype BM(NH) no. 1979102; 40.5 × 8.0 mm.
Discussion: This Australian species is not common in collections. Refer to *T. pertusa* (**77**) for comparison.

80. *Terebra helichrysum* Melvill & Standen, 1903

(Pl. 22, fig. 80)

1903 *Terebra helichrysum* Melvill & Standen, Ann. & Mag. Nat. Hist. 7th ser. 12(69):310.

Description: Shell to 25 mm, very slender; color ivory with subsutural band decorated with white and pale red-brown; outline of whorls slightly convex; protoconch of 3½ whorls; subsutural band with nodes and spiral grooves between, band marked by moderately broad groove; axial sculpture of moderately broad straight ribs, equal to interspaces on early whorls, becoming narrower and more wide spaced; spiral sculpture of about 3 rows of narrow, deep grooves; body whorl with ribs fading at periphery; aperture elongate; columella slightly recurved.
Type locality: "Mussandam, Persian Gulf; 47 fms."
Distribution: Persian Gulf to Gulf of Oman; 60 to 100 m.
Type: Holotype BM(NH) no. 1903.12.15.117; 24.5 mm.
Discussion: This species, like most of those confined to the Persian Gulf area, is uncommon in collections.

81. *Terebra picta* Hinds, 1844

(Pl. 22, figs. 81a, b)
(Color pl. D, fig. 22)

1844 *Terebra picta* Hinds, Proc. Zool. Soc. Lond. pt. 11:156; 1844 Hinds in Sowerby, Thes. Conch. 176, pl. 45, fig. 105; Reeve, Conch Icon. 12, pl. 22, sp. 112b only.

Description: Shell to 30 mm, slender; color shiny gray or tan blotched with brown; outline of whorls flat; protoconch of 2 whorls; subsutural band narrow, noded, convex, defined by shallow groove and deep round punctations between ribs; axial ribs sharp, arcuate, wide-spaced, beginning at nodes on band, about 19 to 23 on penultimate whorl; spiral grooves, about 5, in interspaces; body whorl with narrow white stripe at periphery; aperture elongate; columella straight.
Type locality: "San Nicholas, Cebu Is., Philippines."
Distribution: Philippines to W. Australia.
Type: Holotype BM(NH) no. 1968241; 26.6 mm (top whorls missing).
Discussion: The color of this species gives it a marbled look. It may be separated from *T. campbelli* (**83**) by its longer whorls and convex subsutural band defined by deep round punctations. *T. campbelli* is considerably more straight-sided.

82. *Terebra copula* Hinds, 1844

(Pl. 23, fig. 82)

1844 *Terebra copula* Hinds, Proc. Zool. Soc. Lond. pt. 11:151; 1844 Hinds in Sowerby, Thes. Conch. 1:157, pl. 44, fig. 76; 1860 Reeve, Conch. Icon. 12: pl, 19, figs. 921,b.

Description: Shell to 39 mm; color shiny dark chestnut with black spots between each node on subsutural band, 2 narrow white lines at periphery of body whorl; outline of whorls flat; subsutural band flat with straight ribs slanting to left, marked by narrow deep groove; axial ribs on remainder of whorl curving slightly to left, sometimes closely set, sometimes widely spaced; spiral sculpture absent; aperture moderately quadrate; columella curved.
Type locality: "Guinea, West Africa," error?
Type: Lectotype BM(NH) no. 1968227; 39.4 mm.
Distribution: This species is known only from the type lot.

Plate 22
FIG. **76c**, *Terebra amoena* Deshayes. Lectotype of *T. andamanica* Melvill & Sykes, British Museum (N.H.) No. 1895.4.30.2.; 47.9 mm. **77a,b**, *T. pertusa* (Born). **77a**, Holotype Natural History Museum, Vienna; 64.9 mm. **77b**, Middle whorls. 78, *T. alveolata* Hinds. Illustrated lectotype from Hinds in Sowerby, 1844, pl.45.fig.120. **79a,b**, *T. marmorata* Deshayes. **79a**, Holotype British Museum (N.H.) No. 1979102; 40.5 mm. **79b**, Lower whorls of holotype. **80**, *T. helichrysum* Melvill & Standen. Holotype British Museum (N.H.) No. 1903.12.15.117.; 24.5 mm. **81a,b**, *T. picta* Hinds. **81a**, Holotype British Museum (N.H.) No. 1968241; 26.6+ mm. **81b**, Middle whorls of holotype.

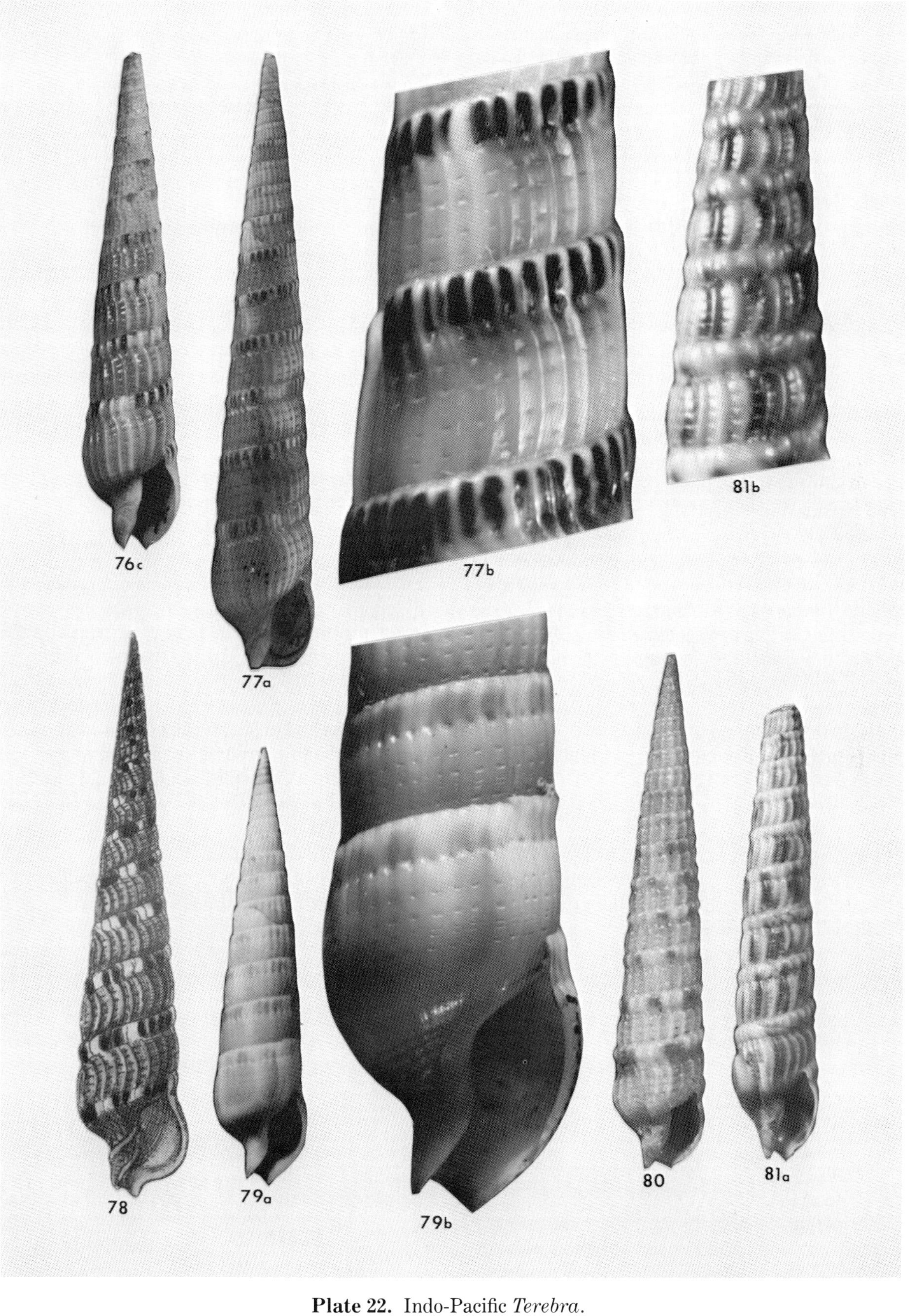

Plate 22. Indo-Pacific *Terebra*.

Discussion: Hinds mentions the color being dark chestnut with black spots on the subsutural band, and his figure shows that coloration as do Reeve's figures. The types now are grayish tan with maroon spots on the band, having faded during the last century, no doubt. Bouchet (1982) concludes that this species probably is not West African in origin.

83. *Terebra campbelli* R. D. Burch, 1965

(Pl. 23, fig. 83)

1965 *Terebra campbelli* R. D. Burch, Veliger 7(4):247, pl. 31, fig. 5.

Description: Shell to 26 mm; color straw color background with random squarish or rectangular brown blotches on subsutural band, occasionally extending anterior to band; outline of whorls extremely flat; protoconch of 1½ rounded whorls; subsutural band flat, occupying about ¼ of whorl, defined by groove, deeper between ribs; axial ribs fine, about 30; spiral striae numerous, impressed, crossing ribs on early whorls, becoming 5 to 7 impressed grooves later; body whorl with ribs becoming obsolete anterior to periphery; aperture moderately elongate; columella straight; apical angle extremely narrow.
Type locality: "Off Tosa Bay, Japan, 33° 25′N,133° 32′E, 70 fms."
Distribution: Japan to New Caledonia, 140 to 400 m.
Type: Holotype MCZ no. 251244; 25.5 × 4.5 mm.
Discussion: This rare shell is extremely slender and straight-sided. This species was named after the Californian terebrid worker, G. Bruce Campbell, who died in 1973 at the age of 39. Refer to *T. picta* (**81**) for comparison.

84. *Terebra suspensa* E. A. Smith, 1904

(Pl. 23, fig. 84)

1904 *Terebra suspensa* E. A. Smith, Journ. Malac. 11(2):30, pl. 2, fig. 2; 1973 Kensley, Sea-shells South Africa p. 216, fig. 879.

1932 *Terebra suspensa affinis* Turton, Mar. Shells Port Alfred p. 11, pl. 3, fig. 91. [non *T. affinis* Gray, 1834]

Description: Shell to 26 mm; color cream with random brownish blotches and brown basalband; outline of whorls slightly convex; protoconch of 2½ whorls; subsutural band flat, scarcely noded, subsutural groove absent but ribs change angle slightly, marking band; axial ribs flat, numerous, curved, almost obsolete, about 32 on penultimate whorl; interspaces filled with rows of punctations, 7 to 12 per whorl; aperture quadrate; columella recurved.
Type locality: *T. suspensa*, "Port Elizabeth, South Africa;" *T. suspensa affinis*, "Port Alfred, S. Africa."
Distribution: South Africa.
Type: *T. suspensa* lectotype BM(NH) no. 1903.12.19.858-862, 19.8 mm; *T. suspensa affinis* holotype Oxford Mus., 18.0 × 4.5 mm.
Discussion: The type specimens of *T. suspensa* are very worn as were the examples we have subsequently examined. Turton's *T. suspensa affinis* (non Gray) is a color form with a dark brown sutural band.

85. *Terebra blanda* Deshayes, 1859

(Pl. 23, fig. 85)

1859 *Terebra blanda* Deshayes, Proc. Zool. Soc. Lond. pt. 27:298; 1860 Reeve, Conch. Icon. 12: pl. 22, sp. 117; 1917 Hirase, Terebridae Jap. Emp. p. 5, pl. 7, fig. 109.

Description: Shell to 30 mm; color whitish with small orange-brown spots on the subsutural band and thin flammules on the remainder of whorl; outline of whorls flat; protoconch of 2 brown whorls; subsutural band almost flat, defined by shallow groove; axial ribs numerous, curved, very weak; spiral sculpture not obvious; body whorl moderately long, aperture elongate; columella straight.
Type locality: "Japan."
Distribution: Japan to Western Australia.
Type: Holotype BM(NH) no. 197984; 29.9 × 8.0 mm
Discussion: The body whorl and aperture are much more elongate in this species than in *T. gotoensis* (**86**) or *T. melanacme* (**87**). Refer to *T. gotoensis* for comparison.

Plate 23
FIG. 82, *Terebra copula* Hinds. Lectotype British Museum (N.H.) No. 1968227; 39.2 mm. **83a-c**, *T. campbelli* Burch. **83a**, Holotype Museum Comparative Zoology, Harvard No. 251244; 25.5 mm. **83b**, New Caledonia; 28.2+ mm. **83c**, Middle whorls of holotype. **84a-c**, *T. suspensa* E. A. Smith. **84a**, Lectotype British Museum (N.H.) No. 1903.12.19.858-862; 19.8 mm. **84b**, Type-figure of *T. suspensa affinis* Turton, from Turton, 1932, pl.3, fig.91. **84c**, Middle whorls. **85a,b**, *T. blanda* Deshayes. **85a**, Holotype British Museum (N.H.) No. 197984; 29.9 mm. **85b**, Middle whorls.

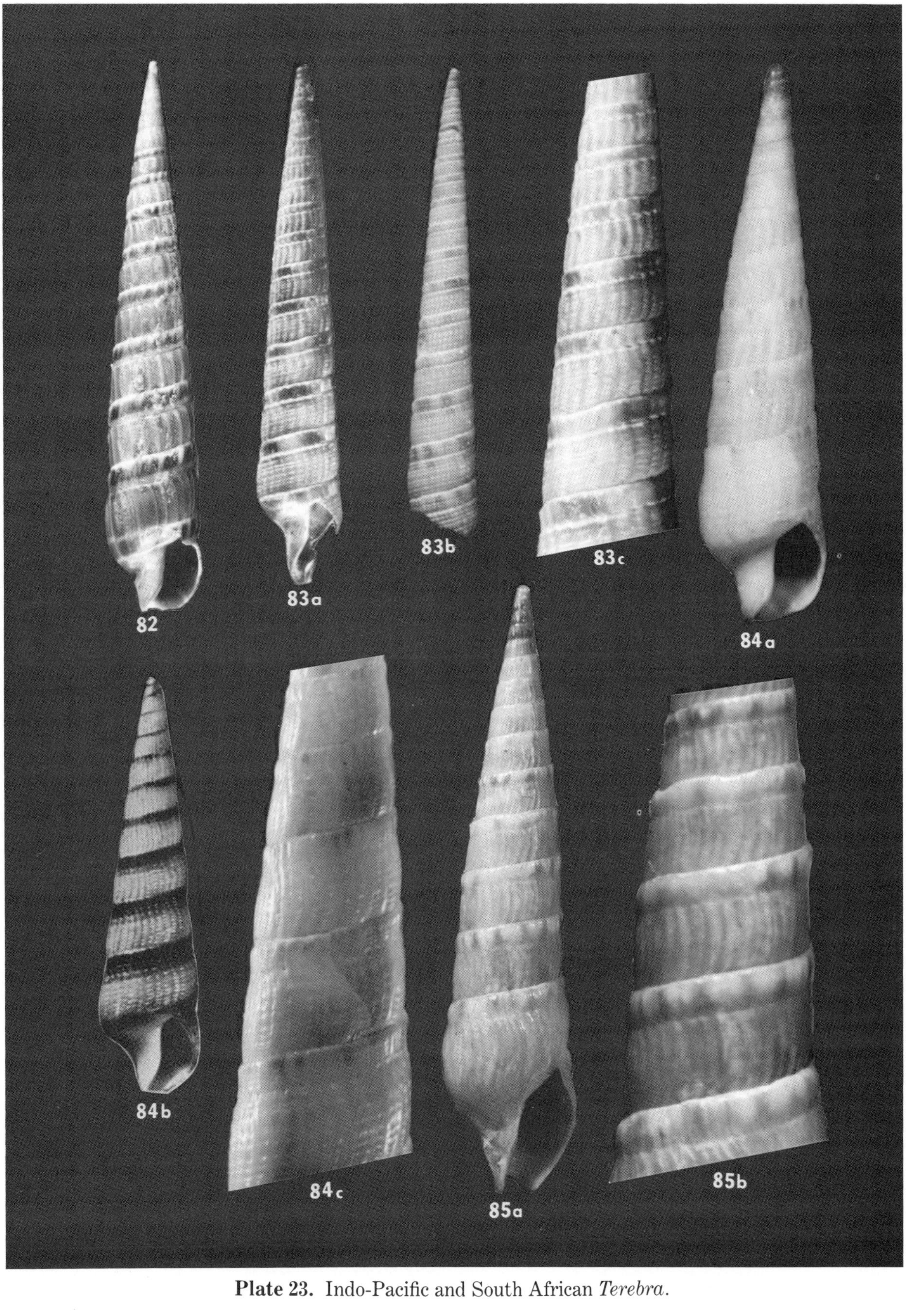

Plate 23. Indo-Pacific and South African *Terebra*.

86. ***Terebra gotoensis*** **E. A. Smith, 1879**

(Pl. 24, figs. 86a-b)

1879 *Terebra gotoensis* E. A. Smith, Proc. Zool. Soc. Lond. p. 183, pl. 19, figs. 1, 1a; 1917 Hirase, Terebridae Jap. Emp. p. 5, pl. 4, figs. 52-54.

1961 *Hastulopsis gotoensis* (E. A. Smith), Oyama & Takemura, Moll. Shells, Resources Exploitation Inst. 5: *Hastulopsis* fig. 4.

Description: Shell to 29 mm; color pale brown or fawn color with a white band spotted with brown, a narrow whitish peripheral band on the body whorl, and often bluish early whorls; outline of whorls almost flat; protoconch of 2 whorls; subsutural band with elongate nodes, defined by grooves between ribs cutting very faintly over ribs on later whorls; axial ribs weak, arcuate, closely spaced on early whorls, moderately spaced later, 24 on the penultimate whorl; spiral grooves fine, scarcely crossing ribs; body whorl short, angled at periphery; aperture quadrate; columella recurved.
Type locality: "Japan."
Distribution: Japan.
Type: Holotype BM(NH) no. 1878.11.7.18; 25.1 × 5.0 mm.
Discussion: This species may be separated from *T. blanda* (85) by its narrower apical angle, short angulate body whorl, quadrate aperture, and recurved columella. *T. blanda* has an elongate gently curved body whorl, elongate aperture, and straight columella. Named after the Goto Islands of southwestern Japan from whence the types probably came.

87. ***Terebra melanacme*** **E. A. Smith, 1873**

(Pl. 24, fig. 87)

1875 *T. melanacme* E. A. Smith, Ann. & Mag. Nat. Hist. Lond. ser. 4, 17:415; 1877 E. A. Smith, Ann. & Mag. Nat. Hist. ser. 4, 19:224; 1917 Hirase, Terebridae Jap. Emp. p. 6, pl. 2, figs. 15, 16.

1961 *Hastulopsis melanacme* (E. A. Smith), Oyama & Takemura, Moll. Shells, Resources. Exploitation Inst. 5: *Hastulopsis* pl., fig. 1.

Description: Shell to 23 mm; color shiny cream splotched with beige, a white subsutural band dotted with chestnut, and a narrow white peripheral band on body whorl; outline of whorls straight; protoconch blackish brown; subsutural band in color only; axial sculpture of straight, evenly spaced ribs from suture to suture on the first 5 or 6 whorls of teleoconch, fading almost completely, later whorls being smooth, shining, with only axial striae faintly visible; body whorl angulate, cutting in sharply anterior to periphery; aperture quadrate; columella curved.
Type locality: "Cape Sima, Japan, 18 fms."
Distribution: Japan.
Type: Holotype BM(NH) no. 873.8.6.11; 17.8 mm.
Discussion: Smith recognized that the holotype of this species is immature. He said (1877), "The above is probably the description only of the young of a largish species; but the characters are so distinctive that it is not, I think, very hazardous to describe it. The black apex, the smooth whorls, the white zone dotted with chestnut, and the narrow white band round the last whorl easily define the species." Two years later he described *T. gotoensis*, which has all those features except the shining almost smooth whorls. It has similar early sculpture but has faint axial ribs and spiral sculpture in later whorls. The types of both have angulate body whorls which cut in anterior to the suture. Additional specimens and study may prove these to be conspecific. *T. melanacme* is the type species of the genus *Hastulopsis* Oyama, 1961.

88. ***Terebra japonica*** **E. A. Smith, 1873**

(Pl. 24, figs. 88a-c)

1873 *Terebra japonica* E. A. Smith, Ann. & Mag. Nat. Hist. 4th ser. 11:265; 1952 Kuroda & Habe, Check list & Bibl. Rec. Mar. Moll. Japan p. 89.

1877 *Terebra lischkeana* Dunker, Malakozool. Blätter, 24:74; 1882 Dunker, Index Moll. Maris Japonica p. 71, pl. 5, figs. 13, 16.

Plate 24
FIGS. **86a,b,** *Terebra gotoensis* E. A. Smith. **86a,** Holotype British Museum (N.H.) No. 1878.11.7.18.; 25.1 mm. **86b,** Middle whorls of holotype. **87,** *T. melanacme* E. A. Smith. Holotype British Museum (N.H.) No. 1873.8.6.11.; 17.8 mm. **88a-c,** *T. japonica* E. A. Smith. **88a,** Lectotype British Museum (N.H.) No. 1979133; 36.2 mm. **88b,** Holotype of *T. lischkeana* Dunker, Zoological Museum Humboldt University, Berlin; 34.1 mm. **88c,** Holotype of *T. awajiensis* Pilsbry, Academy Natural Sciences, Philadelphia No. 86004; 36.5 mm. **89,** *T. isabella* Thiele. Holotype Zoological Museum Humboldt University, Berlin; 13.5 mm. 90, *T. arabella* Thiele. Same Institution; 10.7 mm. **91a,b,** *T. caliginosa* Deshayes. **91a,** Holotype British Museum (N.H.) No. 197987; 29.7 mm. **91b,** Fiji Is.; 28.0 mm.

Plate 24. Indo-Pacific and Japanese *Terebra*.

1904 *Terebra awajiensis* Pilsbry, Proc. Acad. Nat. Sci. Philadelphia 56: 4, pl. 1, figs. 3, 3a.; 1918 Hirase, Terebridae Jap. Emp. p. 34, pl. 2, figs. 17, 18.

1961 *Brevimyurella awajiensis* (Pilsbry), Oyama & Takemura, Moll. Shells, Resources Exploitation Inst. 5: *Brevimyurella* figs. 3, 9.

1961 *Brevimyurella lischkeana* (Dunker), Oyama & Takemura, Moll. Shells, Resources Exploitation Inst. 5: *Brevimyurella* figs. 1, 2, 8; 1962 Kira, Shells W. Pacific Col. 1:111, pl. 39, fig. 14; 1967 Habe & Kosuge, Stand. Book Jap. Shells Col. 3: 102, pl. 40, fig. 18.

1963 *Terebra (Brevimyurella) awajiensis* Pilsbry, Shikama, Sel. Shells World Col. 1:125, fig. 4.

1975 *Brevimyurella japonica* (Smith) Okutani, Mollusks of Japan, Gakken Illus. Nat. Encyclop, vol. 1, pp. 154, 238.

Description: Shell to 47 mm; color straw with a narrow light band at periphery of body whorl and with an interrupted brown band, usually occupying lower half of whorl, sometimes placed midwhorl, edges of ribs whitish; outline of whorls flat; protoconch of $3^1/_2$ conical whorls; subsutural groove sometimes confined to interspaces; axial ribs narrow, sharp, contiguous with ribs on preceding whorl; interspaces wider than ribs; spiral threads or grooves in interspaces fine and numerous on some specimens, completely absent on others, some showing only a few faint grooves under magnification; aperture moderately elongate; columella curved.
Type locality: *T. japonica*, "Japan;" *T. lischkeana*, "Japan," *T. awajiensis*, "Fukura, Awaiji, Japan."
Distribution: Japan; intertidal to 50 m.
Type: *T. japonica* lectotype BM(NH) no. 1979133, 36.2 × 7.7 mm; *T. lischkeana* holotype (pl. 24, fig. 88b) MNHUB, 34.1 mm; *T. awajiensis* holotype (pl. 24, fig. 88c)ANSP no. 86004, 36.5 × 8.2 mm.
Discussion: The chronologically prior taxon, *T. japonica* E. A. Smith, 1873, has replaced *T. lischkeana* Dunker, 1877, in recent Japanese literature. Pilsbry stated that his *T. awajiensis* differed from *T. lischkeana* by having smooth intercostal spaces. The holotype of *T. lischkeana* also lacks spiral sculpture in the intercostal spaces although some of the paratypes are spirally striate. Both types of sculpture may be found in the same populations. *T. lischkeana* Dunker (now *T. japonica*) is the type species of the genus *Brevimyurella* Oyama, 1961.

89. ***Terebra isabella*** Thiele, 1925

(Pl. 24, fig. 89)

1925 *Terebra isabella* Thiele, Wiss. Ergeb. Deut. Tief. Exped. Valdivia 17:345, pl. 29, fig. 23.

Description: Shell to 16 mm; color shiny tan with an indistinct brown band in the center of the whorl and at the periphery of the body whorl; outline of whorls moderately convex; protoconch of $3^1/_2$ whorls; subsutural band indicated only by short grooves between ribs; axial ribs slender, almost straight, widely spaced, from suture to suture; about 4 faint spiral grooves confined to the interspaces, which are wider than ribs; aperture elongate, narrow; columella curved.
Type locality: "Padang," Sumatra, Indonesia.
Distribution: Indonesia.
Type: Holotype MNHUB, 13.5 × 2.5 mm.
Discussion: This species somewhat resembles *T. arabella* **(90)** but is straighter in outline, with fewer, straighter, and more widely-spaced ribs and has a more elongate aperture than that species.

90. ***Terebra arabella*** Thiele, 1925

(Pl. 24, fig. 90)

1925 *Terebra arabella* Thiele, Wiss. Ergeb. Tief. Exped. "Valdivia" 17(2):346, pl. 29, fig. 24; 1984 Trondle, Xenophora 15:8, fig. 4; 1984 Salvat & Rives, Shells of Tahiti, p. 136, fig. lower.

Description: Shell to 22 mm; color shiny tan to brown with lighter axial ribs; outline of whorls convex; protoconch 4 conical whorls; subsutural band marked by moderately deep grooves between ribs; axial ribs well developed, widely-spaced, curved, extending from suture to suture, the same width as interspaces, each rib contiguous with an interspace on preceding whorl; spiral grooves fine, 4 or 5 per whorl; aperture semi-elongate; columella slightly recurved, with moderate parietal callus.
Type locality: "Padang" Sumatra, Indonesia.
Distribution: From Indonesia to French Polynesia.
Type: Lectotype MNHUB; 10.7 mm.
Discussion: This shiny species may be recognized by its arcuate ribs which seem to fall exactly between the ribs of the preceding whorl. For comparison refer to *T. isabella* **(89).**

91. *Terebra caliginosa* Deshayes, 1859

(Pl. 24, figs. 91a-b)

1859 *Terebra caliginosa* Deshayes, Proc. Zool. Soc. Lond. pt. 27: 287; 1860 Reeve, Conch. Icon. 12, pl. 20, fig. 100.

1978 *Hastula caliginosa* (Deshayes), Cernohorsky, Trop. Pacific Mar. Shells, p. 150, pl. 53, fig. 8.

Description: Shell to 30 mm; color dull purplish brown or purplish gray; outline of whorls almost straight except for convex body whorl; subsutural band absent except for grooves between ribs; axial ribs narrow, widely-spaced, straight, becoming sinuous on body whorl; spiral sculpture absent; sutures shallow; body whorl extremely elongate, with ribs fading at the periphery; aperture semi-elongate, with brown interior; columella stained brown, curved, with slight parietal callus; siphonal fasciole with cord-like keel.

Type locality: "Philippine Islands."

Distribution: From Durban, South Africa to the Philippines.

Type: Holotype BM(NH) no. 197987; 29.7 × 6 mm.

Discussion: This is the type species of the genus *Punctoterebra* Bartsch, 1923. It may be separated from *T. nitida* **(92)** by its brown columella and interior of aperture, lack of shine, and very shallow, almost unnoticeable subsutural grooves.

92. *Terebra nitida* Hinds, 1844

(Pl. 25, figs. 92a-c)

1844 *Terebra nitida* Hinds, Proc. Zool. Soc. Lond. pt. 11:152; 1844 Hinds in Sowerby, Thes. Conch. 1:164, pl. 45, fig. 103; 1960 Reeve, Conch. Icon. 12: pl. 22, sp. 115; 1967 Cernohorsky, Mar. Shells Pacific 1:205, pl. 51, fig. 381; 1972 Hinton, Shells New Guinea Cent. Indo-Pacific p. 46, pl. 23, fig. 13; 1975 Salvat & Rives, Coq. Polynésie p. 363, fig. 393; 1984 Trondle, Xenophora 15:12, fig. 23; Aubry, Terebridae pl. 2; 1984 Salvat & Rives, Shells of Tahiti p. 138, pl. 36, fig. 3.

1857 *Terebra plicatella* Deshayes, Journ. Conchyl. 6:96, pl. 3, fig. 5 (Hab. Van Diemen's Land, Tasmania [error]; lectotype BM(NH) no. 197924; 36.7 × 5.8 mm).

1894 *Terebra cernica* Sowerby, Proc. Malac. Soc. London 1 (2):43, pl. 4, fig. 1 (Hab. Mauritius; type lost).

1898 *Terebra nitida* var *sicyodes* Melvill & Sykes, Proc. Malac. Soc. London 3:43, pl. 3, fig. 8 (Hab. Andaman Islands; lectotype BM(NH) no. 1898.4.30.4; 35.0 × 5.5 mm).

1961 *Punctoterebra nitida* (Hinds), Oyama & Takemura, Moll. Shells, Resources Exploitation Inst. 5, *Punctoterebra* fig. 4.

1964 *Terebra (Punctoterebra) nitida* Hinds, Shikama, Sel. Shells World Col. 1:124, pl. 101, fig. 8.

1965 *Hastula nitida* (Hinds), R. D. Burch, Veliger 7(4):252; 1979 Kay, Hawaiian Mar. Shells p. 389, fig. 126; 1984 Abbott & Dance, Compendium of Seashells p. 276.

Description: Shell to 47 mm; color from shiny pale gray to dark brown, monochromatic; outline of whorls flat; protoconch of 4 conical whorls; subsutural band indicated only by deep punctations between ribs; axial ribs straight, slanting, sharp; spiral sculpture usually absent, interspaces, rarely showing a few obsolete spiral grooves; body whorl elongate, with ribs fading at periphery; aperture elongate; columella almost straight, with heavy parietal callus in mature specimens.

Type locality: *T. nitida*, "Marquesas Is., in 7 fms., sandy mud;" *T. plicatella*, "Van Diemen's Land, Tasmania," error; *T. cernica*, "Mauritius;" *T. nitida* var. *sicyodes*, "Andaman Is."

Distribution: Tropical Indo-Pacific; intertidal to 30m.

Type: *T. nitida* holotype BM(NH) no. 1844.6.7.85, 20.5 mm; *T. plicatella* lectotype (pl. 25, fig. 92b) BM(NH) no. 197924, 36.7 × 5.8 mm; *T. cernica* type lost; *T. nitida* var. *sicyodes* lectotype (pl. 25, fig. 92c) BM(NH) no. 1848.4.30.4, 35.0 × 5.5 mm.

Discussion: The elongate aperture with its heavy parietal callus and the smooth, shiny shell with deep subsutural punctations or dashes make this species easy to recognize.

Refer to *T. plumbea* **(93)** for comparison.

93. *Terebra plumbea* Quoy & Gaimard, 1833

(Pl. 25, figs. 93a-e)

1833 *Terebra plumbea* Quoy & Gaimard, Voy. L'Astrolabe Zool. 2:470, pl. 36, figs. 29, 30; 1860 Reeve, Conch. Icon. 12, pl. 22, sp. 111; 1958 Tinker, Pacific Sea Shells ed. 2:192, figs. 2nd row, right and left.

1859 *Terebra bourguignati* Deshayes, Proc. Zool. Soc. Lond. pt. 27:288.

1869 *Terebra assimilis* Pease, American Journ. Conch. 5(2):67 [*non* Angas, 1867].

1871 *Terebra contigua* Pease American. Journ. Conch 7(1):20 [nom nov. pro *Terebra assimilis* Pease, 1869]

1921 *Terebra clappi* Pilsbry, Proc. Acad. Nat. Sci. Philadelphia 72:306 (Hab. Honolulu, Hawaii; lectotype MCZ no. 49962; 17.8 × 4.5 mm).

1925 ?*Terebra castaneofusca* Thiele, Wiss. Ergeb. Deutsch. Tief. Exped. Valdivia 17:345, pl. 29, fig. 21.

1952 *Terebra (Strioterebrum) hoffmeyeri* R. T. Abbott, Nautilus 65(3):78, pl. 5, figs. 5-9

1964 *Hastula plumbea* (Quoy & Gaimard), R. D. Burch, Veliger 6(4): 213; 1966 Cernohorsky & Jennings, Veliger 9(1):6, pl. 7, fig. 52; 1967 Cernohorsky, Mar. Shells Pacific 1:212, pl. 54, fig. 407; 1984 Aubry, Terebridae pl. 8.

Description: Shell to 25 mm; color gray or dark pinkish beige to dark blackish brown, with light area below the suture, ribs often appearing whitish; outline of whorls almost straight; protoconch of 3½ to 4½ whorls; subsutural grooves between ribs rare; axial ribs, 13 to 15, unbroken from suture to suture; spiral striae fine, crowded or wide spaced, often microscopic; aperture moderately quadrate; columella curved.
Type locality: *T. plumbea*, "Moluccas, Indonesia;" *T. bourguignati*, "China Seas;" *T. assimilis*, "Oahu, Hawaii;" *T. clappi*, "Honolulu, Hawaii;" *T. castaneofusca*, "Padang," Sumatra, Indonesia; *T. hoffmeyeri*, "Pasay Beach, Manila Bay, Luzon Is., Philippines."
Distribution: Tropical Indo-West Pacific.
Type: *T. plumbea* holotype MNHNP, 22.9 × 4.8 mm; *T. bourguignati* holotype (pl. 25, fig. 93b) MNHNP, 16.8 × 3.7 mm; *T. assimilis* type lost; *T. clappi* lectotype (pl. 25, fig. 93d) MCZ no. 19962, 17.8 × 4.5 mm; *T. castaneofusca* holotype (pl. 25, fig. 93c) MNHUB, 11.4 × 2.5 mm; *T. hoffmeyeri* holotype (pl. 25, fig. 9593e) USNM 598077, 17.4 × 4.4 mm.
Discussion: The spiral sculpture on this species is fine and ranges from widely-spaced to crowded and sometimes microscopic threads. This variability no doubt led to its being described by several authors. The band of lighter color, sometimes whitish, below the suture appears to be constant. *T. nitida* **(92)** may be separated from this species by its elongate aperture, well-developed parietal callus, deep subsutural punctations, and lack of a light colored subsutural area. The type of *T. castaneofusca* is in bad condition, which is our reason for placing it questionably in synonymy with *T. plumbea*.

94. ***Terebra tiurensis*** Schepman, 1913

(Pl. 25, fig. 94)

1913 *Terebra tiurensis* Schepman, Siboga Exped. 49 1(5):374, pl. 25, fig. 10.

Description: Shell to 32 mm; color ivory with 2 faint small brown blotches on the dorsum midway down shell; outline of whorls slightly convex; protoconch of 1½ conical whorls; subsutural band slightly convex, defined by punctations between the ribs; axial ribs rather weak, unevenly spaced, contiguous with those on band; spiral sculpture absent; aperture semi-quadrate; columella curved.
Type locality: "Near Ruma Luis, north-point of Tiur Island, Indonesia."
Distribution: The holotype is the only specimen of this species examined.
Type: Holotype ZMUA; 14.5 × 3.3 mm.
Discussion: The specimen marked as the type appears to agree with Schepman's figure, but there is a discrepancy between that and his description. None of the "scattered reddish brown spots, just below the subsutural band, in a few cases stretching, though more faintly, down over the whorls" are visible. There are two very small, very faint brown blotches on the dorsum, midway down the whorl.

95. ***Terebra swainsoni*** Deshayes, 1859

(Pl. 25, figs. 95a-c)

1859 *Terebra swainsoni* Deshayes, Proc. Zool. Soc. Lond. pt. 27: 299; Reeve, Conch. Icon. 12: pl. 22, sp. 118 (type fig. of *T. inflexa* Pease).

1859 *Terebra longiscata* Deshayes, Proc. Zool. Soc. Lond. pt. 27: 294; 1967 Cernohorsky, Marine Shells Pacific 1:204: pl. 51, fig. 375; 1978 Salisbury, Hawaiian Shell News 26 (8); 8.

1869 *Terebra swainsonii* Deshayes, *var. inflexa* Pease, American Journ. Conch. 5(2):64.

1958 *Terebra inflexa alta* Tinker, Pacific Sea Shells, p. 198, figs. second row, left and right.

Plate 25

FIGS. **92a-c,** *Terebra nitida* Hinds. **92a,** Holotype British Museum (N.H.) No. 1844.6.7.85.; 20.5 mm. **92b,** Lectotype of *T. plicatella* Deshayes, same Institution No. 197924; 36.7 mm. **92c,** Lectotype of *T. nitida* var. *sicyodes* Melvill & Sykes, same Institution No. 1898.4.30.4.; 35.0 mm. **93a-e,** *T. plumbea* Quoy & Gaimard. **93a,** Holotype Museum National d'Histoire Naturelle, Paris; 22.9. mm. **93b,** !Paratype of *T. bourguignati* Deshayes, British Museum (N.H.); 14.5 mm. **93c,** Immature holotype of *T. castaneofusca* Thiele, Zoological Museum Humboldt University, Berlin; 11.4 mm. **93d,** Lectotype of *T. clappi* Pilsbry, Museum Comparative Zoology, Harvard No. 49962; 17.8 mm. **93e,** Holotype of *T. hoffmeyeri* Abbott, National Museum Natural History, Washington No. USNM-598077; 17.4 mm. **94,** *T. tiurensis* Schepman. Holotype Zoological Museum, University of Amsterdam; 14.5 mm. **95a-c,** *T. swainsoni* Deshayes. **95a,** Lectotype British Museum (N.H.) No. 1979113; 30.0 mm. **95b,** Questionable type of *T. longiscata* Deshayes, same Institution No. 1979101; 9.9 mm. **95c,** Pokai Bay, Hawaiian Is.; 17.0 mm.

Plate 25. Indo-Pacific *Terebra*.

Description: Shell to 30 mm; color variable, pinkish-white, flesh-beige, dark gray or orange-brown; outline of whorls slightly convex; protoconch of 4½ amber whorls; subsutural band defined by deep punctations between ribs and groove which cuts weakly through ribs; axial ribs curved, sharp, narrower than interspaces; spiral threads fine, numerous, not crossing ribs, often almost obsolete; aperture elongate; columella almost straight, with parietal callus.
Type locality: *T. swainsoni*, "Hawaiian Is;" *T. longiscata*, "Philippine Is."
Distribution: From Durban, S. Africa to Tahiti and Hawaii.
Type: *T. swainsoni* lectotype BM(NH): no. 1979113, 30.0 × 6.6 mm; *T. longiscata* questionable type (pl. 25, fig. 95b) BM(NH) no. 1979101, 9.9 mm.
Discussion: The questionable type of *T. longiscata* Deshayes is incomplete, but most characteristics agree with those of *T. swainsoni* with which it is tentatively synonymized. Deshayes named it after the English conchologist, William Swainson (1789-1855).

96. ***Terebra cinctella*** Deshayes, 1859

(Pl. 26, figs. 96a-b)
(Color pl. D, fig. 23)

1859 *Terebra cinctella* Deshayes, Proc. Zool. Soc. Lond. pt. 27: 305.

Description: Shell to 31 mm; color fawn to amber, some with lighter subsutural band and light peripheral stripe on body whorl which shows through outer lip, others with band of lavender hue; outline of whorls convex; protoconch of 2½ small whorls; subsutural band convex, moderately narrow, with elongate nodes, defined by deep suture and constricting subsutural groove, usually with punctations between ribs; axial ribs curved, evenly spaced, well-developed, contiguous with elongate nodes on band; interspaces about equal to ribs; spiral cords numerous, fine, 6 to 12 on penultimate whorl, sometimes also appearing on band; aperture moderately quadrate; columella recurved.
Type locality: "Mouth of the Indus," Pakistan.
Distribution: From the Arabian Sea to the Solomon Is.
Type: Lectotype BM(NH) no. 197988; 27.8 × 5.8 mm.
Discussion: A good diagnostic feature of this species is the constriction both above and below the narrow, convex subsutural band. *T. swainsoni* Deshayes lacks this.

97. ***Terebra livida*** Reeve, 1860

(Pl. 26, fig. 97)

1860 *Terebra livida* Reeve, Conch. Icon. 12: pl. 22, fig. 116.
1961 *Decorihastula livida* (Reeve), Oyama & Takemura, Moll. Shells, Resources Exploitation Inst. 5: *Decorihastula* figs. 7, 13.

Description: Shell to 40 mm; color shiny lead gray; outline of whorls flat; protoconch of 4 whorls; subsutural band defined by deep groove between ribs, faint over summit; axial ribs, about equal to interspaces, slightly swollen at posterior, 14 to 20 on penultimate whorl; spiral grooves, 6 to 10; aperture elongate-quadrate, usually with a narrow light band showing through; columella recurved, with light parietal callus.
Type locality: "Philippine Is."
Distribution: From the Philippines to the Fiji Is.
Type: Holotype BM(NH) no. 1874.12.11.299; 38.4 mm.
Discussion: There are few terebrids with the shiny lead gray color of this species. See *T. succincta* (Gmelin) (**98,** below) for comparison.

98. ***Terebra succincta*** (Gmelin, 1791)

(Pl. 26, figs. 98a-c)
(Color pl. E, fig. 11)

1791 *Buccinum succinctum* Gmelin, Syst. Nat. ed. 13:3502 (Ref. Chemnitz, vol. 4, pl. 154, fig. 1451).
1798 *Epitonium fissum* Röding, Mus. Bolten. p. 95 (Ref. Chemnitz 1780, 4, pl. 154, fig. 1451).

Plate 26
FIGS. **96a, b,** *Terebra cinctella* Deshayes. **96a** Lectotype British Museum (N.H.) No. 197988; 27.8 mm. **96b,** Paralectotype, same Institution No. 197988; 31.0 mm. **97,** *T. livida* Reeve. Holotype British Museum (N.H.) No. 1874.12.11.299.; 38.4 mm. **98a-c,** *T. succincta* (Gmelin). **98a,** Illustrated lectotype from Chemnitz, 1780, vol. 4, pl. 154, fig. 1451. **98b,** Holotype of *T. cancellata* Quoy & Gaimard, Museum National d'Histoire Naturelle, Paris; 31.0 mm. **98c,** Holotype of *T. flavescens* Deshayes, British Museum (N.H.) No. 197994; 44.6 mm. **99a-c,** *T. turrita* (E. A. Smith). **99a,** Middle whorls. **99b,** Holotype British Museum (N.H.) No. 1965137; 23.0 mm. **99c,** Madras, India; 43.1 mm. **100a.** *T. venilia* Tenison-Woods. Holotype Australian Museum, Sydney, No. C-129007; 41.0 mm.

Plate 26. Indo-Pacific *Terebra*.

1807 *Vertagus succinctus* Link, Beschr. Nat.-Sam. lines. Rostock p. 129 (Ref. Chemn. fig. 1451).

1833 *Terebra cancellata* Quoy & Gaimard, Voy. L'Astrolabe, Zool. 2: 471, pl. 36, figs. 27, 28; 1844 Hinds in Sowerby, Thes. Conch. 1:178, pl. 44, fig. 80; 1860 Reeve, Conch. Icon. 12, pl. 14, figs. 58a., b; 1953 Risbec, Bull. Mus. National Hist. Nat. Paris ser. 2,25:578 text figs. 4, 9, 14 (anatomy); [non *Epitonium cancellatum* Röding, 1798 = *Terebra*].

1857 *Terebra bermonti* Lorois, Journ. Conchyl. 6: (2):389, pl. 12, fig. 2.

1859 *Terebra flavescens* Deshayes, Proc. Zool. Soc. Lond. pt. 27: 299

1859 *Terebra undatella* Deshayes, Proc. Zool. Soc. Lond. pt. 27: 300 [nom nov pro *T. cancellata* Hinds, 1844].

1964 *Terebra (Diplomeriza?) cancellata* Quoy & Gaimard, Shikama, Sel. Shells World Col. 2: 128, pl. 69, fig. 16.

1966 *Terebra (Strioterebrum) succincta* (Gmelin), Cernohorsky & Jennings, Veliger 9(1):54, pl. 6, fig. 35, text fig. 10a.

1966 *Terebra (Strioterebrum) flavescens* Deshayes, Cernohorsky & Jennings, Veliger 9 (1):54, pl. 6, fig. 34. text fig. 10b.

1967 *Terebra succincta* (Gmelin) Cernohorsky, Mar. Shells Pacific 1:206, pl. 52, fig. 387; 1972 Hinton, Shells New Guinea & Cent. Indo-Pacific p. 46, pl. 23, fig. 26; 1984 Aubry, Terebridae pl. 12.

Description: Shell to 48 mm; color beige to almost black; outline of whorls slightly convex; protoconch of 1½ whorls; subsutural band with nodes on early whorls, ribs on later ones with about 4 to 8 fine spiral threads in interspaces, band defined by subsutural groove; axial ribs, about 26 to 32, sharp, curved or sinuous, somewhat narrower than interspaces; spiral threads fine, 5 to 10, may or may not cross ribs; body whorl elongate; aperture elongate; columella curved, with moderate parietal callus on mature individuals.

Type locality: *T. succincta*, "Indian Ocean;" *E. fissum*, none given; *T. cancellata*, "Moluccas," Indonesia; *T. bermonti*, "Tahiti;" *T. flavescens*, "Hawaii."

Distribution: Throughout the tropical Indo-Pacific; subtidal to 30m.

Type: *T. succincta*. The specimen illustrated by Chemnitz (1780) on plate 154, fig. 1451, is here designated as the illustrated lectotype; *E. fissum* type is Chemnitz fig. 1780; *T. cancellata* holotype (pl. 26, fig. 98b) MNHNP, 31 mm; *T. bermonti* the type possibly is at MNHN Paris but not segregated [Dr. Bouchet in litt.], original measurements 87 × 12 mm; *T. flavescens* (pl. 26, fig. 98c) holotype BM(NH) no. 1979934, 44.6 × 9 mm.

Discussion: This species has both poison gland and harpoon-shaped radular teeth. (Risbec, 1953). It may be separated from *T. livida* (**97**) by its having a protoconch of 1½ whorls, and axial ribs numbering 26 to 32. *T. livida* has a protoconch of 4 whorls and axial ribs numbering 14 to 20.

99. *Terebra turrita* (E. A. Smith, 1873)

(Pl. 26, figs. 99a-c)
(Color pl. F, fig. 6)

1873 *Myurella turrita* E. A. Smith, Ann. & Mag. Nat. Hist. Lond. ser. 4 11:266.

1966 *Terebra turrita* (E. A. Smith), Cernohorsky & Jennings, Veliger 9(1):55, pl. 6, fig. 40b (only).

Description: Shell to 60 mm; color shiny ivory blotched with brown; outline of whorls turreted; protoconch of 1½ whorls; subsutural band noded, defined by deep punctations between ribs; axial ribs sharp, curved, contiguous with nodes on band; spiral grooves between ribs, usually confined to the anterior portion of the whorls; aperture moderately quadrate; columella recurved.

Type locality: "Torres Straits, northern Australia."

Distribution: From India to Taiwan and to Fiji.

Type: Holotype BM(NH) no. 1965137; 23.0 × 4.0 mm.

Discussion: It is interesting to note that the spiral sculpture on the shell of this species seems to be confined to the lower portion of each whorl. Refer to *T. venilia* (**100**) for comparison.

100. *Terebra venilia* Tenison-Woods, 1880

(pl. 26, fig. 100a;
pl. 27, fig. 100b)

1880 *Terebra venilia* Tenison-Woods, Proc. Linn. Soc. N.S.W. 4: 23, pl. 4, figs. 2, 2a; 1978 Hinton, Guide Austral. Shells pl. 59, fig. 4.

Description: Shell to 45 mm; color creamy white with very faint often interrupted orange spiral bands, outline of whorls turreted; subsutural band defined by punctations between ribs; axial ribs slender, slightly curved, about 21 on penultimate whorl; area anterior to band frequently smooth; spiral striae on remainder of whorl of numerous, crowded, fine, up to 10 on penultimate whorl and up to 30 on body whorl; columella slightly arched centrally, with narrow parietal callus; siphonal fasciole with oblique keel.

Type locality: "Sow and Pigs reef, Port Jackson, New South Wales, Australia, 4 fms."
Distribution: New South Wales, S. E. Australia.
Type: Holotype AMS no. C-129007; 41.0 × 8.6 mm.
Discussion: This species is similar to *T. turrita* **(99)** and can be separated by the spiral sculpture, which consists of numerous finely incised striae rather than deeply incised grooves; the centrally arched columella; and the absence of reddish brown blotches.

101. *Terebra trismacaria* Melvill, 1917

(Pl. 27, figs. 101a-b)

1917 *Terebra trismacaria* Melvill, Journ. Conch. 15(6):188, text fig.

Description: Shell to 24 mm, slender; color ivory with lighter subsutural band; outline of whorls turreted; protoconch of 3½ conical whorls; subsutural band marked by deep grooves between ribs; axial ribs narrow, curved, wide spaced, from suture to suture, about 16 on penultimate whorl; spiral grooves numerous, confined to interspaces, about 2 to 6 on subsutural band, 4 to 6 on remainder or whorl; aperture small, elongate; columella recurved, with heavy parietal callus on mature individuals.
Type locality: "Probably off Charbar, Mekran Coast, Persian Gulf."
Distribution: From the Persian Gulf to Fiji.
Type: Holotype BM(NH) no. 1925.3.12.14; 17.0 mm
Discussion: This species has often been confused with *T. polygyrata* (104), but *T. trismacaria* has a much more slender, turreted shell with a heavy parietal callus.

102. *Terebra exiguoides,* Schepman, 1913

(Pl. 27, figs. 102a-b)

1913 *Terebra (Strioterebrum) exiguoides* Schepman, Siboga Exped. 49(5):369, pl. 25, figs. 5, a, b.

Description: Shell to 18 mm; color shiny brownish beige to dark blackish brown; outline whorls slightly convex; protoconch of 2 large mamillate whorls, beige or dark brown; subsutural band with nodes, marked by deep punctations between the ribs; axial ribs quite sharp and wide spaced, 17 to 20 on penultimate whorl, ribs contiguous with nodes on band; spiral grooves, 3 or 4, deep, evenly spaced, not crossing the ribs; aperture elongate; columella slightly recurved.
Type locality: "Bay of Bima, near South Fort, Indonesia, 55 m."
Distribution: From Indonesia to Western Australia.
Type: Holotype ZMUA; 11.5 × 2.7 mm (incomplete).
Discussion: This is an extremely shiny species which exhibits very little variation except for color.

103. *Terebra textilis* Hinds, 1844

(Pl. 27, figs. 103a-e)

1844 *Terebra textilis* Hinds, Proc. Zool. Soc. London. pt. 11:156; 1844 Hinds in Sowerby, Thes. Conch. 1:177, pl. 44, fig. 73; 1860 Reeve, Conch. Icon. 12: pl. 24, sp. 130; 1885 Tryon, Man. Conch. 1st ser. vol. 7:20, pl. 5, figs. 75, 76; 1917 Hirase, Terebridae Jap. Emp. 7: pl. 6, fig. 95; 1967 Cernohorsky, Mar. Shells Pacific 1:206, pl. 52, fig. 388; 1984 Aubry, Terebridae pl. 6.

1850 *Terebra roseata* Adams & Reeve, Zool. Voy. H. M. S. Samarang p. 30, pl. 10, fig. 24; 1960 Reeve, Conch. Icon. 12: pl. 21, fig. 104; 1978 Cernohorsky, Trop. Pacific Mar. Shells p. 148, pl. 53, fig. 2.

1859 *Terebra exigua* Deshayes, Proc. Zool. Soc. Lond. pt. 27:301 1967 Cernohorsky, Mar. Shells Pacific 1:199, pl. 50, fig. 363.

1859 *Terebra soulyeti* Deshayes, Proc. Zool. Soc. Lond. pt. 27: 303.

1873 *?Myurella macgillivrayi* E. A. Smith, Ann. & Mag. Nat. Hist. 4th ser. vol. 11:267 Guinea, 35 fms.;

1879 *Terebra subtextilis* E. A. Smith; Proc. Zool. Soc. Lond. p. 185, pl. 19, fig. 3; 1917 Hirase, Terebridae Jap. Emp. p. 8, figs. 25, 26; 1961 Oyama & Takemura, Moll. Shells, Resources Exploitation Inst. 5, Strioterebrum fig. 2; 1967 Habe & Kosuge, Stand. Book Jap. Shells Col. 3:102, fig. 13.

1904 *Terebra hizenensis* Pilsbry, Proc. Acad. Nat. Sci. Philadelphia vol. 56:4, pl. 1, figs. 2, 2a.

1966 *Terebra (Strioterebrum) exigua* Deshayes, Cernohorsky & Jennings, Veliger 9(1):53, pl. 6, fig. 38.

1966 *Terebra (Strioterebrum) textilis* Hinds, Cernohorsky & Jennings, Veliger 9(1):55, pl. 6, fig. 36.

Description: Shell to 40 mm; color white, rose, lavender, orange-tan or brown; outline of whorls almost flat; protoconch 3½ slender conical whorls; subsutural band defined by deep punctations or grooves between ribs, grooves sometimes coalescing into a faint groove over ribs, but usually confined to interspaces; axial ribs from suture to suture, slender, sharp, usually enlarged into nodes below suture, about 13 to 21 on penultimate whorl; interspaces wider than ribs; spiral cords rarely crossing ribs, 5 to 11 on

penultimate whorl, including those which, sometimes occur on subsutural band, aperture elongate-quadrate; columella almost straight, with parietal callus on some mature specimens.
Type locality: *T. textilis*, "Sorgoson, Bay of Manila, Philippines and Straits of Macassar; in 6 to 13 fms." The former is here selected as the type locality; *T. roseata*, "Sooloo Is.;" *T. exigua*, "E. coast of Australia;" *T. soulyeti*, "Gulf of Mexico", error; *Myurella macgillivrayi*, "Bruinie Is., S. Coast of New Guinea; *T. subtextilis*, "Japan," *T. hizenensis*, "Hirado, Hizen, Japan."
Distribution: Tropical Indo-Pacific; intertidal to 30 m.
Type: *T. textilis* lectotype BM(NH) no. 1844.6.7.80, 25.7 mm; *T. roseata* questionable type, (pl. 27, fig. 103a) BM(NH) no. 1979131, 29.9 mm; *T. exigua* holotype (pl. 27, fig. 103b BM(NH) 197993, 20.8 × 3.6 mm; *T. soulyeti* probable type (pl. 27, fig. 103c) lectotype MNHNP, 40.6 × 7.5 mm; *T. macgillivrayi* holotype (pl. 27, fig. 103d) BM(NH) no. 1856.12.8.93; *T. subtextilis* holotype (pl. 28, fig. 103f) BM(NH) no. 1878.10.16.28, 35 mm; *T. hizenensis* holotype (pl. 28, fig. ANSP no. 85,993, 26.0 × 6.0 mm.
Discussion: This is an extremely variable species in color and sculpture. The spiral sculpture of some whorls on some specimens, especially worn ones, tends to look more like spiral grooves than cords. The parietal callus will vary also. Of the three specimens in the type lot of *T. roseata* Adams & Reeve, one has a heavy callus, one light, one none at all. Of the type lot of *T. soulyeti* Deshayes, the largest measures only 40.6 × 7.5 mm, but Deshayes, whose measurements have proved quite accurate, gave the measurements as 49 × 8 mm. Although some of his description fits the possible type lot, some of it does not, especially his remark than the species most nearly approaching it is *T. larvaeformis* (**183**). The body whorl of the type of *T. macgillivrayi* E. A. Smith, appears longer and the aperture more elongate than is expected of specimens of this species, but there is a bad break in the early part of the shell which could change the form. The size, color, and sculpture fit the species, but it is included here as being questionably conspecific.

104. *Terebra polygyrata* Deshayes, 1859

(Pl. 28, figs. 104a-b)

1859 *Terebra polygyrata* Deshayes, Proc. Zool. Soc. Lond. 27:301.
1912 *Terebra ambrosia* Melvill, Proc. Malac. Soc. Lond. 10(3):250, pl. 11, fig. 10.

Description: Shell to 18 mm; color from ivory to light beige with subsutural band lighter on most individuals; outline of whorls convex, with early whorls of teleoconch quite inflated; protoconch of $1^{1}/_{2}$ dome-shaped whorls; subsutural band ribbed, defined by groove; axial ribs contiguous with those on band, more widely spaced on early whorls, 16 to 21 on penultimate whorl; spiral grooves occasionally cutting ribs; body whorl having more numerous and deeper spiral grooves; aperture moderately quadrate; columella slightly curved, with moderate parietal callus.
Type locality: *T. polygyrata*, "Philippine Is;" *T. ambrosia*, "Charbar, Mekran Coast," 5 fms.
Distribution: From the Gulf of Oman to the Fiji Is.
Type: *T. polygyrata* holotype BM(NH) no. 1979101, 12.6 × 4.9 mm; *T. ambrosia* lectotype (pl. 28, fig. 104b) BM(NH) no. 1912.9.17.28; 15.9, 5.9 × 4.9 mm.
Discussion: This is a somewhat broad species. It may be distinguished from *T. textilis* (**103**) by its broader, more convex outline, and by its protoconch of $1^{1}/_{2}$ whorls.

Plate 27
FIG. **100b,** *Terebra venilia* Tenison-Woods. Middle whorls of holotype. **101a, b,** *T. trismacaria* Melvill. **101a,** Holotype British Museum (N.H.) No. 1925.3.12.14.; 17.0 mm. **101b,** Momi, Fiji Is.; 22.5 mm. **102a, b,** *T. exiguoides* Schepman. **102a,** Holotype Zoological Museum University Amsterdam; 11.5+ mm. **102b,** Port Hedland, West Australia; 16.0 mm. **103a-e,** *T. textilis* Hinds. **103a,** Questionable type of *T. roseata* Adams & Reeve, British Museum (N.H.) No. 1979131; 29.9 mm. **103b,** Holotype of *T. exigua* Deshayes, same Institution No. 197993; 20.8 mm. **103c,** Probable type of *T. souleyeti* Deshayes, Museum National d'Histoire Naturelle, Paris; 40.6 mm. **103d,** Holotype of *Myurella macgillivrayi* E. A. Smith, British Museum (N.H) No. 1856.12.8.93.; 22.0 mm. **103e,** Lectotype of *T. textilis* Hinds, Brit. Mus. (N.H.) No. 1844.6.7.80.; 25.7 mm.

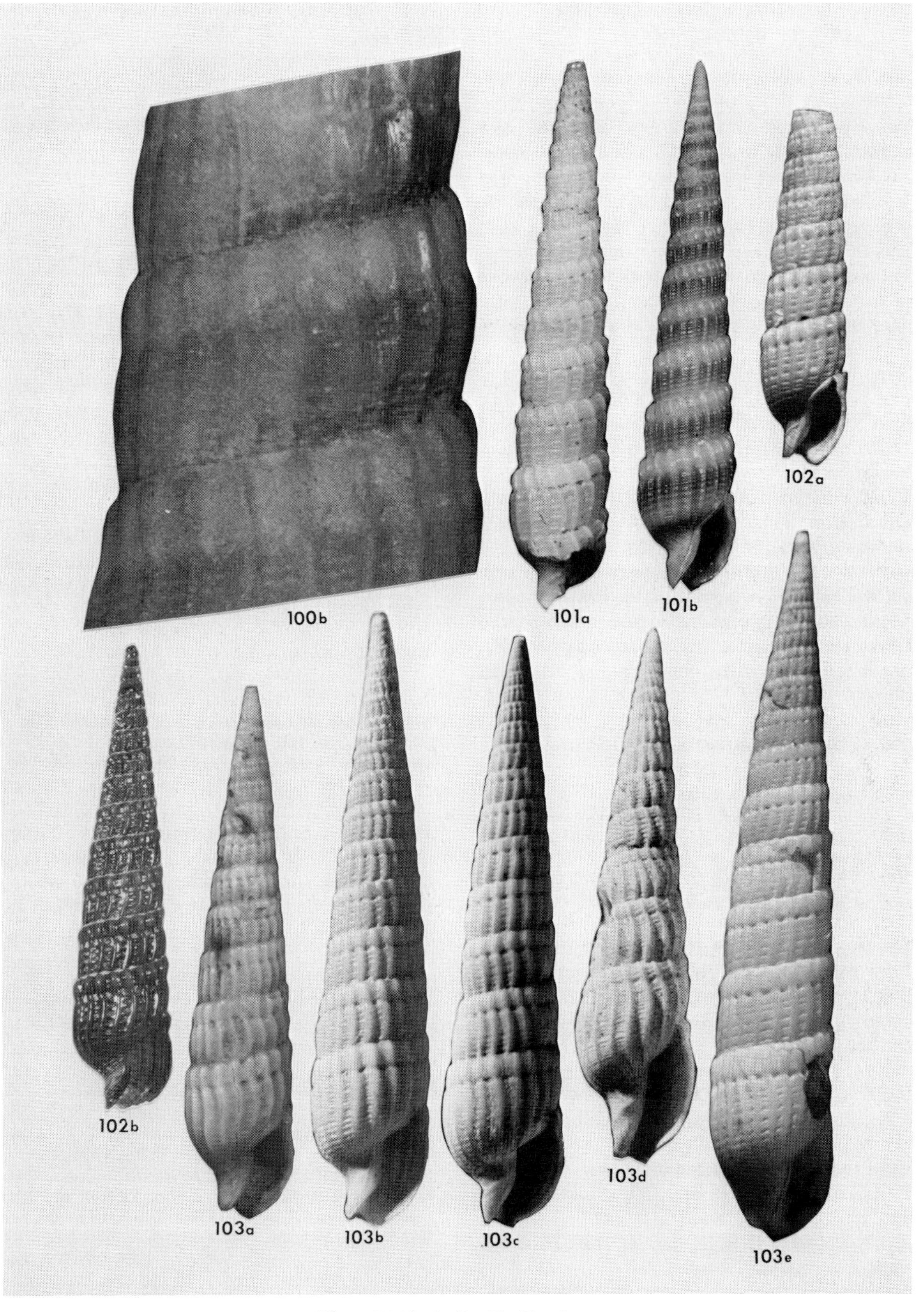

Plate 27. Indo-Pacific *Terebra*.

105. *Terebra turschi* Bratcher, 1981
(Pl. 28, figs. 105a-b)

1981 *Terebra turschi* Bratcher, Veliger 23(4):331, figs. 7, 8.

Description: Shell to 23 mm, slender; color somewhat shiny brown with a light band below suture; outline of whorls turreted; protoconch of 4½ slender whorls; subsutural band defined by grooves between ribs, rarely crossing ribs; axial ribs thin, sharp, curved, from suture to suture, the posterior ends forming weak elongate nodes on band; spiral threads fine, faintly crossing ribs; aperture elongate; columella curved, with heavy parietal callus.
Type locality: "Hansa Bay, North coast of Papua New Guinea, 04°06′S,144°06′E, in 30 m., mud bottom."
Distribution: From Papua New Guinea to Indonesia and Fiji.
Type: Holotype LACM no. 1191; 14.1 × 3.0 mm.
Discussion: This species somewhat resembles the eastern Pacific *T. iola* **(182)** which lacks the parietal callus and the groove marking the subsutural band. *T. polygyrata* **(104)** has a broader, heavier shell, a quadrate aperture, and lacks the heavy parietal callus. This species was named in honor of the Belgian malacologist, Bernard Tursch.

106. *Terebra contracta* (E. A. Smith, 1873)
(Pl. 28, fig. 106)

1873 *Myurella contracta* E. A. Smith, Ann. Mag. Nat. Hist. ser. 4 11:268.
1901 *Terebra contracta* (E. A. Smith), Melvill & Standen, Proc. Zool. Soc. Lond. 2:428.

Description: Shell to 16 mm; color gray with brownish early whorls and brown aperture; outline of whorls turreted; subsutural band indicated only by a spiral row of grooves anterior to the suture; axial ribs curved, from suture to suture; spiral cords evenly spaced, faintly crossing ribs, about 10 per whorl; body whorl elongate; aperture elongate; columella curved, brown, with parietal callus.
Type locality: None given.
Distribution: Arabian Sea. (Melvill & Standen, 1901)
Type: Holotype BM(NH) no. 1979137; 16 × 3.1 mm.
Discussion: This species was subsequently reported by Melvill & Standen from Pakistan.

107. *Terebra subangulata* Deshayes, 1859
(Pl. 29, figs. 107a-b)

1859 *Terebra subangulata* Deshayes, Proc. Zool. Soc. Lond. pt. 26: pt. 26:300; 1860 Reeve, Conch. Icon. 12: pl. 18, sp. 87; 1977 Bratcher, Nautilus 91(2):42, fig. 11; Aubry, Terebridae pl. 7.

Description: Shell to 37 mm; color yellowish; outline of whorls slightly convex, somewhat turreted; subsutural band angled inward from remainder of whorl, defined by groove, deeper between ribs, cutting lightly through ribs; axial ribs strong, well developed, widely spaced, about 16 on penultimate whorl; interspaces, including those on band, filled with fine spiral threads, about 25 on penultimate whorl; body whorl elongate; aperture narrow and elongate; columella almost straight.
Type locality: None given.
Distribution: Unknown.
Type: Holotype MNHN Paris; 33.6 × 6.8 mm.
Discussion: The type locality and distribution are unknown, and we have seen no specimens other than the holotype.

108. *Terebra cognata* E. A. Smith, 1877
(Pl. 29, fig. 108)

1877 *Terebra (Myurella) cognata* E. A. Smith, Ann. & Mag. Nat. Hist. ser. 4, 19:229.
1901 *Terebra cognata* E. A. Smith, Melvill & Standen, Proc. Zool. Soc. Lond. 2:428, pl. 21, fig. 9.

Description: Shell to 14 mm; color white; outline of early whorls extremely convex, later ones only slightly convex; protoconch of 1½ whorls; subsutural band of elongate nodes, defined by deep grooves between ribs; axial ribs narrower

---

Plate 28
FIGS. **103f-h,** *Terebra textilis* Hinds. **103f,** Holotype of *T. subtextilis* E. A. Smith, British Museum (N.H.) No. 1878.10.16.28.; 35.0 mm. **103g,** Holotype of *T. hizenensis* Pilsbry, Academy Natural Sciences, Philadelphia; 26.0 mm. **103h,** Middle whorls, **104a-c,** *T. polygyrata* Deshayes. **104a,** Holotype British Museum (N.H.) No. 1979101; 12.6 mm. **104b,** Lectotype of *T. ambrosia* Melvill, same Institution No. 1912.9.17.28.; 15.9 mm. **104c,** Schouten I., West Irian; 16.0 mm. **105a, b,** *T. turschi* Bratcher. **105a,** Holotype Los Angeles County Museum No. 1191; 14.1 mm. **105b,** Middle whorls. **106,** *T. contracta* (E. A. Smith). Holotype British Museum (N.H.) No. 1979137; 16.0 mm.

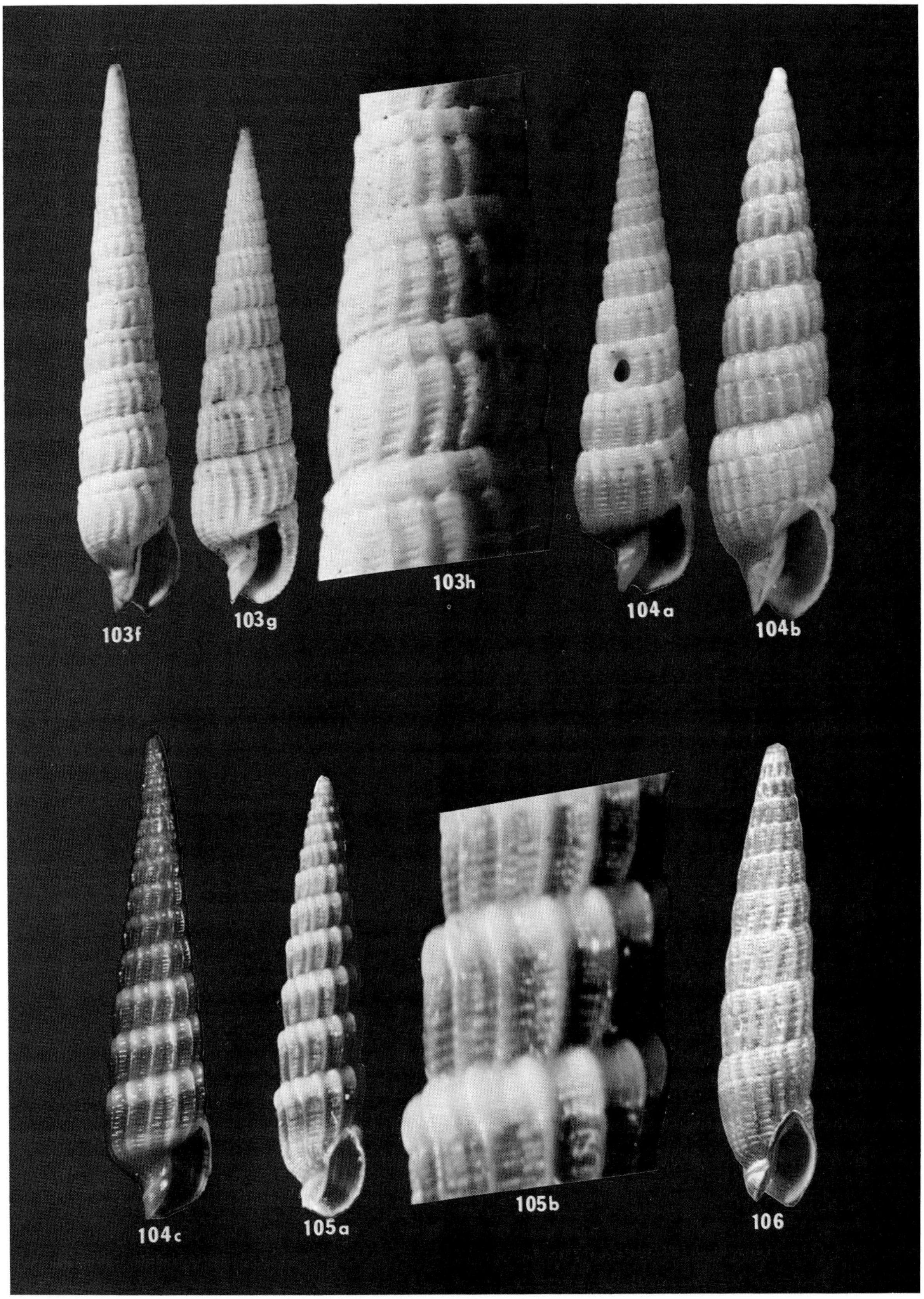

Plate 28. Indo-Pacific *Terebra*.

than interspaces; spiral cords, about 6 per whorl; aperture quadrate; columella curved.
Type locality: "Persian Gulf."
Distribution: Persian Gulf to Karachi, Pakistan; 6 to 14 m.
Type: Holotype BM(NH) no. 1873.7.5.7; 13.3 × 3.2 mm.
Discussion: Melvill & Standen mentioned more than once that this species and *T. ambrosia* Melvill, which is a junior synonym of *T. polygrata* **(109)** are very closely allied. They have the same range and may prove to be conspecific. The types of *T. polygrata* and *T. ambrosia* are pale beige with somewhat lighter subsutural bands.

109. *Terebra polygonia* Reeve, 1860

(Pl. 29, fig. 109)

1860 *Terebra polygonia* Reeve, Conch. Icon. 12: pl. 27, sp. 154.

Description: Shell to 11 mm; color light fulvous-brown with darker subsutural band and darker base of body whorl; outline of whorls convex, protoconch of 3½ slender whorls; subsutural band narrow, somewhat constricted, with weak nodes, band defined by groove; axial ribs rather sharp, wide spaced, 13 on penultimate whorl; spiral sculpture not evident; body whorl with groove at periphery; aperture moderately elongate; columella slightly recurved.
Type locality: None given.
Distribution: Unknown.
Type: Holotype BM(NH) no. 1874.12.11.302; 10.7 mm.
Discussion: The type locality and distribution are unknown, and no other specimen is known except the holotype.

110. *Terebra fijiensis* (E. A. Smith, 1873)

(Pl. 29, figs. 110a-c)

1873 *Myurella fijiensis* E. A. Smith, Ann. Mag. Nat. Hist., 4th Ser. 11; 266; 1967 Cernohorsky, Mar. Shells Pacific 1:200, pl. I:200, pl. 50, fig. 366.
1967 *Terebra (Clathroterebra) fijiensis* E. A. Smith, Cernohorsky & Jennings, Veliger 9 (1): . 55. , pl. 6, fig. 42.
1984 *Terebra fijiensis* E. A. Smith, Aubry, Terebridae pl. 9.

Description: Shell to 29 mm; color usually white, occasionally amber with white subsutural and peripheral bands; outline of whorls slightly convex, somewhat turreted; protoconch of 1¾ to 2 conical whorls; subsutural band indicated only by a spiral row of subsutural grooves between ribs; axial ribs numerous, narrow, arcuate from suture to suture; interspaces wider than ribs; spiral cords numerous, evenly spaced, well-developed; aperture elongate; columella recurved.
Type locality: "Ovalau Is., Fiji Is."
Distribution: From the Philippines to Fiji Is.
Type: Holotype BM(NH) no. 1856.10.27.8; 19.8 mm.
Discussion: Although the sculpture of this species is similar to that of *T. fortunei* **(111)**, the shape of the whorls and of the shell is quite different, this species being broader and shorter and the whorls being more turreted and squared off rather than being gently rounded. Refer to *T. fortunei* for comparison.

111. *Terebra fortunei* Deshayes, 1857

(Pl. 29, figs. 111a-c)
(Color pl. E, fig. 1)

1857 *Terebra fortunei* Deshayes, Journ. Conchyl. 6:79, pl. 4, fig. 1; 1860 Reeve, Conch. Icon. 12, pl. 13, sp. 53; 1978 Cernohorsky, Trop. Pacific Mar. Shells p. 148, pl. 52, fig. 8; 1984 Aubry, Terebridae pl. 14.
1913 *Terebra (Strioterebrum) multistriata* Schepman, Siboga Exped. 49I(5):371, pl. 25, fig. 8.
1961 *Clathroterebra fortunei* (Deshayes), Oyama & Takemura, Moll. Shells, Resources Exploitation Inst. 5: *Clathroterebra* fig. 9; 1962 Kira, Shells W. Pacific Col. 1:208, pl. 71, fig. 5 a, b.
1964 *Terebra (Clathroterebra) fortunei* Deshayes, Shikama, Shells World Col. 1:125, fig. 9.

Description: Shell to 68 mm, slender; color white, usually with broad brownish yellow area on base of each whorl; outline of whorls convex; subsutural band indicated only by a spiral row of inconspicuous grooves; axial ribs sharp, arcuate,

---

Plate 29
FIGS. **107a, b,** *Terebra subangulata* Deshayes. **107a,** Holotype Museum National d'Histoire Naturelle, Paris; 33.6 mm. **107b,** Lower whorls. **108,** *T. cognata* E. A. Smith. Holotype British Museum (N.H.) No. 1873.7.5.7.; 13.3 mm. **109,** *T. polygonia* Reeve. Holotype British Museum (N.H.) No. 1874.12.11.302.; 10.7 mm. **110a-c** *T. fijiensis* (E. A. Smith). **110a,** Holotype British Museum (N.H.) No. 1856.10.27.8,; 19.8 mm. **110b,** Zamboanga, Philippines, on *Xenophora* shell; 18.0 mm. **110c,** Rabaul, Papua New Guinea; 20.8 mm. **111a-c,** *T. fortunei* Deshayes. **111a,** Holotype British Museum (N.H.) No. 197961; 68.1 mm. **111b,** Holotype of *T. multistriata* Schepman, Zoological Museum University Amsterdam; 34.2 mm. **111c,** Middle whorls.

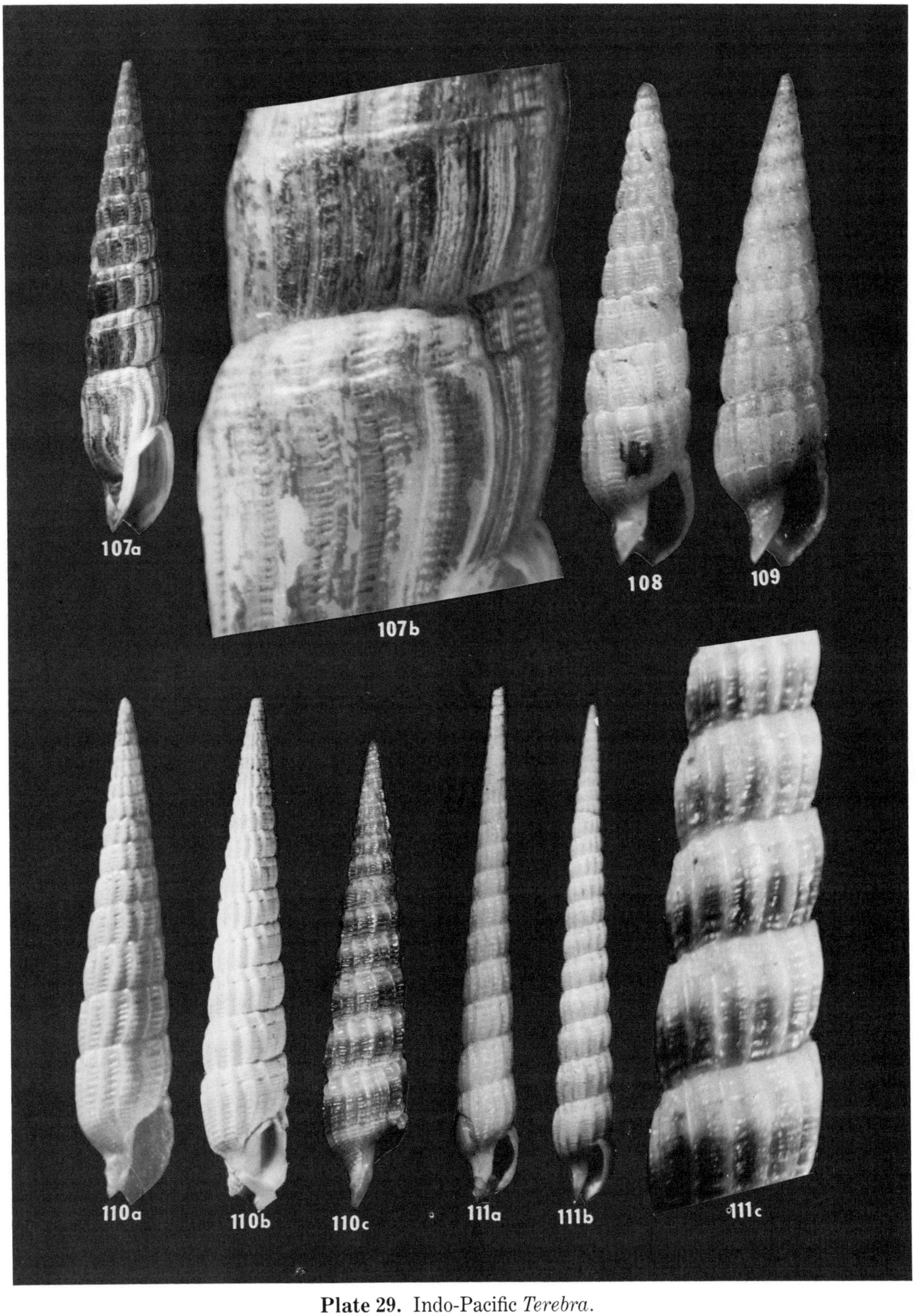

Plate 29. Indo-Pacific *Terebra*.

from suture to suture, narrower than interspaces, about 14 to 19 on penultimate whorl; spiral cords numerous, unevenly spaced, of varying strengths, faintly crossing ribs, 16 to 23 on penultimate whorl; body whorl and aperture elongate; columella recurved.

Type locality: *T. fortunei*, "China Seas"; *T. multistriata*, "Kwandang Bay entrance," Indonesia.

Distribution: From China and Japan to the Philippine Islands, northern Queensland, Australia, and Melanesia; subtidal.

Type: *T. fortunei* holotype BM(NH) no. 197961, 68.1 mm. *T. multistriata* holotype (pl. 29, fig. 111b) UZMC, 34.2 × 4.2 mm.

Discussion: This is the type species of the genus *Clathroterebra* Oyama, 1961. It has gently rounded whorls, curving outward to the center of the whorl, then inward. The shell of *T. fijiensis* **(110)** is slightly turreted, with more "squared-off" whorls, is less slender, lacks the elongate body whorl and is a smaller species, not attaining the length of *T. fortunei*. Named after an English collector, Riley Fortune, who travelled throughout China in the 1840s.

112. *Terebra mactanensis* Bratcher & Cernohorsky, 1982

(Pl. 30, figs. 112a-b)
(Color pl. E, figs. 2, 3)

1982 *Terebra mactanensis* Bratcher & Cernohorsky, Nautilus 96(2):63, figs. 1, 10; 1986 Springsteen & Leobrera, Shells of the Philippines p. 260, pl. 74, fig. 13.

Description: Shell to 62 mm, extremely slender; color light orange-brown or dark brown with white early whorls and occasional white spots between ribs below the suture and small white dots at the periphery of the body whorl, or white with similarly located yellowish brown spots, peripheral spots sometimes absent on white specimens; outline of whorls somewhat convex, slightly turreted; protoconch of 3 white conical whorls; subsutural band defined by punctations or short grooves between ribs; axial ribs straight, from suture to suture, narrower than interspaces, more widely-spaced on some individuals than others, from about 13 to 19 on penultimate whorl; spiral grooves, about 8 to 18, confined to interspaces; ribs on body whorl terminating at periphery in a white spot; aperture elongate; columella recurved, white; siphonal fasciole extremely long.

Type locality: "Punta Engano, Mactan Island, Cebu, Philippines at 200 m."

Distribution: Known only from the type locality.

Type: Holotype LACM no. 1968; 54.4 × 7.2 mm.

Discussion: Although this species is known only from the type locality, the tangle nets have been bringing up a number of specimens. The white form may be compared with *T. fortunei* **(111)** which has more inflated whorls and more arcuate ribs, but lacks the brown dots and somewhat turreted outline.

113. *Terebra fuscotaeniata* Thiele, 1925

(Pl. 30, fig. 113)

1925 *Terebra fuscotaeniata* Thiele, Wiss. Ergeb. Deut. Tief. Exped. Valdiva 17:345. pl. 29, fig. 22.

Description: Shell to 15 mm; color beige with a nebulous dark line at subsutural area and at periphery of body whorl; outline of whorls moderately convex; protoconch of $1^{1/2}$ whorls; subsutural band absent; axial ribs from suture to suture, widely spaced on early whorls becoming less widely spaced later; spiral lines numerous, fine, evenly spaced; aperture quadrate; columella curved.

Type locality: "Padang," Sumatra, Indonesia.

Distribution: Indonesia.

Type: Lectotype MNHUB; 14.7 mm.

Discussion: All in the type lot were collected dead and were fragmented. The species is known only from the type lot and has not been subsequently reported.

Plate 30

FIGS. **112a, b,** *Terebra mactanensis* Bratcher & Cernohorsky. **112a,** Holotype Los Angeles County Museum No. 1968; 54.4 mm. **112b,** Middle whorls. **113,** *T. fuscotaeniata* Thiele. Lectotype Zoological Museum Humboldt University, Berlin; 14.7 mm. **114a, b,** *T. castaneostriata* Kosuge. **114a,** Middle whorls. **114b,** Holotype Institute of Malacology, Tokyo No. 79-20.; 33.1 mm. **115a-c,** *T. lauretanae* Tenison-Woods. **115a,** Immature holotype Australian Museum, Sydney No. C-2464; 19.7 mm. **115b** *Holotype of T. lauretanae tabifica* Iredale, same Institution No. C-24436; 41.7 mm. **115c,** Port Hacking, New South Wales, Australia; 33.3 mm. **116,** *T assecla* (Iredale). Holotype Australian Museum, Sydney No. C-102823; 28.6 mm.

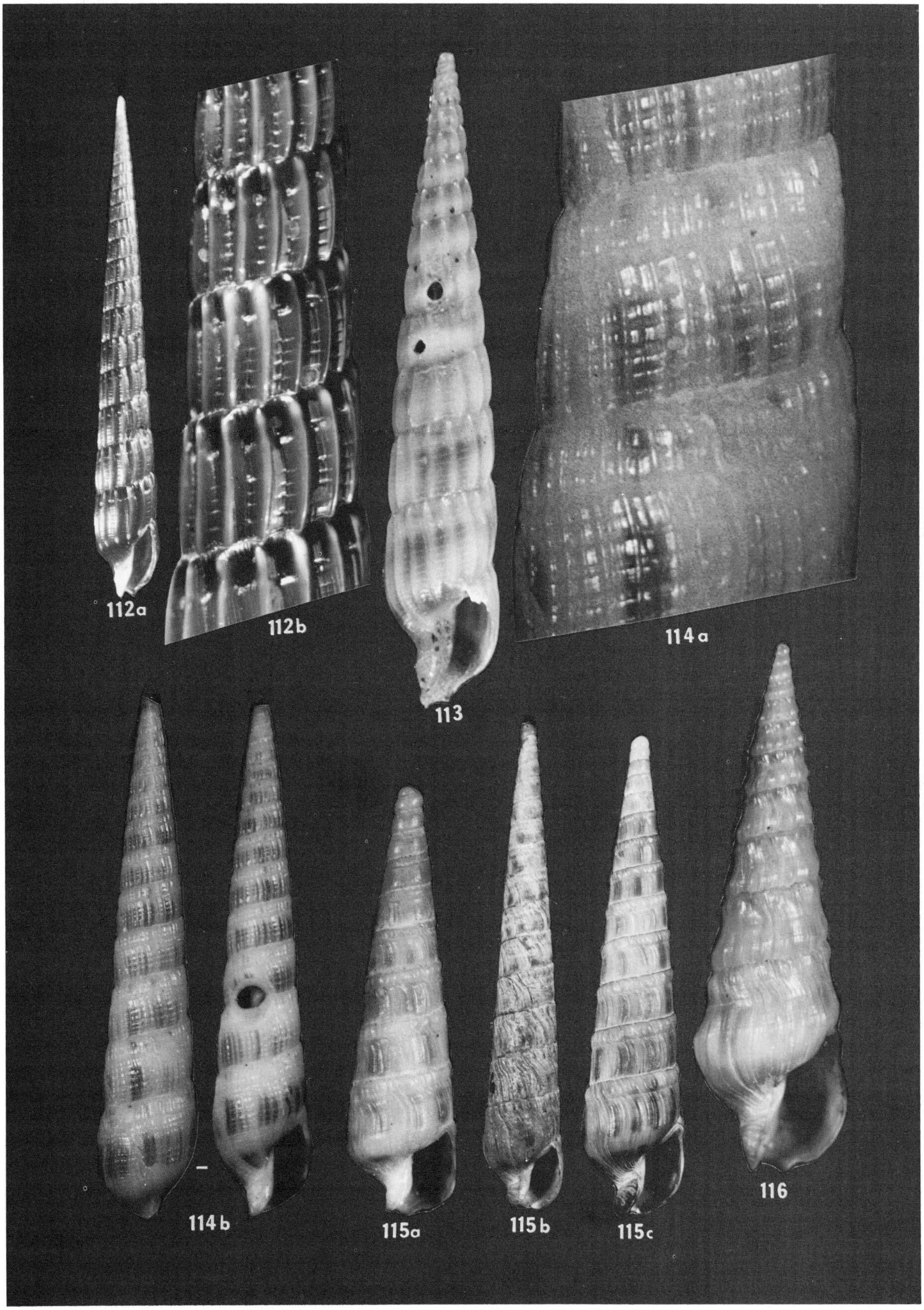

Plate 30. Indo-Pacific and Australian *Terebra.*

114. *Terebra castaneostriata* Kosuge, 1979

(Pl. 30, figs. 114a-b)

1979 *Terebra castaneostriata* Kosuge, Bull. Inst. Mal. Tokyo 1(2):32, pl. 5, fig. 7.

Description: Shell to 37 mm; color white, anterior half of whorls streaked with orange brown or reddish brown; outline of whorls slightly convex; subsutural band flat, with 2 spiral grooves between axial riblets, band defined by subsutural groove; axial sculpture of very slender, low, irregular riblets, 41 on penultimate whorl; spiral grooves, about 7 on penultimate whorl; aperture quadrate; columella straight.
Type locality: "Off Midway Island (28°33.1′N, 176°39.4′W); 84 m."
Distribution: Midway Island and possibly Hawaii.
Type: Holotype I. M. Tokyo no. IMT–79–20; 33.1 × 6.7 mm.
Discussion: One dead-collected specimen, which appears to be the same species, was dredged from deep water at Honolulu, Hawaii, by E. R. Cross.

115. *Terebra lauretanae* Tenison-Woods, 1878

(Pl. 30, figs. 115a-c)

1878 *Terebra lauretanae* Tenison-Woods, Proc. Linn. Soc. New South Wales 2:262; 1906 Hedley & Petterd, Rec. Australian Mus. 6(3):222, pl. 37, fig. 9; 1923 May, Illust. Index Tasmanian Shells, pl. 34, fig. 12.
1925 *Terebra lauretanae tabifica* Iredale, Rec. Australian Mus. 14(4):268.
1978 *Nototerbra (sic) lauretanae* (Tenison-Woods), Hinton Guide Austral. Shells, pl. 59, fig. 2, 2a.

Description: Shell to 48 mm; color yellowish beige with a double row of brown spots which occasionally coalesce into axial rectangles in center of whorl; outline of whorls flat; protoconch of 1½ mamillate whorls; subsutural band weakly noded, defined by shallow groove; axial ribs weak, curved, striae-like; spiral sculpture of inconspicuous, microscopic striae; aperture moderately elongate; columella slightly curved.
Type locality: *T. lauretanae*, "Off Port Jackson Heads, 45 fms," New South Wales; *T. lauretannae tabifica*, "Off Sydney, N.S.W., Australia, 300 fms."
Distribution: Eastern Australia, New South Wales to Tasmania; to 600 m.
Type: *T. lauretanae* holotype AMS no. C-2464; 19.7 × 4.0 mm (immature specimen); *T. lauretanae tabifica* holotype (pl. 30, fig. 115b) AMS no. C-24435, 41.7 × 7.0 mm.
Discussion: The holotype of this species is an immature individual with only eight whorls of the teleoconch. The shape of the mature aperture has not yet developed, and it is not so slender as more mature ones. The width of the shell seems quite variable. The sculpture with the subsutural band being more pronounced in some specimens than in others.

116. *Terebra assecla* (Iredale, 1924)

(Pl. 30, fig. 116)

1924 *Pervicacia assecla* Iredale, Proc. Linn. Soc. N.S.W. 49(3):263, pl. 36, fig. 16; 1976 Garrard, Jour. Malac. Soc. Austral. 3(3–4):181, fig. 3, figd. holotype; 1978 Hinton, Guide Austral. Shells, pl. 59, fig. 8.

Description: Shell to 28 mm; color shiny pinkish white to light fawn with fulvous spots between subsutural nodes, light brown anterior to periphery of body whorl; outline of whorls moderately bulging posterior to the suture, remainder of whorl concave; protoconch 1½ whorls; subsutural band with a row of nodes, defined by a shallow, broad depression, another row of nodes occurring posterior to suture; axial striae becoming almost obsolete ribs, crossed by microscopic spiral striae; aperture quadrate; columella straight.
Type locality: "Twofold Bay, New South Wales, 18–46 m." Australia.
Distribution: New South Wales to South Australia and Tasmania; 20 to 238 m.
Type: Holotype AMS no. C-102823; 28.6 × 7.4 mm.
Discussion: It is easy to separate this species from other Australian terebrids, none of which have nodes both anterior and posterior to the suture, causing the outline to be concave.

117. *Terebra ballina* (Hedley, 1915)

(Pl. 31, figs. 117a-b)

1915 *Duplicaria ballina* Hedley, Proc. Linn. Soc. New South Wales 39(4):729, pl. 84, fig. 86.
1952 *Diplomeriza ballina* (Hedley), Cotton, Trans. Royal Soc. Austral. 75:42.
1976 *Terebra (Punctoterebra) ballina* (Hedley), Garrard, Austral. Shell News. No. 13:3. Text fig.
1978 *Punctoterebra ballina* (Hedley), Hinton, Guide Austral. Shells, pl. 59, fig. 5.

Description: Shell to 34 mm; color from pale cream or grayish mauve to dark reddish brown; outline of whorls convex; protoconch of 3 amber whorls; subsutural band indicated only by deep punctations between ribs; axial ribs, 15 to 22, sharp, slanting; spiral grooves, 12 to 16, filling moderately wide interspaces; aperture quadrate; columella curved, with parietal callus.
Type locality: "Trial Bay, N.S.W." Australia.
Distribution: Southern Queensland to Port Jackson, N.S.W., interridal to 30 m.
Type: Holotype AMS no. C-28471; 23.5 × 5.3 mm.
Discussion: This species suffers from a weak outer lip which frequently is found broken or repaired. Live-taken specimens are rare.

118. *Terebra jacksoniana* (Garrard, 1976)

(Pl. 31, fig. 118)

1900 *Terebra fictilis* Hedley, Proc. Linn. Soc. N.S.W. 25:509, pl. 26, fig. 14. [non Hinds, 1844]
1976 *Pervicacia jacksoniana* Garrard, Journ. Malac. Soc. Australia 3(3–4):183, fig. 6.

Description: Shell to 15 mm; color shiny pale cream with light brown spots between nodes and suture, light brown or orange flamules and streaks on most whorls, and light brown anterior to periphery of body whorl; outline of whorls almost straight, turreted; protoconch of 1½ whorls; subsutural band broad, of large blunt nodes defined by a thin depressed groove; axial ribs of thick, rounded almost noded posteriorly and almost obsolete anteriorly, macroscopic growth striae evident; spiral sculpture absent; aperture quadrate; columella straight, with slight parietal callus.
Type locality: "Post Jackson, New South Wales," Australia.
Distribution: Northern N.W.S. to Victoria, and South Australia; subtidal to 25 m.
Type: Holotype AMS no. C8097; 15.2 × 4.3 mm. (very worn specimen).
Discussion: The few large coarse nodes below the suture, with a small brown spot above each and the short, wide, blunt ribs below the subsutural band make this species easily recognizable.

119. *Terebra scalariformis* (Cotton & Godfrey, 1932)

(Pl. 31, fig. 119)

1932 *Gradaterebra scalariformis* Cotton & Godfrey, S. Australian Naturalist 13(2):66, pl. 13, fig. 9.

Description: Shell to 12 mm; color brown; outline of whorls extremely convex; protoconch of 1½ whorls; subsutural band indicated by weak constiction of ribs; axial ribs fine, slender, from suture to suture, 28 on penultimate whorl; spiral sculpture absent; aperture of holotype broken; columella slightly curved.
Type locality: "Newland Head, South Australia, 20 fms."
Distribution: South Australia; to 40 m.
Type: Holotype SAM no. D-110176; 12.1 × 3.2 mm.
Discussion: Some specimens are translucent with a blackish purple protoconch. This is the type species of the genus *Gradaterebra* Cotton & Godfrey, 1932. Their diagnosis of the genus is, "Tapering, narrow; axial ribs numerous; flexous; protoconch blunt, two turns, smooth, round; last whorl cylindrically oblong, contracted at the base; columella slightly convex; anterior notch a little everted. Closely related to *Euterebra* but distinguished by the development of the axials and the convex whorls."

120. *Terebra tristis* Deshayes, 1859

(Pl. 31, figs. 120a-f, pl. 32, fig. 120g-k)

1859 *Terebra tristis* Deshayes, Proc. Zool. Soc. Lond. pt. 27:306; 1860 Reeve, Conch. Icon. 12, pl. 21, sp. 102; 1885 Tryon, Man. Conch. 7:36, pl. 10, fig. 100; 1913 Suter, Man. N. Z. Moll. p. 515; 1915 Suter, Atlas, pl. 46, fig. 29; 1984 Aubry, Terebridae pl. 10.
1867 *Acus (Abretia) bicolor* Angas, Proc. Zool. Soc. Lond. p. 111, pl. 13, fig. 7.
1867 *Acus (Abretia) assimilis* Angas, *ibid.* p. 111, pl. 13, fig. 8.
1873 *Acus antarctica* E. A. Smith, Ann. & Mag. Nat. Hist. Lond. 4(11):270.
1873 *Cerithium kirki* Hutton, Cat. Mar. Moll. New Zealand p. 27.
1902 *Terebra inconspicua* Pritchard & Gatliff, Proc. Roy. Soc. Victoria 14(2):181, pl. 9, fig. 2.
1909 *Terebra flexicostata* Suter, Proc. Mal. Soc. 8:255, pl. 11, fig. 8; 1913 Suter, Man. N. Z. Moll. p. 514; 1915 Suter Atlas, pl. 22, fig. 20.
1909 *T. tristis crassicostata* Suter, Proc. Mal. Soc. 8:255.
1940 *Pervicacia mariae* Powell, Trans. Royal Soc. New Zealand 70:246, pl. 33, fig. 15; 1961 Powell, Shells N. Z. p. 203, pl. 1, fig. 103.

1952 *Pervicacia subplicata* Cotton, Trans. Roy. Soc. S. Australia 75:40, pl. 3, fig. 3.
1962 *Pervicacia bicolor* (Angas), Macpherson & Gabriel, Mar. Moll. Victoria, p. 240, fig. 280; 1978 Hinton, Guide Austral. Shells pl. 59, figs. 11, 11a, b.
1968 *Pervicacia propelevis* Ponder, Rec. Dom. Mus. 6(4):46, pl. 4, fig. 58
1969 *Pervicacia tristis* (Deshayes), Rudman, Veliger 12(1):53, text figs., animal & anatomy.
1976 *Pervicacia bicolor bicolor* (Angas), Garrard, Journ. Malac. Soc. Australia 3(3–4):181, figs. 15, 17.
1976 *Pervicacia bicolor subplicata* (Cotton) Garrard, *ibid.*, (3–4):182.
1979 *Duplicaria (Pervicacia) tristis* (Deshayes), Powell, New Zealand Moll. p. 246, pl. 47, figs. 16, 17.
1979 *Duplicaria (Pervicacia) flexicostata* (Suter), Powell, *ibid.*, p. 246, pl. 47, fig. 15.
1979 *D. (Pervicacia) propelevis* (Ponder), Powell, *ibid.*, p. 246.

Description: Shell to 24 mm, variable in shape, color and sculpture; color cream to brown, sometimes streaked or mottled with brown, usually dark anterior to periphery of body whorl; outline of whorls convex, occasionally faintly indented anterior to sutures; protoconch of 1½ to 2 whorls; axial sculpture of feeble riblets, or indistinct ribs which become obsolete on last 3 or 4 whorls or which persist to the body whorl; spiral sculpture absent to naked eye, some individuals showing very fine macrostriae; aperture quadrate; columella weakly or distinctly recurved.
Type locality: *T. tristis*, "Japan," error. Ohope Beach, New Zealand is here designated as the type locality; *A. bicolor*, "Middle Harbour, Port Jackson," NSW, Australia; *A. assimilis*, "Port Jackson," NSW, Australia; *A. antarctica*, "Antarctic region," New Zealand; *C. kirki* "Auckland, New Zealand"; *T. inconspicua*, "off Rhyll Is., Phillip Is., Westernport, Victoria, 5–6m"; *T. flexicostata*, *"Cape Maria, van Diemen," New Zealand; T. tristis crassicostata*, "Lyall Bay," New Zealand; *P. mariae*, "Cape Maria, van Diemen," New Zealand; *P. subplicata*, "Backstairs Passage," S. Australia, 2 fms.; *P. propelevis*, "Waikanae Beach, W. Coast, Wellington," New Zealand.
Distribution: Throughout New Zealand and from southern Queensland to Tasmania and South Australia; from shallow water to 183 m.
Type: *T. tristis* holotype BM(NH) no. 1979115; 17.7 × 5.3 mm, *A. bicolor* lectotype (pl. 32, fig. 120h) BM(NH) no. 1870.10.26.52., 15.0 mm; *A. assimilis* lectotype (pl. 31, fig. 120b) BM(NH) no. 1870.10.26.53, 11.7 mm; *A. antarctica* lectotype (pl. 31, fig. 120c) BM(NH) no. 1880.7.2.5, 13.3 × 4.1 mm; *T. kirki* holotype Canterbury Mus., New Zealand; *T. inconspicua* holotype (pl. 32, fig. 120j) NMVM no. F-541, 10.8 × 3.3 mm; *T. flexicostata* lectotype (pl. 32, fig. 120i) New Zealand Geol. Survey no. 4145, 18.9 × 5.1 mm; *T. tristis crassicostata* lectotype (pl. 31, fig. 120e) N. Z. Geol. Surv. no. 4089, 9.5 × 3.1 mm; *P. mariae* holotype (pl. 31, fig. 120f) AIM no. TM-617, 22.3 × 7.7 mm; *T. subplicata* holotype (pl. 32, fig. 120g) SAM no. D-10177, 10.1 × 3.3 mm; *P. propelevis* paratype (pl. 32, fig. 120k) Nat. Mus. N. Z., Wellington, 9.7 mm.
Discussion: In a large sample from Ohope Beach, New Zealand, which we have examined, one could observe a complete gradation between large and small, broad and narrow, weakly and strongly sculptured individuals. Mr. B. Marshall, National Museum of New Zealand, Wellington *(in litt.)* expressed his opinion that *T. crassicostata* (Suter), *T. flexicostata* (Suter), *T. mariae* (Powell), and *T. propelevis* (Ponder) are all forms of *T. tristis*. We concur with Mr. Marshall's findings, and we also include the Australian populations in this species complex. In the Australian populations one can find a similar gradation of characters and a similar number of forms described. Garrard (1976) also realized that *T. bicolor* Angas, *T. assimilis* (Angas), and *T. subplicata* (Cotton) are forms of the same species. *T. tristis* is only one of several species which have a common S. E. Australian-New Zealand distribution. *T. inconspicua* Pritchard & Gatliff (now *T. tristis*) is the type species of the genus *Euterebra*.

Plate 31
FIGS. **117a, b,** *Terebra ballina* (Hedley). **117a,** Holotype Australian Museum, Sydney, No. C-28471; 23.5 mm. **117b,** Lennox Head, New South Wales, Australia; 30.9 mm. **118,** *T. jacksoniana* Garrard. Holotype Australian Museum, Sydney, No. C-8097; 15.2 mm. **119,** *T. scalariformis* (Cotton & Godfrey). Holotype South Australian Museum, Adelaide No. D-10176; 12.1 mm. **120a-f,** *T. tristis* Deshayes. **120a,** Holotype British Museum (N.H.) No. 1979115; 17.7 mm. **120b,** Lectotype of *Acus assimilis* Angas, same Institution No. 1870.10.26.53.; 11.7 mm. **120c,** Lectotype of *A. antarctica* E. A. Smith, same Institution No. 1880.7.2.5.; 13.3 mm. **120d,** Te Ao Inlet, Bay of Islands, New Zealand; 14.0 mm. (broad form). **120e,** Lectotype of *T. tristis crassicostata* Suter, New Zealand Geological Survey, Lower Hutt No. 4089; 9.5 mm. **120f,** Holotype of *Pervicacia mariae* Powell, Auckland Institute and Museum No. TM-617; 22.3 mm.

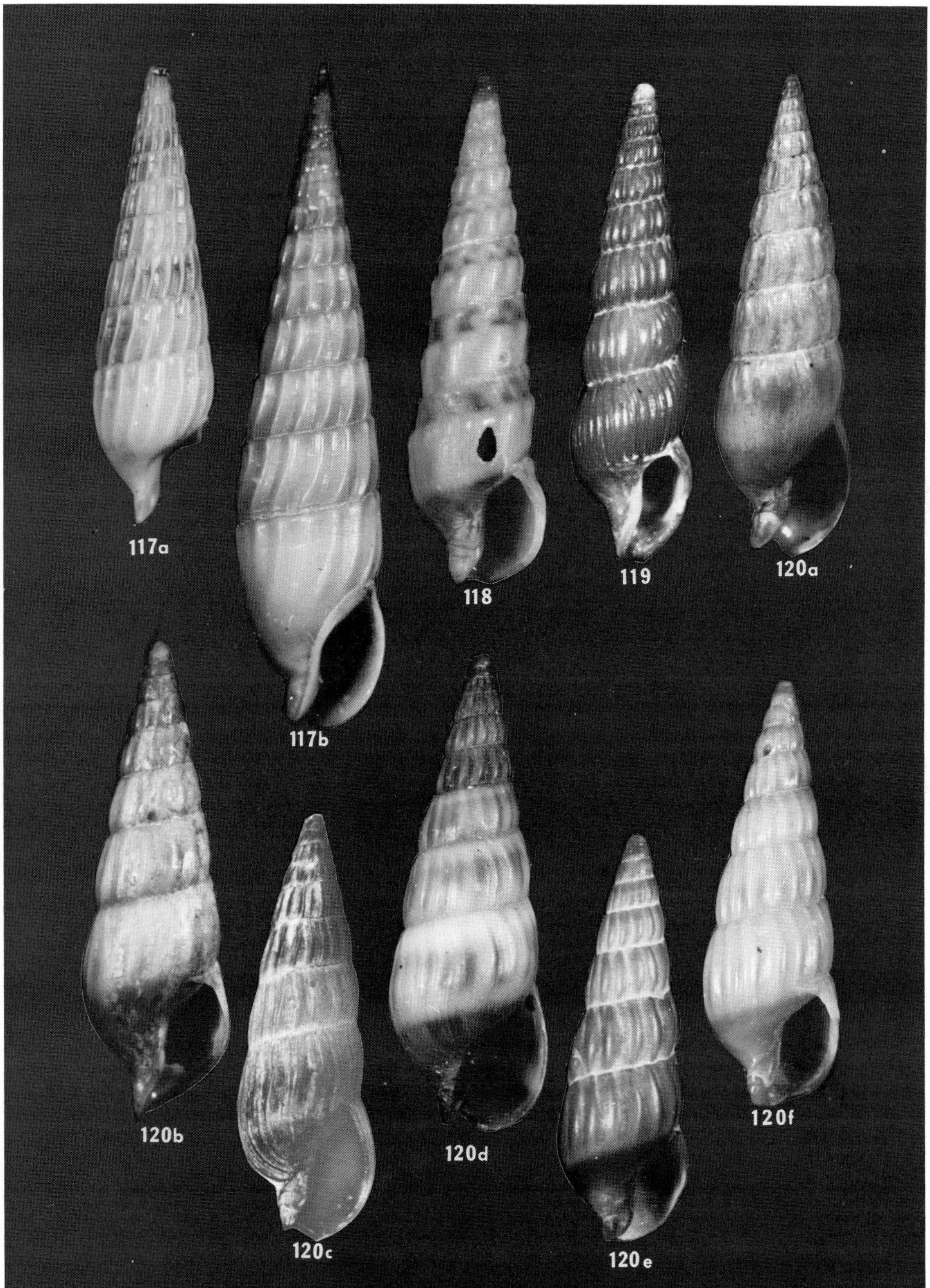

Plate 31. Australian and New Zealand *Terebra*.

121. ***Terebra capensis*** **E. A. Smith, 1873)**

(Pl. 32, fig. 121)
(Color pl. D, fig. 24)

1873 *Myurella capensis* E. A. Smith, Ann. & Mag. Nat. Hist. ser. 4 11:269. 1959 *Terebra capensis* (E. A. Smith), Barnard, Ann. S. African Mus. 44(4):79, fig. 1e.
1984 *Duplicaria capensis* E. A. Smith, Aubry, Terebridae pl. 9.

Description: Shell to 17 mm; color pale buff with darker subsutural, peripheral, and basal bands; outline of whorls convex; protoconch of $1^1/_2$ to 2 whorls; subsutural band and groove absent, but darker coloration and change in angle of ribs give the appearance of a band; axial ribs slightly wider than interspaces; spiral grooves numerous and fine on fresh specimens; aperture quadrate; columella straight.
Type locality: "Port Elizabeth, Cape of Good Hope, South Africa."
Distribution: South Africa.
Type: Lectotype BM(NH) no. 1979139; 17.0 × 4.7 mm.
Discussion: The shell of this species has the appearance of carved from ivory with caramel-colored stain beneath the sutures and periphery. Even worn specimens tend to retain that soft ivory-like sheen. It is a delicate looking little shell.

122. ***Terebra lightfooti*** **E. A. Smith, 1899**

(Pl. 32, fig. 122)

1899 *Terebra (Abretia) lightfooti* E. A. Smith, Journ. Conch. 9(8):247, pl. 5, fig. 1.
1958 *Terebra lightfooti* E. A. Smith, Barnard, Ann. S. African Mus. 44(4):79, fig. 1c; 1973 Kensley, Sea-shells S. Africa p. 216, fig. 871.

Description: Shell to 20 mm, solid; color dark cream with a lighter area around the columella; outline of whorls convex, turreted; appearing to have a protoconch of 2 whorls (most specimens with eroded protoconchs); subsutural band absent; axial ribs wide spaced, rounded, from suture to suture; sutures deep, crenulated by ribs; body whorl long; aperture elongate; columella almost straight.
Type locality: "Table Bay," South Africa; 10 fms.
Distribution: South Africa; 20 to 193 m.
Type: Holotype BM(NH) no. 1899.9.9.2; 19.5 × 6.0 mm.
Discussion: In his original description, Smith mentioned that this species is very closely allied to *T. tristis* **(120)**, as it is indeed, but there is little chance that they are conspecific as the range of *T. tristis* is limited to New Zealand and Australia, while this species is found only in South Africa. Refer to *T. taylori* for comparison. The species was named for a local 19th century South African collector, Mr. Lightfoot.

123. ***Terebra taylori*** **Reeve, 1860**

(Pl. 32, fig. 123)

1860 *Terebra taylori* Reeve, Conch. Icon. 12: pl. 23, sp. 124; 1885 Tryon, Man. Conch. 7:25, pl. 7, fig. 22.

Description: Shell to 20 mm; color gray or light brown; sometimes with a bluish purple hue; outline of whorls convex; protoconch small, mamillate, of $1^1/_2$ to $1^3/_4$ whorls; subsutural band of moderately heavy elongate nodes, defined by a space rather than a subsutural groove; axial ribs thick but short, aligned with the nodes on band; inconspicuous spiral grooves rarely occurring in the interspaces; axial and spiral sculpture fading anterior to subsutural nodes on mature whorls of some individuals; body whorl with ribs becoming weak and with only axial striae on some specimens, others with wide-spaced ribs; aperture quadrate; columella short and somewhat curved.
Type locality: "Torres Strait, Australia."
Distribution: From Tanzania, East Africa, to Japan and the Solomon Is.

Plate 32
FIGS. **120g-k,** *Terebra tristis* Deshayes. **120g,** Holotype of *T. subplicata* Cotton, South Australian Museum, Adelaide No. D-10177; 10.0 mm. **120h,** Lectotype of *Acus bicolor* Angas, British Museum (N.H.) No. 1870.10.26.52.; 15.0 mm (slender, elongate form). **120i,** Lectotype of *Terebra flexicostata* Suter, New Zealand Geological Survey, Lower Hutt No. 445; 18.9 mm. **120j,** Holotype of *T. inconspicua* Pritchard & Gatliff, National Museum of Victoria No. F-541; 10.8 mm. **120k,** Paratype of *Pervicacia propelevis* Ponder, National Museum of New Zealand, Wellington; 9.7 mm. **121,** *Terebra capensis* (E. A. Smith). Lectotype British Museum (N.H.) No. 1979139; 17.0 mm. **122,** *T. lightfooti* E. A. Smith. Holotype British Museum (N.H.) No. 1899.9.9.2.; 19.5 mm. **123,** *T. taylori* Reeve. Holotype British Museum (N.H.) No. 1874.12.11.300.; 19.3 mm. **124a-c,** *T. tantilla* (E. A. Smith). 124a, Holotype British Museum (N.H.) No. 1873.7.5.2.; 7.5 mm. **124b,** Holotype of *Myurella pumilio* E. A. Smith, same Institution No. 1979141; 11.0 mm. **124c,** Type-figure of *Terebra asiatica* Ray, from Ray 1968, textfig.27.

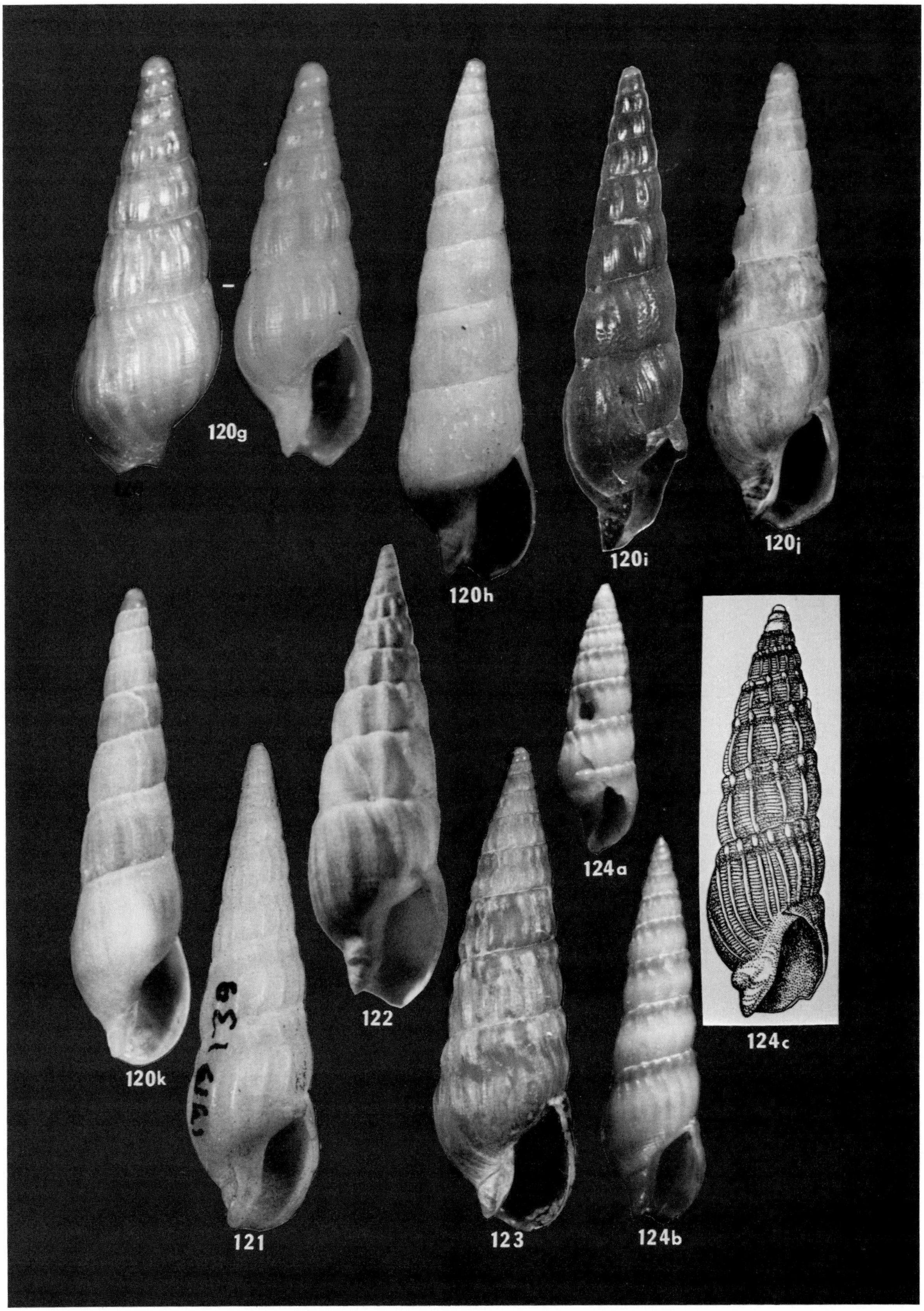

Plate 32. Australian and South African *Terebra*.

Type: Holotype BM(NH) no. 1874.12.11.300; 19.3 mm.
Discussion: This species usually appears sturdy for its size. It differs from *T. lightfooti* **(122)** by having a noded subsutural band and gray, light brown or bluish color. *T. lightfooti* has no subsutural band and is cream color. The species was named for Thomas Lombe Taylor, a well-known English collector of the last half of the 19th century.

124. *Terebra tantilla* E. A. Smith, 1873

(Pl. 32, figs. 124a-c)

1873 *Myurella tantilla* E. A. Smith, Ann. & Mag. Nat. Hist. ser. 4, 11:270.
1873 *Myurella pumilio* E. A. Smith, Ann. Mag. Nat. Hist. 4th ser. 11:269.
1879 *Terebra tantilla* E. A. Smith, Proc. Zool. Soc. Lond. p. 185, pl. 19, fig. 4; 1885 Tryon, Man. Conch. 7:35, pl. 11, fig. 4; 1917 Melvill & Standen, Journ. Conch. 15(7):212.
1968 *Terebra asiatica* Ray, Indian Mus. Bull. Calcutta 3(1/2):28, text fig. 27.

Description: Shell to 11 mm, slender; color cream with a light orange-brown band below the suture and the same color on the base of the body whorl; outline of whorls moderately convex; protoconch of 3 conical whorls; subsutural band narrow, with small rounded nodes defined by a faint depression rather than a subsutural groove; axial ribs weak, curved left; interspaces, including those on band, filled with numerous fine spiral threads; aperture somewhat elongate; columella curved, with a light parietal callus.
Type locality: *T. tantilla*, "Japan," error (?); *T. pumilio*, none; *T. asiatica*, off Puri on the Orissa coast of Bay of Bengal, India."
Distribution: From the Persian Gulf to eastern India, subtidal.
Type: *T. tantilla* holotype BM(NH) no. 1873.7.5.2, 7.5 × 2.3 mm; *M. pumilio* holotype (pl. 32, fig. 124b) BM(NH) no. 1979141, 11.0 × 2.4 mm; *T. asiatica* type figure (pl. 32, fig. 124c).
Discussion: E. A. Smith (1879) was the first to recognize that *T. tantilla* and *T. pumilio* are conspecific. *T. padangensis* **(126)** is similar in size and color but may be separated by its extremely convex outline and by its sharper and fewer axial ribs and more translucent look. However, additional material may prove these two to be conspecific. *T. macandrewii* **(125)**, another small species from the same area, has a broader apical angle, broader subsutural band, and brown markings which are below the subsutural band rather than on it.

125. *Terebra macandrewii* E. A. Smith, 1877

(Pl. 33, fig. 125a-b)

1877 *Terebra (Myurella) macandrewi* E. A. Smith, Ann. & Mag. Nat. Hist. Ser. 4, 19:228, 1901 Melvill & Standen, Proc. Zool. Soc. Lond. 2:428, pl. 21, fig. 6.

Description: Shell to 22 mm; color cream or tan with brown below the periphery of the body whorl and in spots between ribs at the subsutural depression, sometimes coalescing into a stripe; also spots of the same color are present posterior to the suture; outline of whorls convex, with elongate nodes, defined by depression but no groove; axial ribs widely spaced, contiguous with nodes on band; spiral threads fine, from suture to suture; aperture quadrate; columella recurved.
Type locality: "Persian Gulf."
Distribution: Persian Gulf; intertidal to 10 m.
Type: Holotype BM(NH) no. 1873.7.5.5; 13.4 mm.
Discussion: This species has extremely inflated whorls and is sculptured with numerous fine threads. *T. nassoides* **(130)** also commonly has a pattern of spots between the axial ribs, but it has less inflated whorls and a smaller apical angle. It also lacks the spiral threads seen on this species. See *T. tantilla* **(124)** for comparison.

Plate 33
FIGS. **125a, b,** *Terebra macandrewi* E. A. Smith. **125a,** Holotype British Museum (N.H.) No. 1873.7.5.5.; 13.4 mm. **125b,** Masirah I., Sultanate of Oman; 11.4 mm. **126a, b,** *T. padangensis* Thiele. **126a,** Lectotype Zoological Museum Humboldt University, Berlin; 6.7 mm. **126b,** Type-figure of *T. durgella* Ray, from Ray, 1968, textfig.26. **127,** *T. pellyi* E. A. Smith. Lectotype British Museum (N.H.) No. 1873.7.5.1.; 12.4 mm. **128a-c,** *T. fuscobasis* E. A. Smith. **128a,** Holotype British Museum (N.H.) No. 1873.7.5.8.; 11.3 mm. **128b,** Holotype of *T. wilkinsi* Dance & Eames, same Institution No. 1893.12.15.74.; 9.3 mm. **128c,** Paratype of *T. wilkinsi* Dance & Eames, same Institution; 7.8+ mm. **129,** *T. fuscocincta* E. A. Smith. Lectotype British Museum (N.H.) No. 1873.7.5.3.; 7.7 mm. **130a, b,** *T. nassoides* Hinds. **130a,** Lectotype British Museum (N.H.) No. 1968251; 13.5 mm. **130b,** Masirah I., Sultanate of Oman; 18.4 mm.

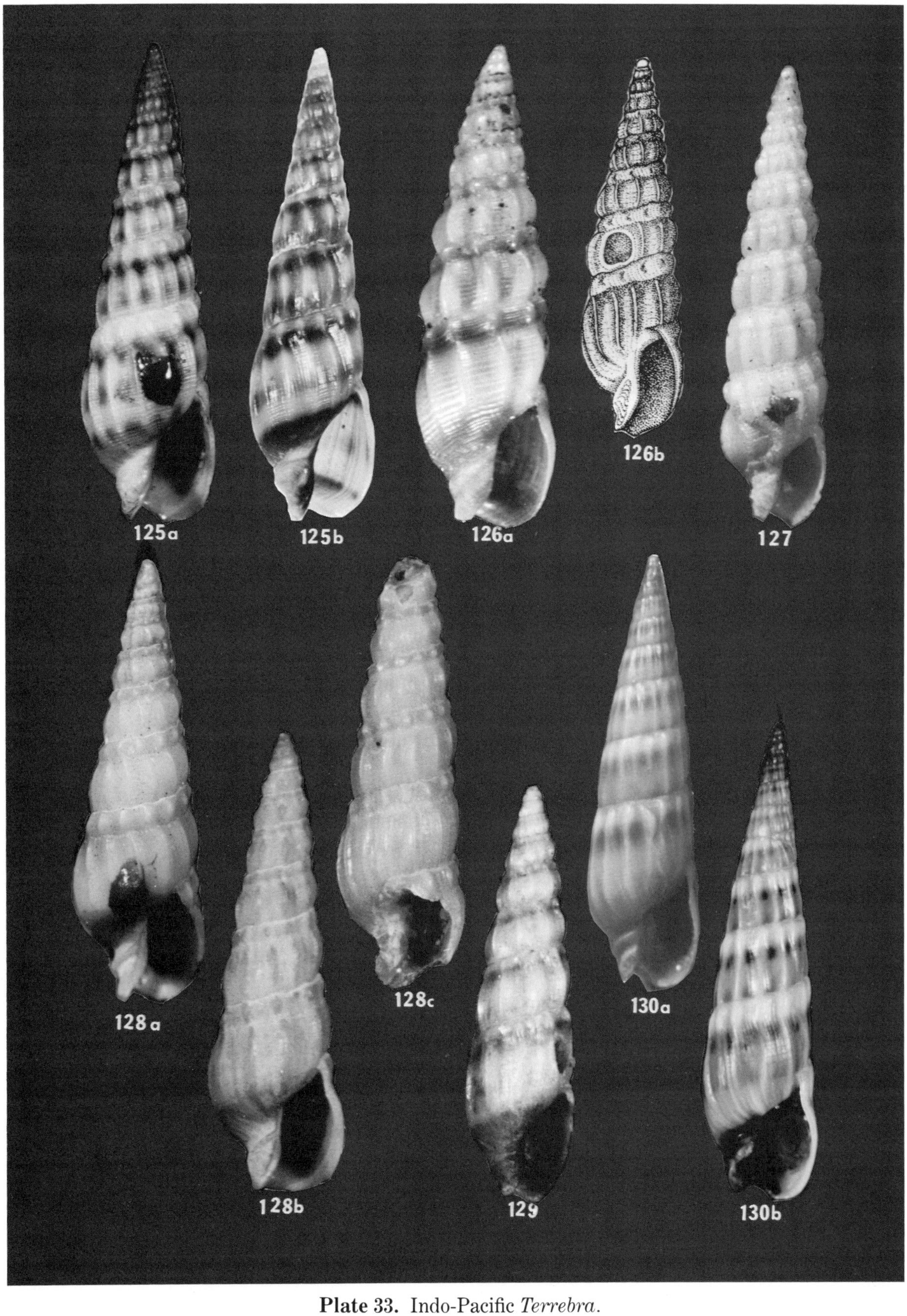

Plate 33. Indo-Pacific *Terrebra.*

126. ***Terebra padangensis*** Thiele, 1925

(Pl. 33, fig. 126a-b)

1925 *Terebra padangensis* Thiele, Wiss. Ergeb. Tief. Exped. Valdiva 17(2):346, pl. 42, fig. 2.

1968 *T. durgella* Ray, Indian Mus. Bull. Calcutta 3(1/2):28, text fig. 26.

Description: Shell to 17 mm; color cream with fulvous subsutural stripe and broader stripe below periphery of body whorl; outline of whorls convex; protoconch of 3 conical whorls; subsutural band with well-developed round nodes defined by subsutural depression but with no subsutural groove; remainder of whorl sculptured with broadly spaced, arcuate ribs crossed by many fine spiral lines; aperture quadrate; columella curved.
Type locality: *T. padangensis*, "Padang," Sumatra, Indonesia, *T. durgella*, "off Puri, Orissa coast, Bay of Bengal."
Distribution: Eastern India to Indonesia.
Type: *T. padangensis* lectotype MNHU, 6.7 mm; *T. durgella* holotype (pl. 33, fig. 126) Zool. Mus. Calcutta no. M-1806/3, 16.5 × 4.5 mm.
Discussion: The type lot consists of many specimens, of which the lectotype of 6.7 mm is the largest. See *T. pellyi* **(127)** for comparison.

127. ***Terebra pellyi*** E. A. Smith, 1877

(Pl. 33, fig. 127)

1877 *Terebra pellyi* E. A. Smith, Ann. & Mag. Nat. Hist. ser. 4, 19:226; 1901 Melvill & Standen, Proc. Zool. Soc. Lond. 2:428, pl. 21, fig. 10; 1917, Jour. Conch. 15(7):210.

Description: Shell to 15 mm; color ivory to ash-gray; outline of early whorls extremely convex, later ones moderately convex; protoconch of 1½ whorls; subsutural band marked by groove cutting through ribs; axial ribs slightly curved; spiral grooves fine, crossing ribs on some individuals only, another narrow spiral groove occurring immediately posterior to the sutures on some specimens including the lectotype, aperture moderately quadrate, columella curved.
Type locality: "Persian Gulf."
Distribution: Persian Gulf to Arabian Sea; 6 to 14 m.
Type: Lectotype BM(BH) no. 1873.7.5.1; 12.4 mm.
Discussion: All eight specimens in the type lot have very thin shells and appear to be juvenile. They were collected by Colonel Pelly. This species may be separated from *T. cognata* **(108)** by its inflated whorls and lack of numerous fine spiral lines. *T. padangensis* **(126)** differs by having a noded, constricting subsutural band.

128. ***Terebra fuscobasis*** E. A. Smith, 1877

(Pl. 33, fig. 128a-c)

1877 *Terebra (Myurella) fuscobasis* E. A. Smith, Ann. & Mag. Nat. Hist. ser. 4, 19:227.

1966 *Strioterberum (Partecosta) wilkinsi* Dance & Eames, Proc. Malac. Soc. Lond. 37(2):42, pl. 3, fig. 4, and pl. 5, fig. 5.

Description: Shell to 11 mm; color white with a fulvous brown zone at or below the subsutural band and below the periphery of the body whorl; outline of whorls angulate on early whorls, convex on later ones; protoconch of 2½ conical whorls; subsutural band with widely spaced nodes; band defined by shallow or distinct, smooth or macroscopically punctate groove; axial ribs arcuate, contiguous with nodes on band; spiral sculpture absent except for faint microstriae; aperture quadrate; columella recurved.
Type locality: *T. fuscobasis*, "Persian Gulf"; *T. wilkinsi*, "Fao, Persian Gulf."
Distribution: Persian Gulf.
Type: *T. fuscobasis* holotype BM(NH) no. 1873.7.5.8, 11.3 mm; *S. wilkinsi* holotype BM(NH) no. 1893.12.15.74, 9.2 × 2.8 mm.
Discussion: This species is variable in the strength of the subsutural groove and the prominence of the subsutural nodes. It is the type species of the genus *Partecosta* Dance & Eames, 1966. Some syntype specimens in the type lot of *T. wilkinsi* Dance & Eames also have a brown-stained base and are more similar to *T. fuscobasis* than to the holotype. Refer to *T. nassoides* **(130)** and *T. fuscocincta* **(129)** for comparisons.

129. ***Terebra fuscocincta*** E. A. Smith, 1877

(Pl. 33, fig. 129)

1877 *Terebra (Myurella) fuscocincta* E. A. Smith, Ann. & Mag. Nat. Hist. ser. 4, 19:228.

Description: Shell to 8 mm; color beige with a fine brown subsutural line and light brown base of body whorl; outline of whorls convex; protoconch of 3 smooth whorls; subsutural band narrow, noded, defined by depression but no

groove; axial ribs widely spaced, about 12 on subsutural whorl, beginning with node on band; spiral sculpture absent; aperture elongate; columella curved.
Type locality: "Persian Gulf."
Distribution: Persian Gulf.
Type: Lectotype here selected BM(NH): no. 1873.7.5.3; 7.7 × 2.0 mm.
Discussion: All five specimens in the type lot were collected dead, and the species has been reported only once since the original description. *T. fuscobasis* **(128)**, another small Persian Gulf species, is distinctly constricted at the base of the body whorl and lacks the interrupted brown bands and brown base of *T. fuscocincta*. See *T. nassoides* **(130)** for comparison.

130. *Terebra nassoides* Hinds, 1844
(Pl. 33, fig. 130a-b)

1844 *Terebra nassoides* Hinds, Proc. Zool. Soc. Lond. pt. 11:158; 1844 Hinds in Sowerby, Thes. Conch. 1:182, pl. 45, fig. 115. 1860 Reeve, Conch. Icon. 12, pl. 26, sp. 144; 1982, K. Smythe, Seashells of the Arabian Gulf 72, pl. 13, fig. d.
1901 *Terebra (Euryta) nassoides* Hinds, Melvill & Standen, Proc. Zool. Soc. Lond. pt. 11. 429.

Description: Shell to 14 mm; color light-brown with a pale gray subsutural area followed by a narrow dark-brown band; outline of whorls faintly convex; protoconch of 2½ conical whorls; subsutural band of color only; axial ribs, from suture to suture, narrower than interspaces, later whorls with ribs sometimes becoming obsolete; spiral sculpture, seen only under magnification, consisting of about 5 evenly-spaced lines; aperture quadrate; columella curved; siphonal fasciole with posterior keel.
Type locality: None given. The Gulf of Oman is here designated as the type locality.
Distribution: Persian Gulf and Gulf of Oman; intertidal.
Type: Lectotype BM(NH) no. 1968251; 13.5 mm.
Discussion: Melvill & Standen said that this species is found in a "great variety of form and coloration, at low tide, just below the surface of the water." It may be separated from the two other small, smooth Persian Gulf species, *T. fuscobasis* **(128)** and *T. fuscocincta* **(129)** by its almost straight outline and by the complete lack of a subsutural band.

131. *Terebra tenera* Hinds, 1844
(Pl. 34, fig. 131)

1844 *Terebra tenera* Hinds, Proc. Zool. Soc. Lond. pt. 11:158; 1944 Hinds in Sowerby, Thes. Conch. 1:184, pl. 45, fig. 111.

Description: Shell to 9 mm, delicate; color almost translucent yellowish beige with reddish brown anterior to suture and the same color on base of body whorl; outline of whorls slightly convex; protoconch of 2½ to 3 conical whorls; subsutural band absent; axial ribs widely spaced, unbroken from suture to suture, 9 to 15 on penultimate whorl; spiral sculpture absent, but with a narrow brown spiral stripe one fourth way down the whorl; aperture semi-elongate; columella curved.
Type locality: "Straits of Malacca, 17 fms."
Distribution: From India to Indonesia.
Type: Lectotype BM(NH) no. 1844.6.7.86–7; 6.2 mm.
Discussion: This tiny species varies in the number of axial ribs and in the apical angle, some individuals having a very broad apex while others are more slender.

132. *Terebra kowiensis* Turton, 1932
(Pl. 34, fig. 132)

1932 *Terebra kowiensis* Turton, Mar. Shells Port Alfred, S. Africa p. 11, pl. 2, fig. 92; 1958 Barnard, Ann. S. African Mus. 44(4):82.

Description: Shell to 18 mm; color shiny pale brown blotched with rich reddish brown, giving the appearance of interrupted brown lines in middle of whorls; base of columella deep mauve; outline of whorls almost straight, slightly turreted, narrowly ledged at suture; protoconch 1½ bulbous whorls; subsutural band absent; axial ribs rather flat, unbroken from suture to suture, about 20 on body whorl; interspaces a mere groove between ribs; spiral sculpture absent; aperture moderately elongate; columella recurved.
Type locality: "Port Alfred, South Africa."
Distribution: South Africa.
Type: Oxford Univ.; 18.0 × 4.5 mm.
Discussion: *T. planecosta* **(133)** may be separated from this species by its smaller size, half the length in the same number of whorls; less numerous axial ribs; brown color, and body whorl which cuts in abruptly below the periph-

ery while that of this species slants in a weak curve almost like that of a *Hastula*.

133. *Terebra planecosta* Barnard, 1958

(Pl. 34, figs. 133a-c)

1958 *Terebra planecosta*, Barnard, Ann. S. African Mus. 44(4):81, figs. 1, 1a, b.

Description: Shell to 9 mm; color shiny cream to brown, darker on base of body whorl; protoconch of 1½ mamillate whorls; subsutural band and groove absent, but ribs may develop small inconspicuous nodes immediately below suture on about first 5 whorls; axial sculpture of straight, broad flat ribs, 17 on penultimate whorl; interspaces mere grooves; spiral sculpture absent; columella recurved, somewhat angled; aperture quadrate.
Type locality: "Off Cove Rock, East London area, 4½ miles, 22 fms," South Africa.
Distribution: South Africa.
Type: Holotype South African Museum no. A8659; 8.7 × 2.2 mm.
Discussion: The ribs on this species are very flat and broad with only a narrow groove between them. Consult *T. kowiensis* Turton **(132)** for comparison.

134. *Terebra castigata* A. H. Cooke, 1885

(Pl. 34, figs. 134a-b))

1885 *Terebra castigata* A. H. Cooke, Ann. & Mag. Nat. Hist. Lond. ser. 5, p. 329; 1978 Bratcher & Mienis, Argamon 6(3–4):21, figs. 1, 2, figured lectotype.

Description: Shell to 20 mm; color shiny warm beige, white at posterior end, light subsutural band with occasional small round fulvous dots scattered at random; outline of whorls concave; protoconch of 2 bulbous translucent beige whorls; subsutural band of inconspicuous elongate nodes, becoming less pronounced in later whorls, defined by groove and sometimes punctations between ribs; axial ribs curved, somewhat swollen at posterior, interspaces equal to ribs; spiral grooves, 2 to 3, widely spaced, finely punctate under magnification; body whorl with ribs fading at periphery; aperture semi-quadrate; columella curved and stained brown; siphonal fasciole quite smooth, with rounded keel.
Type locality: "Gulf of Suez, Jubal Island, southern end of Gulf."
Distribution: The Red Sea, Gulf of Suez and Gulf of Eilat.
Type: Lectotype Zoological Department of the University of Cambridge; 18.5 × 4.9 mm.
Discussion: Specimens of this species show little variation except that the whorls of some are slightly more concave than others. The brown stain on the inside of the columella appears to be constant.

135. *Terebra severa* Melvill, 1897

(Pl. 34, fig. 135)

1897 *Terebra severa* Melvill, Mem. Proc. Manchester Lit. Phil. Soc. 41(7):9, pl. 6, fig. 8; 1917 Melvill & Standen, Journ. Conch. 15(7):211.

Description: Shell to 15 mm; color grayish beige, base somewhat darker; outline of early whorls angulate, later ones moderately convex; protoconch of 1½ large mamillate whorls; subsutural band scarcely evident, a shallow, weak groove sometimes visible on a few whorls; axial ribs on early whorls few, widely spaced, with a node at center causing the angulate outline, later becoming numerous axial striae and axial rows of weak nodes, about 3 per whorl, sometimes coalescing into obsolete ribs; spiral sculpture not evident on some individuals, occasionally with 3 strong cords; body whorl elongate; aperture elongate; columella curved.
Type locality: "Mekran Coast," Arabian Sea.
Distribution: Endemic to the Persian Gulf, the Gulf of Oman, and the Arabian Sea.
Type: Holotype BM(NH) no. 1897.7.30.103; 14.8 × 4.0 mm.
Discussion: Melvill & Standen stated that there are three strong spiral lirae, which are not evident in the holotype.

Plate 34
FIG. **131**, *Terebra tenera* Hinds. Lectotype British Museum (N.H.) No. 1844.6.7.86.; 6.2 mm. **132**, *T. kowiensis* Turton. Holotype Oxford University; 18.0 mm. **133a-c**, *T. planecosta* Barnard. **133a**, Holotype South African Museum No. A-8659; 8.7 mm. **133b**, Cape St. Blaize, Port Elizabeth, Sth. Africa; 19.8 mm. **133c**, Middle whorls. **134a, b**, *T. castigata* Cooke. **134a**, Lectotype Zoological Department, University of Cambridge; 18.5 mm. **134b**, Ras Muhammad, Red Sea; 20.0 mm. **135**, *T. severa* Melvill. Holotype British Museum (N.H.) No. 1897.7.30.103.; 14.8 mm. **136a-c**, *T. walkeri* E. A. Smith. **136a**, Holotype British Museum (N.H.) No. 1891.11.21.326.; 26.2 mm. **136b**, Lower whorls. **136c**, Upper whorls of holotype.

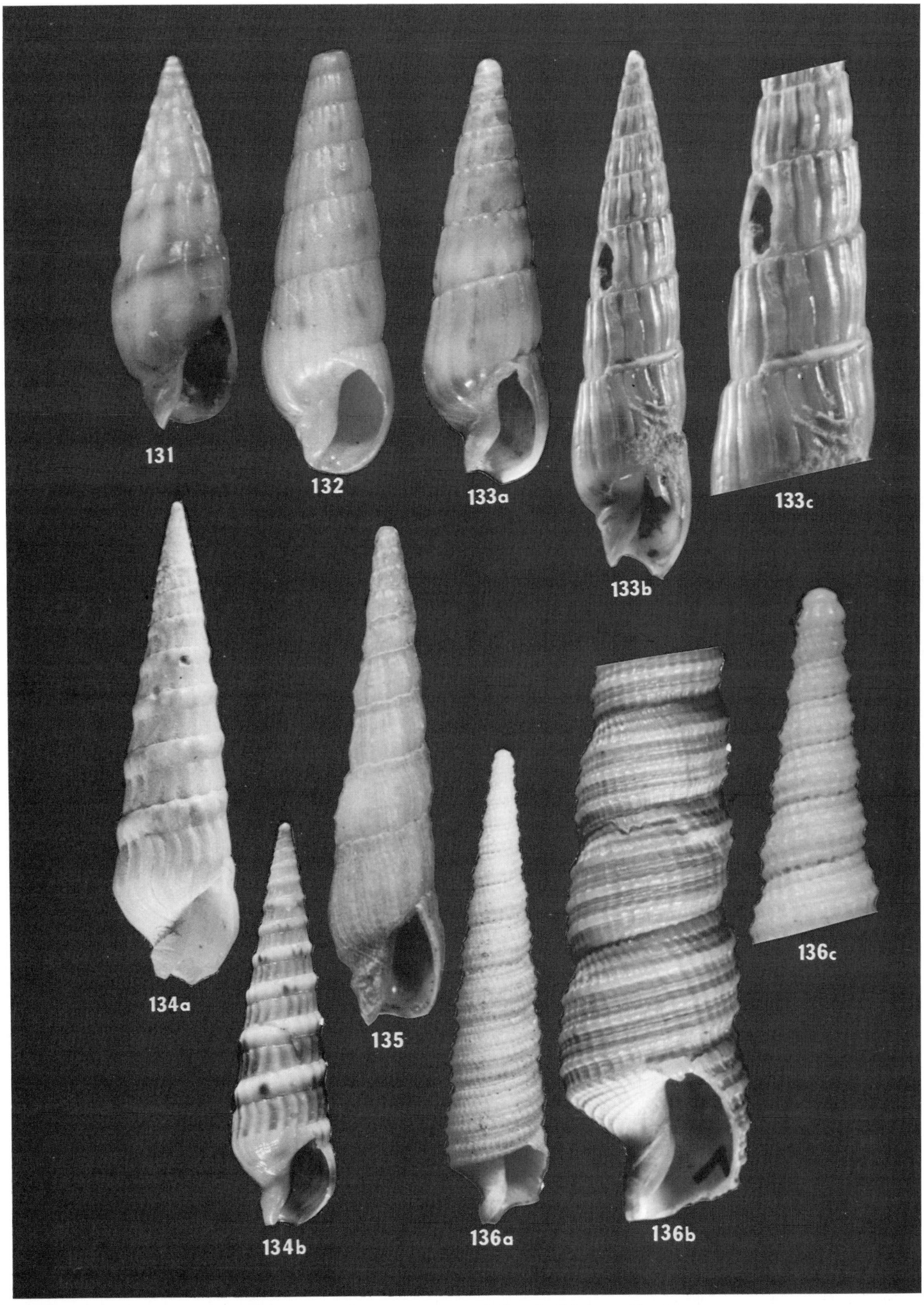

Plate 34. Indian Ocean and South African *Terebra.*

136. ***Terebra walkeri*** **E. A. Smith, 1899**

(Pl. 34, figs. 136a-c)

1899 *Terebra walkeri* E. A. Smith, Proc. Mal. Soc. Lond. 3(6):312, fig. 1.

Description: Shell to 26 mm; color white; outline of whorls convex; protoconch of 2 whorls, the first one flattened, the second one with about 10 sturdy ribs jutting out beyond the following whorl; entire teleoconch with fine cancellate sculpture; subsutural area indented, followed by 3 enlarged sharp cords of small nodes formed by 36 axial cords crossing over about 12 smaller spiral cords; body whorl with 4 keels, the anterior one at the periphery; aperture quadrate; columella curved.

Type locality: "Holothuria Banks, N. W. Australia; 10–30 fms."

Distribution: Endemic to N. W. Australia.

Type: Holotype BM(NH) no. 1891.11.21.326; 26.2 × 6.0 mm.

Discussion: The unusual protoconch and sculpture make the placement of this species in the Terebridae questionable. It may be in the Cerithiacea. It was named for its Australian collector, J. J. Walker.

EASTERN PACIFIC AND ATLANTIC TEREBRA

137. ***Terebra robusta*** **Hinds, 1844**

(Pl. 35, figs. 137a-d)
(Color pl. B, fig. 5)

1844 *Terebra robusta* Hinds, Proc. Zool. Soc. Lond. pt. 11:149; 1844 Hinds in Sowerby, Thes. Conch. 1:152, pl. 42, fig. 35; 1860 Reeve, Conch. Icon. 12, pl. 3, sp. 10; 1944 M. Smith, Panamic Mar. Shells p. 36, fig. 471; 1958 Keen, Seashells Trop. W. America ed. 1:498, fig. 948; 1961 Hanna & Hertlein, Pro. Calif. Acad. Sci. 4th ser. 30(3):71, pl. 6, figs. 3, 7, 8; pl. 7, fig. 1; 1962 Abbott, Sea Shells of the World, p. 125, fig. 2; 1964 Shikama, Sel. Shells World Col. 1:126, pl. 101, fig. 21; 1966 Morris, Field Guide to Shells p. 200, pl. 61, fig. 23; 1971 Bratcher & Burch, Proc. Calif. Acad. Sci. 37 (21) 554, fig. 25; 1971 Bratcher & Burch in Keen, Seashells Trop. W. America, ed, 2:682, fig. 1560; 1982 Abbott & Dance, Compendium of Seashells p. 277, pl. 1; 1984 Aubry, pl. 1.

1844 *Terebra lingualis* Hinds, Proc. Zool. Soc. London pt. 11:153; 1844 Hinds in Sowerby, Thes. Conch. 1:167, pl. 43, fig. 50; 1860 Reeve, Conch. Icon. 12: pl. 5, sp. 15; 1944 M. Smith, Panamic Mar. Shells p. 35, fig. 463; 1958 Keen, Seashells Trop. W. America ed. 1:488, fig. 946; 1961 Hanna & Hertlein, Proc. Calif. Acad. Sci. 30(3):69, pl. 6, figs. 4–6; 1963 Campbell, Veliger 5(6):102, pl. 12, fig. 1; pl. 13, fig. 1; 1964 Shikama, Sel. Shells World Col. 1:125, pl. 101, fig. 19.

1854 *Terebra loroisii* Guérin-Méneville, Rev. Mag. Zool. ser. 2, 6:218, pl. 4, fig. 5. [pers. comm. Dr. Philippe Bouchet].

1857 *Terebra insignis* Deshay Journ. Conchyl. 6:70, pl. 3, fig. 2.

1961 *Terebra dumbauldi* Hanna & Hertlein, Proc. Calif. Acad. Sci. 30(3):77, pl. 6, fig. 2, pl. 7, figs. 2–5.

Description: Shell to 140 mm; color cream to light beige with brown spots, often coalescing into axial stripes; outline of whorls concave on early whorls, straight to convex on later ones; protoconch of 3½ to 4½ whorls; subsutural band on early whorls consisting of double row of nodes divided by a groove, in some individuals the anterior row being smaller or almost obsolete, band becoming a single row of rather flat nodes defined by a narrow shallow groove, then becoming entirely smooth and defined by an almost completely obsolete groove; axial ribs well-developed on early whorls, each beginning at node on band, all sculpture fading except for axial growth striae and obsolete subsutural groove; aperture moderately elongate; columella recurved; operculum small and thick.

Type locality: *T. robusta*, "Panama, Gulf of Nicoya, Gulf of Papagayo and San Blas, 4 to 18 fms." The type locality is here restricted to Kobee Beach, Canal Zone, Panama; *T. lingualis*, "Gulf of Papagayo, Bay of Montejo, west coast of

---

Plate 35

FIGS. **137a-d,** *Terebra robusta* Hinds. 137a, Lectotype British Museum (N.H.) No. 1968171; 117.4mm. 137b, Lectotype of *T. lingualis* Hinds, same Institution No. 1968232; 68.1 mm. **137c,** Holotype of *T. insignis* Deshayes, same Institution No. 197963; 77.0 mm. **137d,** Holotype of *T. dumbauldi* Hanna & Hertlein, California Academy Sciences No. 12333; 106.6 mm. **138a-d,** *T. formosa* Deshayes. **138a,** Middle whorls of holotype. **138b,** Holotype British Museum (N.H.) No. 197960; 70.9 mm. **138c,** Holotype of *T. incomparabilis* Deshayes, same Institution No. 197997; 84.6 mm. **138d,** Worn lectotype of *T. pachyzona* Mörch, University Zoological Museum, Copenhagen; 51.0 mm. **139,** *T. ornata* Gray. Manzanillo, Mexico; 61.2 mm. **140,** *T. argosyia* Olsson. Holotype National Museum of Natural History, Washington No. USNM-701160; 71.0 mm. **141a, b,** *T. strigata* Sowerby. **141a,** Isabela (Albemarle) I., Galapagos Is.; 87.3 mm. **141b,** Off Guaymas, Mexico; 79.2 mm (brown axial stripes missing).

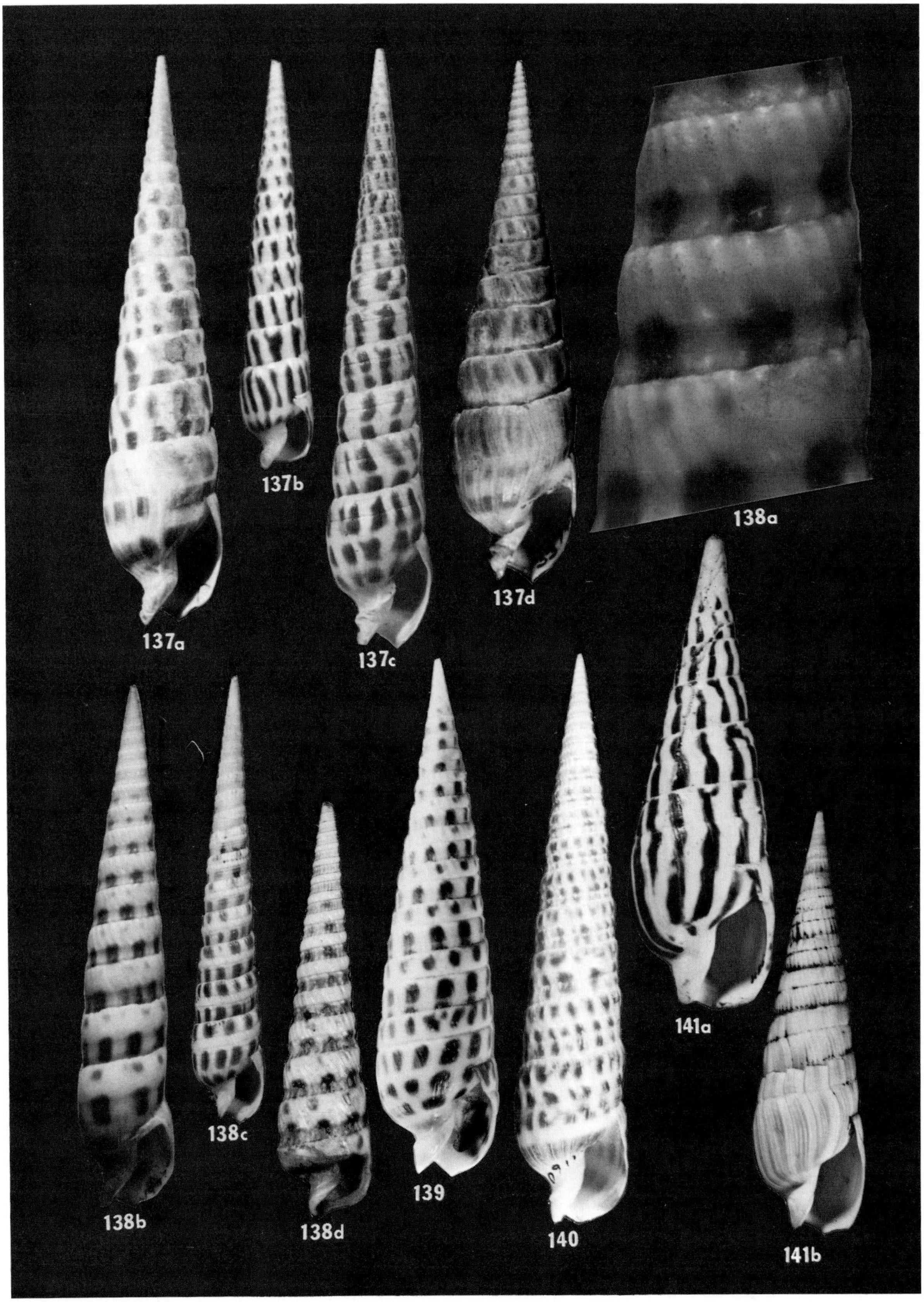

Plate 35. Eastern Pacific *Terebra*.

America, 0–17 fms;" *T. loroisii*, none; *T. insignis*, "Panama"; *T. dumbauldi*, "Panama."

Distribution: From the outer coast of Baja California, Mexico, to the Galapagos Is.; intertidal to 110 m.

Type: *T. robusta* lectotype BM(NH) no. 1968171, 117.4 mm; *T. lingualis* lectotype BM(NH) no. 1968232, 68.1 mm; *T. loroisii* type not located [pers. comm. Dr. Philippe Bouchet], pub. size 120 × 36 mm; *T. dumbauldi* holotype CAS no. 12333, 106.6 × 27 mm.

Discussion: This large species is extremely variable both in sculpture and outline. Some individuals are slender, others obese. The broad form, described as *T. dumbauldi* seems more common in the southern part of the range. Hanna & Hertlein separated *T. robusta*, *T. lingualis*, and *T. dumbauldi* on the apical angle and the diameter of the later whorls of mature specimens. They stated that *T. robusta* and *T. lingualis* cannot be separated when they are as young as the type of *T. lingualis*. There is a gradual transition from one form to another, from slender to inflated, making it impossible to place the intergrades under one of the above species or form names.

George Cummings, Jr. (Hawaiian Shell News 22(7):4) states that over a two-year period of collecting in Panama, the more slender *T. lingualis* form was found only in areas of fine, clean sand in calm water while the more obese form was in muddy sand where the water was opaque on an ebbing tide.

T. robusta is the type species of the genus *Panaterebra* Olsson, 1967. The latter diagnosed the genus thus: "Like *Paraterebra* in general appearance, shell large, robust; sculpture of three, axially ribbed or noded, encircling bands but restricted to a relatively short stretch, the sculpture becomes obsolete on the middle or the final sections of the shell; columella with a single fold."

To distinguish this species from the other two large, spotted Panamic terebrids see *T. ornata* **(139)**.

138. *Terebra formosa* Deshayes, 1857

(Pl. 35, figs. 138a-d)
(Color pl. B, fig. 12)

1857 *Terebra formosa* Deshayes, Journ. Conchyl. 6:65, pl. 3, fig. 6; 1860 Reeve, Conch. Icon. 12, pl. 12, sp. 49; 1963 Campbell, Veliger, 5(3):101; pl. 12, figs. 8–12, pl. 13, figs. 3–6; 1971 Bratcher & Burch in Keen, Seashells Trop. W. America ed. 2:676, fig. 1536; 1984 Aubry, Terebridae pl. 2.

1859 *Terebra incomparabilis* Deshayes, Proc. Zool. Soc. Lond. pt. 27:307.

1861 *Terebra pachyzona* Mörch, Malakazool. Blätter. 7:105; 1966 Keen, Occas. Pap. Calif. Acad. Sci. no. 59: 29, figs. 37, 37a.

1958 *Terebra (Terebra) formosa* Deshayes, Keen, Seashells Trop. W. America 1st ed.: 488, fig. 945.

Description: Shell to 102 mm; color white with 2 rows of squarish brown spots per whorl, 3 on body whorl; outline with extremely broad convex subsutural band and remainder of whorl short, concave; protoconch of 3½ whorls; subsutural band on early whorls ornamented with nodes becoming ribs progressively, then fading completely; band on later whorls occupying ⅔ of whorl; subsutural groove which borders the band becomes obsolete in later whorls; axial ribs on early whorls numerous, closely spaced, well-developed, swollen into small nodes at posterior end; later whorls having no sculpture except axial striae and swollen band; aperture moderately quadrate; columella short, recurved, with heavy plication.

Type locality: *T. formosa*, "Panama"; *T. incomparabilis*, "Panama"; *T. pachyzona*, "Realejo, Central America."

Distribution: From Guaymas, Mexico to Peru; intertidal to 92 m.

Type: *T. formosa* holotype BM(NH) no. 197960, 70.9 mm; *T. incomparabilis* holotype BM(NH) no. 197997, 84.6 × 13.0 mm; *T. pachyzona* worn lectotype, here selected, UZMC, 51.0 × 12.0 mm.

Discussion: The subsutural band, which occupies two thirds of the whorl, separates this from all other spotted Panamic species.

139. *Terebra ornata* Gray, 1834

(Pl. 35, fig. 139)
(Color pl. B, fig. 13)

1834 *Terebra ornata* Gray, Proc. Zool. Soc. Lond. pt. 2:62; 1844 Hinds in Sowerby, Thes. Conch. 1:152, pl. 42, fig. 34; 1860 Reeve, Conch. Icon. 12, pl. 4, sp. 14; 1958 Keen, Seashells Trop. W. America ed. 1:489, fig. 947; 1961 Hanna & Hertlein, Proc. Calif. Acad. Sci. 30(3):71, pl. 6, fig. 1; 1963 Campbell, Veliger 5(3):103, pl. 12, fig. 2; 1971 Bratcher & Burch, Proc. Calif. Acad. Sci. 4th ser., 37(21):553, fig. 24; 1971 Bratcher & Burch in Keen, Seashells Trop. America 2nd ed.: 680, fig. 1554; 1982 Abbott & Dance, Compendium of Seashells p. 276; 1984 Aubry, Terebridae pl. 2.

Description: Shell to 82 mm, rather obese; color, first few whorls of teleoconch white, remainder ivory to beige, with a row of dark brown spots on subsutural band and 2 rows, which occasionally coalesce into a stripe, on remainder of whorl, 3 on body whorl; outline of whorls turreted; protoconch of 1½ brown whorls; subsutural band convex and moderately noded in early whorls, becoming flat later, band marked by deep subsutural groove; axial ribs low, curved, evenly spaced, inconspicuous, becoming obsolete and fading completely, mature whorls sculptured only with macroscopic axial striae and subsutural groove; apical angle broad; body whorl short; aperture quadrate; columella short, extremely recurved; siphonal fasciole long, striate, with sharp keel.
Type locality: None given. Manzanillo, Mexico, is here designated as the type locality.
Distribution: From San Luis Gonzaga Bay, Baja California, Mexico, to the Galapagos Islands; intertidal to 82 m.
Type: The type has not been located.
Discussion: Although Gray did not illustrate his species, and his type is lost, it has been elucidated by subsequent authors such as Hinds (1844) and Reeve (1860). The name *T. ornata* is well-entrenched in malacological literature. This is the type species of the genus *Myurellina* Bartsch, 1923. He diagnosed the genus as follows: "Columella with two folds. Spiral sculpture of subsutural groove only. Axial ribs not strongly developed on all whorls. Axial ribs evanescent on the later turns."

This species may be separated from the other large spotted Panamic terebrids by its lack of axial ribs on the early whorls. It also differs from *T. robusta* (**137**) by having a short body whorl, a deep subsutural groove in later whorls, and regularly placed round brown spots. *T. formosa* (**138**) always has a white instead of ivory or beige background, and a subsutural band that occupies ⅔ of the whorl. *T. argosyia* (**140**) has a groove below the subsutural groove which divides the remainder of the whorl in half.

140. *Terebra argosyia* Olsson, 1971

(Pl. 35, fig. 140)

1971 *Terebra argosyia* Olsson, Stud. Trop. American Moll. p. 41, figs. 28–30; 1982 Abbott & Dance, Compendium of Seashells p. 277, holotype fig'd; 1984 Aubry, Terebridae pl. 7.

Description: Size to 71 mm; color cream with 3 rows of brown spots per whorl, 4 on body whorl; outline of whorls almost straight; protoconch unknown; subsutural band faintly noded, defined by deep groove on early and middle whorls, becoming progressively weaker; followed by a second, weaker band; axial ribs strong on early whorls, becoming progressively obsolete; mature whorls divided into 3 slightly convex bands separated by spiral grooves; body whorl with 3 rows of brown dots and a somewhat broad brown band anterior to the periphery; aperture quadrate; columella recurved, with 2 plications; siphonal fasciole with heavy striations.
Type locality: "St. 72, Isla La Plata, Ecuador, 37 m."
Distribution: Ecuador; to 37 m.
Type: Holotype USNM no. 701160; 71.0 × 14.5 mm.
Discussion: This rare species has a flatter outline and a groove below the subsutural groove which divides the remainder of the whorl in half, distinguishing it from the other large Panamic terebrids.

141. *Terebra strigata* Sowerby, 1825

(Pl. 35, figs. 141a-b)
(Color pl. C, figs. 10, 16, 17)

1825 *Terebra strigata* Sowerby, Cat. Shells Tankerville, App. xxiii; 1844 Hinds in Sowerby, Thes. Conch. 1:151, pl. 41, fig. 10; 1961 Hanna & Hertlein, Proc. Cal. Acad. Sci., 4th ser. 30(3):75, pl. 7, figs. 6, 7; 1962 Abbott, Sea Shells of the World, p. 125, fig. 3; 1963 Campbell, Veliger 5(3):102, pl. 12, fig. 7; 1964 Shikama, Selected Shells World Col. p. 124, pl. 101, fig. 20; 1971 Bratcher & Burch, Proc. Calif. Acad. Sci. 4th Ser. 37(21):551, figs. 21–23; 1971 Bratcher & Burch in Keen, Seashells Trop. W. America ed. 2:694, fig. 1566; 1974 Dance, Encycl. Shells p. 219, top left fig. center column; 1982 Abbott & Dance, Compendium of Seashells p. 277; 1984 Aubry, Terebridae pl. 1.

1828 *Buccinum elogatum* Wood, Suppl. Index Test. p. 13, pl. 4, fig. 25d (Hab: India, error) [non Sowerby, 1815].

1832 *Terebra flammea* Lesson, Illust. Zool. pl. 48 [non Lamarck, 1822].

1839 *Terebra zebra* Kiener, Spéc, Gén. Icon. Coq. Viv. 9:5, pl. 3, fig. 5.

1958 *Terebra (Terebra) strigata* Sowerby, Keen, Seashells Trop. W. America ed. 1:489, fig. 949.

Description: Shell to 130 mm, heavy, sturdy; color ivory, with dark brown wavy axial zebra-like stripes; outline of whorls slightly convex; protoconch of 2½ whorls; subsutural band on early whorls defined by groove, both band and groove becoming obsolete later in some individuals; spiral ribs curved on early whorls, later becoming more shallow and numerous, up to 55 on body whorl, or fading completely; spiral sculpture absent; body whorl exceptionally long; aperture long but moderately quadrate; columella recurved.

Type locality: *T. strigata*, "Panama"; *T. zebra*, "Gulf of Antilles and coast of the Isthmus of Panama."

Type: *T. strigata* type lost; *T. zebra* type lost.

Distribution: From the Gulf of California to Payta, Peru, and the Galapagos Is.; intertidal to 55 m.

Discussion: Although the type is lost, Sowerby's description is applicable only to this particular species from Panama. Kiener's name, *T. zebra*, would have been an apt name for the species. It may be separated from the other American species of terebrids by its extremely long body whorl and usually by the definite dark brown stripes, although several cream-colored specimens without brown stripes have been examined. This species may be obese or slender, but always has a very heavy shell.

142. *Terebra taurina* (Lightfoot, 1786)

(Pl. 35, figs. 142a-d)
(Color pl. B, fig. 8)

1786 *Buccinum taurinum* Lightfoot. Cat. Portland Mus. p. 142, no. 3158 (ref. Lister, pl. 841, fig. 69).

1798 *Epitonium feldmanni* Röding. Mus. Bolten. p. 94, no. 1216 (ref. Chemnitz, 4: pl. 154, fig. 1446).

1807 *Terebra undata* Fischer von Waldheim, Mus. Demidoff: 180 (ref. to Chemnitz, 4: pl. 154, fig. 1446).

1822 *Terebra flammea* Lamarck, Hist. Nat. Anim. s. Vert. 7:284.

1898 *Terebra texana* Dall, Nautilus 12(4):45.

1961 *Terebra taurinus* Solander, Warmke & Abbott, Caribb. Seashells p. 132, pl. 3, fig. I; 1974, American Seashells ed. 2:259, col. pl. 10, fig. 2835; 1982 Abbott & Dance, Compendium of Seashells p. 276; 1984 Aubry, Terebridae pl. 2.

1967 *Terebra taurinum* (Lightfoot), Rehder, Proc. U. S. Nat. Mus. 121:25.

1971 *Terebra (Strioterebrum) taurinus* (Solander) [Lightfoot], Andrews, Sea Shells of the Texas Coast p. 125, text fig.

1975 *Terebra taurina* (Lightfoot), Rios, Brazilian Mar. Mol. Icon. p. 126, pl. 38, fig. 551.

1975 *Terebra taurinus* (Lightfoot), Humfrey, Sea Shells West Indies p. 179, pl. 20, figs. 8, 8a; 1985 Abbott, Sea Shells World, revised ed., p. 124, fig. 4; 1986 Sutty, Seashell Treasures of the Caribbean pp. 94, 96.

Description: Shell to 167 mm, sturdy; color cream with brown flammules on subsutural band and a spiral row of nebulous brown spots on remainder of whorl, 2 on body whorl; outline of whorls straight; protoconch of $2^1/_2$ whorls; subsutural band convex, with double row of elongate nodes or ribs on early whorls, the posterior being larger, becoming progressively flatter, band on mature whorls unnoded, occupying two thirds of whorl, divided in half by a very shallow groove; axial ribs evenly spaced, slanted about equal to interspaces, on early whorls, becoming axial striae later; aperture elongate; columella strongly recurved.

Type locality: *T. taurina*, "China," error. Lake Worth, Florida, is here designated as the type locality; *E. feldmanni*, none; *T. undata*, none; *T. flammea*, "Indian Ocean," error; *texana*, "Gulf Coast of Matagorda Is.," Mexico.

Distribution: From southwest Florida and the Gulf of Mexico to the West Indies and Brazil; 3 to 80 m., usually in deeper water.

Type: *T. taurina*, the two Portland specimens and the Lister specimen can no longer be traced. We here select the specimen illustrated by Lister, 1770, pl. 841, fig. 60, as the illustrated lectotype; *E. feldmanni*, ref. Chemnitz 4, pl. 154, fig. 1446 is here selected as the illustrated lectotype; *T. undata*, ref. Chemnitz 4, pl. 154, fig. 1446 is here selected as the illustrated lectotype; *T. flammea* holotype MHNG no. 1101/16/1, 136.0 mm; *T. texana* holotype (pl. 36, fig. 142b) USNM no. 107373, 114.6 mm.

Discussion: The animal is yellow and the operculum corneus brown. *T. texana* Dall (now *T. taurina*) is the type species of the genus *Paraterebra* Woodring. His diagnosis of the genus is: "Shell large, generally moderately slender; aperture relatively narrow; outer lip inflected forward at about a third of distance from suture to base of anterior canal; anterior canal with wide, deep notch; columella with thick parietal callus; sculpture of early whorls consisting of a subsutural band, below which lies a narrower band, and of fine axial ribs curbed forward; sculpture disappearing on later whorls."

The sculpture and color pattern closely resemble those of the Panamic *T. robusta* (**137**) but can be separated by the broad convex subsutural band which occupies a third of the whorls. *T. formosa* (**138**), another spotted Panamic terebrid, has such a subsutural band but always has pure white rather than a cream or beige background, has smooth later whorls, and does not have as large a shell.

143. *Terebra floridana* (Dall, 1889)

(Pl. 36, figs. 143a-e)

1889 *Subula floridana* Dall, Bull. Mus. Comp. Zool. Harvard 18(2):63.

1902 *Terebra (Subula) floridana* Dall, Proc. U.S. Nat. Mus. 24; 503, pl. 29, fig. 9; 1903 Dall, Bull. U.S. Nat. Mus. new ed. 37: pl. 75, fig. 9.

1954 *Terebra (Myurella) floridana* (Dall), R. T. Abbott, Nautilus 68(2):38, pl. 2, fig. 7; 1974 R. T. Abbott, American Seashells 2nd ed.: 259, fig. 2837; 1985 R. T. Abbott, Seashells of the World: 125, fig. 3.

1954 *Terebra (Myurella) floridana stegeri* R. T. Abbott, Nautilus 68(2):37, pl. 2, figs. 5, 6 (Hab.: 50 miles W. of Campeche, Yucatan Peninsula, Mexico, 12 fms.)

Description: Shell to 75 mm; color from yellowish white to orange, occasionally with faint brownish spots; outline of whorls concave; protoconch large, mamillate; subsutural band with double row of nodes, separated by an incised groove on early whorls, both rows equal in size, anterior band becoming weaker and less prominent in later whorls, nodes on band later becoming elongate (17 to 23); remainder of whorl sculptured with evenly spaced axial ribs on early whorls, becoming almost obsolete later; spiral cords 3 or 4; aperture quadrate; columella recurved.

Type locality: *T. floridana*, "Straits of Florida, 56 fms, sand"; *T. floridana stegeri*, "50 miles W. of Campeche, Yucatan Peninsula, Mexico, 12 fms."

Distribution: South Carolina to northern Brazil; 10 to 215 m.

Type: *T. floridana* holotype USNM no. 87222, 67.7 × 10.9 mm; *T. floridana stegeri* holotype (pl. 36, fig. 143c) USNM no. 613884, 68.0 × 13.0 mm.

Discussion: This species has an attractive concave-whorled shell. Very faint brownish spots on the whorls are rare. The broader form, *T. floridana stegeri*, appears to be a more southern form or subspecies.

144. *Terebra corrugata* Lamarck, 1822

(Pl. 36, figs. 144a-b;
pl. 37, figs. 144c-d)
(Color pl. B, fig. 14)

1822 *Terebra corrugata* Lamarck, Hist. Nat. Anim. S. Vert. 7:287; 1839 Kiener, Spéc. Gén. Icon. Coq. Viv. 9:25, pl. 13, fig. 31, 31a; 1860 Reeve, Conch. Icon. 12: pl. 6, figs. 19a, b; 1885 Tryon, Man. Conch. 7:26, pl. 7, figs. 21, 28, 32; 1952 Knudsen, Vidensk. Medd. Dansk. Naturh. Foren. 114:182, extensive synonmy; 1983 Bouchet, Boll. Malac. Milano 18(9–12):188, figs. 17–18; 1984 Aubry, Terebridae pl. 11.

1857 *Terebra regina* Deshayes, Journ. Conchyl. 6:67, pl. 3, fig. 7.

1859 *Terebra bitorquata* Deshayes, Proc. Zool. Soc. Lond. pt. 27:313 [nom. nov. pro *T. corrugata* Kiener, 1839, pl. 13, fig. 31a].

Description: Shell to 87 mm; color ivory to light beige with small brown spots between nodes of the subsutural band, and 2 rows of larger spots which may coalesce into a rectangle on remainder of whorl; an extra row of spots below periphery of body whorl; outline of whorls straight, somewhat turreted; protoconch of 3 whorls; subsutural band with round bead-like nodes on early whorls, less prominent ones later; band defined by deep groove on early whorls, becoming progressively more shallow; axial sculpture absent except for striae; spiral sculpture of a row of small, round nodes below the subsutural band, becoming obsolete or forming a narrow raised cord, sometimes scarcely noticeable; remainder of whorl smooth except for axial striae; body whorl short; aperture small, quadrate; columella extremely recurved, short.

Type locality: *T. corrugata*, none given. Goree Bay, Dakar, Senegal, W. Africa is here designated as the type locality; *T. regina*, "Senegal," W. Africa.

Distribution: West Africa from Senegal and the Cape Verde Islands to the Gulf of Guinea; subtidal to 80 m.

Type: *T. corrugata* lectotype MHN Geneva no. 1101/23, 63.5 × 11.5 mm; *T. regina* holotype (pl. 36, fig. 144b) BM(NH) no. 197971, 76.3 × 12.8 mm.

Discussion: *T. bitorquata* Deshayes was described from a specimen in the Crosse collection which matched the specimen figured by Kiener as a junior variety of *T. corrugata*. It was a specimen with well-developed nodes on and below the subsutural band. *T. corrugata* differs from the west Atlantic *T. floridana* (**143**) by its lack of ribs, its somewhat turreted whorls and its brown spots.

145. *Terebra guineensis* Bouchet, 1983

(Pl. 36, fig. 145)

1983 *Terebra guineensis* Bouchet, Boll. Malac. Milano 18(9–12):189, figs. 2, 19.

Description: Shell to 30 mm; color brown-red

flammules on a white background; outline of whorls concave; protoconch of $1\frac{1}{2}$ mamillate whorls; subsutural band consisting of a double row of nodes, the posterior being larger on mature whorls, broader and flatter on early whorls; axial sculpture absent; spiral sculpture of about 4 heavy cords divided by deep, moderately wide grooves; aperture missing from unique holotype; interior columella with 2 heavy plaits.
Type locality: "West of Los Island, Guinea, West Africa; 09°20′N; 14°13′W; 32m; Atlantide Expedition sta. 145."
Distribution: Known only from type locality.
Type: Holotype UZMC, 29.7 mm.
Discussion: This species may be distinguished from *T. corrugata* (**144**) by its mamillate protoconch, indicating direct development, and by the brown flammules which reach from suture to suture in contrast to *T. corrugata*'s multispiral protoconch and flammules which are confined to the area anterior to the subsutural band.

146. *Terebra tuberculosa* Hinds, 1844

(Pl. 37, figs. 146a-c)
(Color pl. E, fig. 18)

1844 *Terebra tuberculosa* Hinds, Proc. Zool. Soc. Lond. pt. 11:154; 1844 Hinds in Sowerby, Thes. Conch. 1:175, pl. 43, fig. 48; 1860 Reeve, Conch. Icon. 12: pl. 14, sp. 61; 1958 Keen, Seashells Trop. W. America ed. 1:494, fig. 979; 1961 Campbell, Veliger 4(1):28, pl. 5, fig. 14 protoconch fig'd.; 1964 Campbell, Veliger 6(3):133, pl. 17, figs. 6, 7; 1966 Keen, Veliger 8(4):274; pl. 47, fig. 13 figd. syntype; 1971 Bratcher & Burch in Keen, Seashells Trop. W. America ed. 2:686, fig. 1570; 1982 Abbott & Dance, Compendium of Seashells p. 277; 1984 Aubry, Terebridae pl. 13.

1930 *Terebra cracilenta* Li, Bull. Geol. Soc. China 9(3):274 pl. 8, fig. 67; 1944 M. Smith, Panamic Mar. Shells p. 35, fig. 457; 1971 Bratcher & Burch in Keen, Seashells Trop. W. America ed. 2:674, fig. 1530; 1984 Aubry, Terebridae pl. 8.

1930 *Terebra tenuis* Li, Bull. Geol. Soc. China 9(3):274, pl. 8, fig. 68.

1931 *Terebra (Strioterebrum) cracilenta* Li, Pilsbry, Proc. Acad. Sci. Philadelphia 83:439, text figs. 1, 1a figd. types: 1958 Keen, Seashells Trop. W. America ed. 1:491, fig. 961; 1961 Campbell, Veliger 4(1)28, pl. 5, fig. 12 protoconch fig'd.; 1964 Campbell, Veliger 6(3):133, pl. 17, figs. 8, 9.

Description: Shell to 68 mm; color shiny gray or peach to dark beige, the early whorls often having a bluish color; outline of whorls faintly convex; protoconch of 3 to 4 whorls; subsutural band convex, noded, defined by a narrow groove; sculpture of 1 to 5 rows of nodes, axially and spirally aligned, nodes often being the only sculpture, but sometimes being connected by low axial ribs, spiral cords, or both, with the nodes occurring at the intersections; body whorl with a row of nodes at the periphery, some specimens having up to 4 rows of smaller nodes posterior to the periphery and/or 4 rows anterior, some with no additional rows; aperture elongate on mature individuals; columella recurved.
Type locality: *T. tuberculosa*, "Panama, Gulf of Papagayo, and San Blas, 4 to 11 fms." The type locality is here restricted to San Blas, Mexico; *T. cracilenta*, "Gatun formation, Miocene, Panama Bay"; *T. tenuis*, "Gatun formation, Miocene, Panama Bay."
Distribution: From Bahia Santa Maria, Baja California, Mexico, to Ecuador; intertidal to 110 m.
Type: *T. tuberculosa* holotype BM(NH) no. 1844.6.7.82, 55.7 mm; *T. cracilenta* holotype AMNH, dept. fossil invert. no. 22/27A, 45.0 × 10.0 mm; *T. tenuis* holotype AMNH dept. fossil invert. no. 22/28A, 46.0 × 10.00 mm.
Discussion: The color of this species varies, but it is always monochromatic. It may be separated from *T. glauca* (**147**), a species with a brown shell, by *tuberculosa*'s much wider apical angle, sturdier shell, and wider subsutural band with larger nodes.

Plate 36
FIGS. **142a-d,** *Terebra taurina* (Lightfoot). **142a,** Illustrated lectotype from Lister, 1770, pl. 841, FIG. **142b,** Holotype of *T. texana* Dall, National Museum Natural History, Washington No. USNM-107373; 114.6+ mm. **142c,** Lake Worth, Florida, E. coast U.S.A.; 137.7 mm. **142d,** Middle whorls. **143a-e,** *T. floridana* Dall. **143a,** Holotype National Museum Natural History, Washington No. USNM-87222; 67.7 mm. **143b,** Pensacola, Florida; 66.0 mm. **143c,** Holotype of *T. floridana stegeri* Abbott, National Museum Natural History, Washington No. USNM-613884. **143d,** Lower and middle whorls. **143e,** Lower whorls. **144a,b,** *T. corrugata* Lamarck. **144a,** Lectotype Museum d'Historie Naturelle, Geneva No. 1102/23; 63.5 mm. **144b,** Holotype of *T. regina* Deshayes, British Museum (N.H.) No. 197971; 76.3 mm.

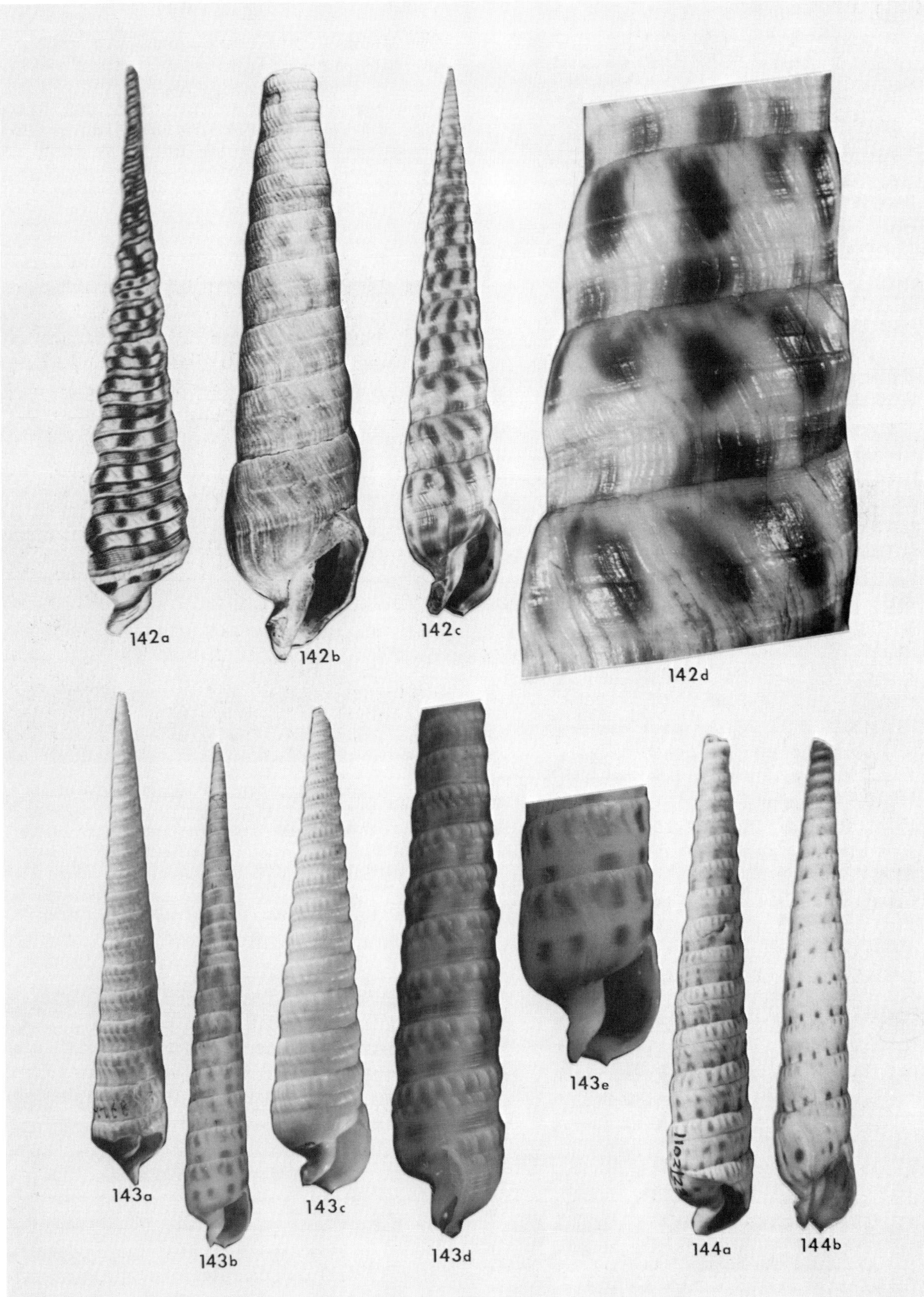

Plate 36. Western Atlantic and West African *Terebra.*

147. *Terebra glauca* Hinds, 1844

(Pl. 37, figs. 147a-b;
pl. 38, figs. 147c-d)

1844 *Terebra glauca* Hinds, Proc. Zool. Soc. Lond. pt. 11:155; 1844 Hinds in Sowerby, Thes. Conch. I:175, pl. 44, fig. 85; 1966 Keen, Veliger 8(4):273, pl. 47, fig. 16; 1979 Bratcher, Veliger 11(1):62, fig. 3; 1984 Aubry, Terebridae pl. 8 (upper left fig. only).

1844 *Terebra radula* Hinds, Proc. Zool. Soc. Lond. pt. 11:155; 1844 Hinds in Sowerby, Thes. Conch. 1:174, pl. 44, fig. 95 [non Gravenhorst, 1807]; 1966 Keen, Veliger 8(4):274, pl. 47, fig. 15.

1970 *Terebra dorothyae* Bratcher & Burch, Veliger 12(3):297, pl. 44, figs. 7, 8; 1971 Bratcher & Burch in Keen, ed. 2: 676, fig. 1634.

Description: Shell to 47 mm, delicate; color brown, portions of subsutural band and nodes slightly lighter; outline of whorls somewhat convex; protoconch of 3 slender whorls, subsutural band with round nodes, becoming elongate later, band defined by shallow groove; sculpture of rows of pustules in both axial and spiral rows, occasionally connected by slender cords; aperture elongate; columella slightly curved.
Type locality: *T. glauca*, none given. Manzanillo, Mexico, is here designated as the type locality; *T. radula*, "Puerto Potrero, W. Coast of America"; *T. dorothyae*, "Near San José, Guatemala."
Distribution: From the Gulf of California, Mexico, to Ecuador and the Galapagos Is.
Type: *T. glauca* probable holotype BM(NH) no. 1968268, 27.8 mm; *T. radula* holotype (pl. 38, fig. 147c) BM(NH) no. 1968236, 40.8 mm; *T. dorothyae* holotype (pl. 38, fig. 147d) LACM no. AHF 1250, 29.9 × 6.5 mm.
Discussion: This species has a thin, delicate shell with small widely spaced nodes. *T. petiveriana* (**148**) has a heavier, broader shell with more file-like, coarser sculpture.

148. *Terebra petiveriana* Deshayes, 1857

(Pl. 38, figs. 148a-c)
(Color pl. E, fig. 19)

1844 *Terebra aspera* Hinds, Proc. Zool. Soc. Lond. pt. 11:154; 1844 Hinds in Sowerby, Thes. Conch. 1:174, pl. 43, fig. 44. [non Bosc, 1801]

1857 *Terebra petiveriana* Deshayes, Journ. Conchyl. 6:85, pl. 5, fig. 10; 1979 Bratcher, Veliger 22(1):63, fig. 4.

1944 *Terebra glauca* Hinds, M. Smith, Panamic Shells p. 35, fig. 466; 1971 Bratcher & Burch in Keen, Seashells Trop. W. America ed. 2:676, fig. 1538. [non Hinds, 1844]

1981 *Strioterebrum gatunensis kugleri* Rutsch, Petuch, Malacologia 20(2):336, figs. 99, 100. [non Rutsch, 1934]

Plate 37
FIGS. **144c,d,** *Terebra corrugata* Lamarck. **144c,** Cape Verde Is.; 85.6 mm. **144d,** Middle whorls. **145,** *T. guineensis* Bouchet. Holotype University Zoological Museum, Copenhagen; 29.7 mm. **146a-c,** *T. tuberculosa* Hinds. **146a,** Holotype British Museum (N.H.) No. 1844.6.7.82.; 55.7 mm. **146b,** Panama Bay, Panama; 54.1 mm. **146c,** Middle whorls. **147a, b,** *T. glauca* Hinds. **147a,** Probable holotype British Museum (N.H.) No. 1968268; 27.8 mm. **147b,** Middle whorls.

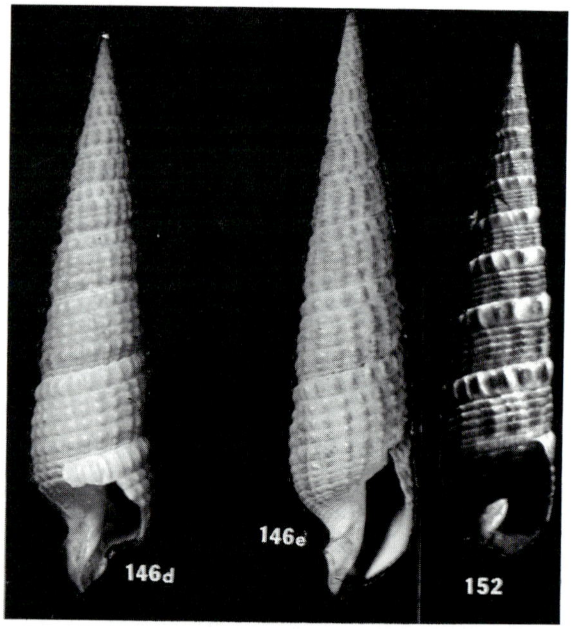

Plate 37a. Miocene specimens. FIGS. **146c,d,** *Terebra tuberculosa* Hinds. FIG. **146c:** Holotype of *T. cracilenta* Li, American Museum of Natural History, fossil invertebrates no.; 22/27a; 45.0 mm. FIG. **146d:** Holotype of *T. tenuis* Li, American Museum of Natural History, fossil invertebrates no. 22/28a; 46.0 mm. FIG. **152.** Holotype of *Terebra africana* Gray, 1834 (see sp. **152,** *Terebra variegata* Gray, 1834).

Description: Shell to 46 mm; color usually gray to brown, with whitish subsutural band, rarely monochromatic flesh to beige; outline of whorls convex with convex band; protoconch of 3 to 3½ slender glassy whorls; subsutural band thick, convex, nodded, defined by deep groove; axial ribs strong, about equal to interspaces; spiral cords, sometimes weak, rarely absent, often as strong as ribs, usually crossing ribs creating a file-like appearance, frequently developing small nodes at interspaces; outer lip moderately sturdy; aperture quadrate; columella recurved.
Type locality: *T. petiveriana*, "Panama"; *T. aspera*, "Panama, Monte Cristi, St. Elna, W. Coast of America."

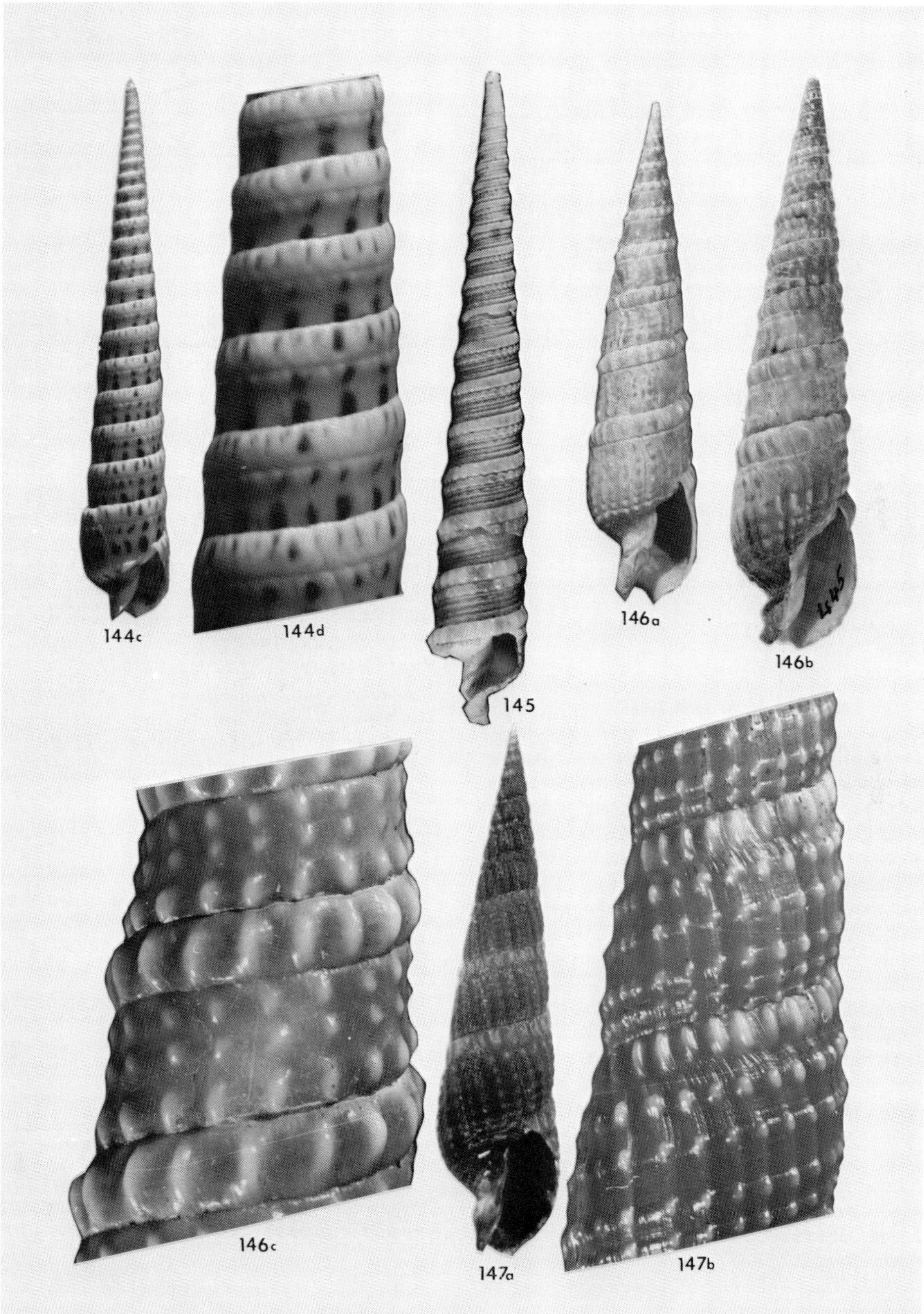

Plate 37. West African and Eastern Pacific *Terebra*.

Distribution: From the Gulf of California, Mexico, to Ecuador, Cartegena, Colombia, and Santa Marta, West Indies; intertidal to 46 m.
Type: *T. petiveriana* lectotype BM(NH) no. 197968, 43.0 × 10.0 mm; *T. aspera* lectotype (pl. 38, fig. 148b) BM(NH) no. 979125, 54.4 mm.
Discussion: It is interesting to note that after Reeve figured a specimen of *T. petiveriana* as *T. glauca* **(147)** reviewers followed his lead and figured *T. petiveriana* as *T. glauca* or synonymized *T. glauca* with *T. aspera*, a synonym of *T. petiveriana*. One specimen of this species lacking spiral sculpture has been examined. Refer to *T. glauca* for comparison. Specimens collected in the Caribbean are impossible to separate from those collected on the coast of Baja California, Mexico. Named after James Petiver (1658–1718), a wealthy English apothecary and ardent naturalist.

149. *Terebra lucana* Dall, 1908

(Pl. 38, figs. 149a-c)
(Color pl. E, fig. 17)

1908 *Terebra (Strioterebrum) lucana* Dall, Bull. Mus. Comp. Zool. Harvard, 43(6):252; 1958 Keen, Seashells Trop. W. America ed. 1:492, fig. 969; 1984 Aubry, Terebridae pl. 13.
1971 *Terebra lucana* Dall, Bratcher & Burch in Keen, Seashells W. America ed. 2:680, fig. 1548.

Description: Shell to 86 mm; straw color; outline of whorls flat; protoconch of $1^{1}/_{2}$ whorls; subsutural band flat, defined by inconspicuous punctations between ribs; axial ribs, 26 to 32, thin, sometimes flexuous, cordlike; spiral cords, 6 or 7, plus another 3 or 4 on subsutural band; axial and spiral cords about equal in strength; aperture elongate; columella straight; internal columella with 2 plications, 1 exterior plication faintly visible.
Type locality: "West of southern extreme of Lower California, 66 fms, at 74°F."
Distribution: From Cedros Island, Baja California, Mexico to Peru; 11 to 275 m.
Type: Lectotype USNM no. 96567; 34.9 × 7.6 mm. (broken apex)
Discussion: This species has an extremely elongate aperture and is the largest of the Panamic cancellate-sculptured terebrids. Although it has a wide distribution, the majority of the specimens examined have come from the outer coast of Baja California. See *T. specillata* **(151)** to separate from that species.

150. *Terebra hancocki* Bratcher & Burch, 1970

(Pl. 38, figs. 150a-b)

1970 *Terebra hancocki* Bratcher & Burch, Veliger 12(3):299, pl. 44, figs. 11, 12; 1971 Bratcher & Burch in Keen, Seashells Trop. W. America ed. 2:678, fig. 1540; 1982 Abbott & Dance, Compendium of Seashells p. 277; 1984 Aubry, Terebridae pl. 7.

Description: Size to 76 mm; color shiny pale beige background with irregular blotches of reddish brown; outline of whorls moderately convex; protoconch of $3^{1}/_{2}$ somewhat inflated whorls; subsutural band with rounded nodes becoming elongate later, band defined by deep subsutural groove; axial sculpture of narrow ribs, much narrower than interspaces, about 16 on penultimate whorl; spiral sculpture of heavy strap-like cords, crossing ribs and producing small nodes, giving the sculpture a file-like appearance; aperture moderately quadrate; columella extremely recurved.
Type locality: "Off La Libertad, Ecuador (2° 08′S, 81°00′W)."
Distribution: From the Tres Marias Is., Mexico, to Ecuador; intertidal to 90 m.
Type: Holotype LACM no. AHF1255; 75.3 × 15.3 mm.
Discussion: This is one of the larger Panamic terebrids and has rather coarse sculpture. *T. specillata* **(151)** has a narrower subsutural band and more elongate aperture. See *T. variegata* **(152)** for comparison with that species.

Plate 38
FIGS. **147c,d,** *T. glauca* Hinds. **147c,** Holotype of *T. radula* Hinds, British Museum (N.H.) No. 1968236; 40.8 mm. **147d,** Holotype of *T. dorothyae* Bratcher & Burch, Los Angeles County Museum No. AHF-1250; 29.9 mm. **148a-c,** *T. petiveriana* Deshayes. **148a,** Lectotype British Museum (N.H.) No. 197968; 43.0 mm. **148b,** Lectotype of *T. aspera* Hinds, British Museum (N.H.) No. 1979125; 54.4 mm. **148c,** San Filipe; 32.9 mm (specimen lacking spiral sculpture). **149a-c,** *T. lucana* Dall. **149a,** Lectotype National Museum Natural History, Washington No. USNM-96567; 34.9+ mm. **149b,** San Lucas, Baja California; 34.9 mm. **149c,** Middle whorls. **150a,b,** *T. hancocki* Bratcher & Burch. **150a,** Holotype Los Angeles County Museum No. AHF-1255; 75.3 mm. **150b,** Middle whorls.

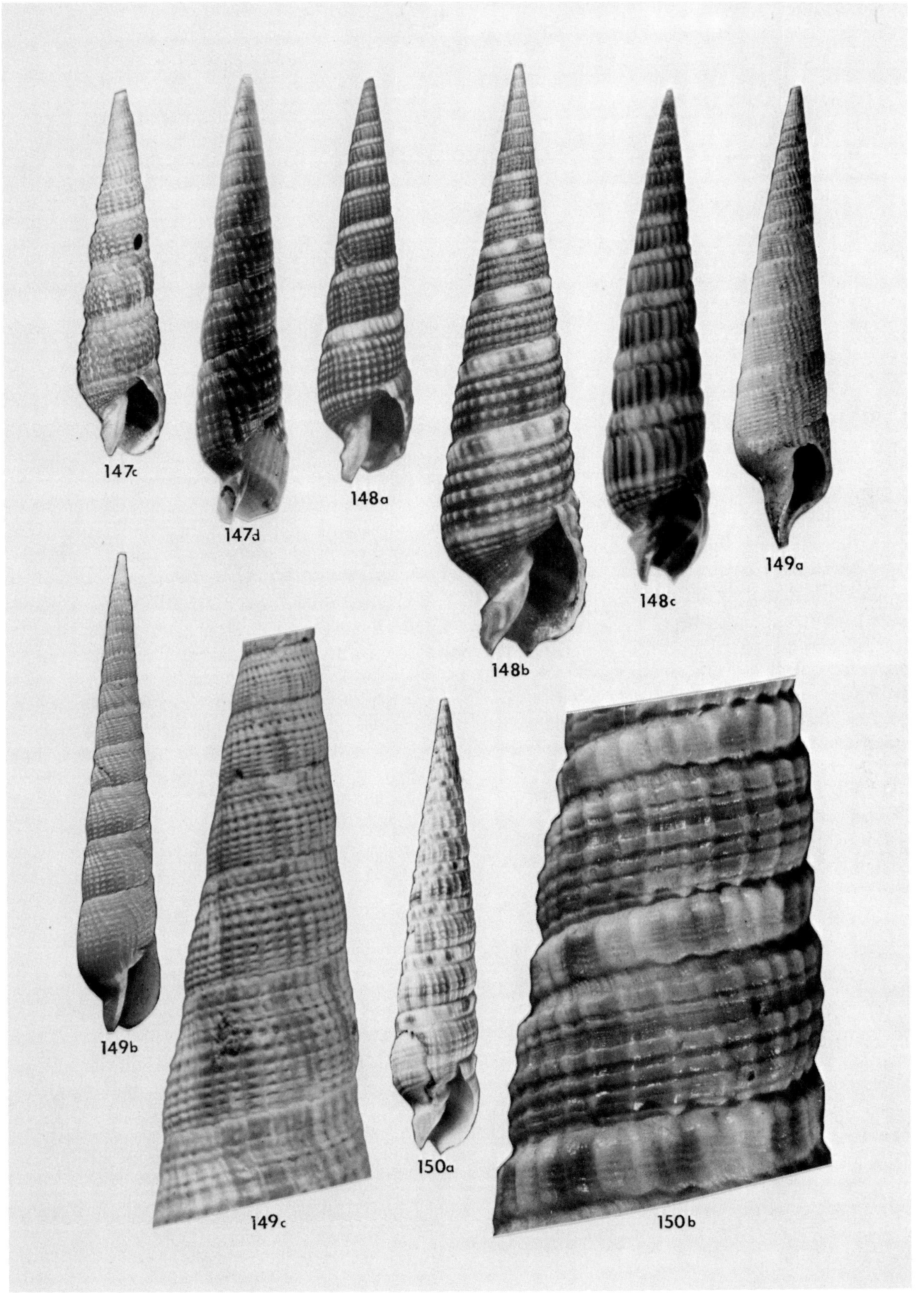

Plate 38. Eastern Pacific *Terebra*.

151. *Terebra specillata* Hinds, 1844

(Pl. 39, fig. 151a-d)

1844 *Terebra specillata* Hinds, Proc. Zool. Soc. Lond. pt. 11:155; 1844 Hinds in Sowerby, Thes. Conch. 1:163, pl. 44, fig. 96 and pl. 45, fig. 116; 1966 Keen, Veliger 8(4):274, pl. 47, fig. 17; 1971 Bratcher & Burch in Keen, Seashells Trop. W. America ed. 2:684, fig. 1565; 1984 Aubry pl. 7.

1857 *Terebra pulchella* Deshayes, Journ. Conchyl. 6:94, pl. 5, fig. 4.

1908 *Terebra (Strioterebrum) balaenorum* Dall, 1908, Bull. MCZ, Harvard 43(6):252; 1958 Keen, Seashells Trop. America ed. 1:490, fig. 957.

1958 *Terebra (Strioterebrum) specillata* Hinds, Keen, Seashells Trop. W. America, ed. 1:494, fig. 978.

Description: Shell to 57 mm; color white, beige, or light brown, often with blotches of cinnamon-brown; outline of whorls somewhat concave when subsutural nodes are large, otherwise flat with a slightly convex subsutural band; protoconch of 1½ to 2 small dome-shaped whorls; subsutural band of elongate nodes, defined by punctations in early whorls, a shallow groove later; sculpture extremely variable; axial ribs usually predominating on early whorls, sometimes continuing, sometimes becoming obsolete; spiral cords, obsolete or strong, crossing ribs; body whorl elongate, often with a cord, several cords, or a few obsolete nodes at the periphery; aperture elongate; columella straight.

Type locality: *T. specillata*, "San Blas, 7 fms.," W. Mexico; *T. pulchella*, "China Seas," error; *T. balaenorum*, "off Ballenas Bay, W. coast of Lower California."

Distribution: From Santa Maria Bay, outer coast of Baja California, Mexico, to Ecuador; 9 to 90 m.

Type: *T. specillata* lectotype, here designated, BM(NH) no. 1844.6.7.84; 39.3 mm; *T. pulchella* lectotype BM(NH) no. 197969, 39.2 × 8.0 mm; *T. balaenorum* holotype (pl. 39, fig. 151b) USNM no. 110559, 27.0 × 6.0 mm.

Distribution: The concave whorls exhibited by some individuals of this species are caused by the sometimes enlarged subsutural nodes on the otherwise flat-sided shell. *T. lucana* **(149)** also has a beige, flat-sided shell with an elongate body whorl and aperture, but it has numerous closely set, cord-like axial ribs and a flat subsutural band to distinguish it. Comparison of the types of *T. specillata* and *T. balaenorum* show little difference.

152. *Terebra variegata* Gray, 1834

(Pl. 39, figs. 152a-b))
(Color pl. B, fig. 4)

1834 *Terebra africana* Gray in Griffith & Pidgeon, Anim. Kingdom of Cuvier, Moll. & Radiata 12:600, pl. 23, fig. 5. [*nomen oblitum* ICZN Opinion 1156].

1834 *Terebra variegata* Gray, Proc. Zool. Soc. Lond. pt. 2:61; 1860 Reeve, Conch. Icon. 12: pl. 16, sp. 12; 1962 Abbott, Sea Shells of the World p. 125, fig. 1; 1971 Bratcher & Burch in Keen ed. 2:686, fig. 1571; 1982 Abbott & Dance, Compendium of Seashells p. 277; 1984 Aubry, Terebridae pl. 5.

1857 *Terebra hupei* Lorois, Journ. Conchyl. 6:388, pl. 12, fig. 1.

1931 *Terebra (Strioterebrum) melia* Pilsbry, Proc. Acad. Sci. Philadelphia 83:437, text figs. 3, 4.

1958 *Terebra (Strioterebrum) variegata* Gray, Keen, Seashells Trop. W. America ed. 1:494, fig. 980; 1963 Shikama, Select Shells World Col. 1:125, pl. 102, fig. 4.

Description: Shell to 87 mm; color mottled gray or tan with a whitish subsutural band spotted with brown; outline of whorls flat and somewhat turreted with a thick, squared-off subsutural band; protoconch of 3 conical whorls; axial ribs well-developed on some individuals, obsolete on others, with only an axial row of inconspicuous nodes to note their existence; spiral grooves 3 to 5, deep, with the crossing axial ribs often forming small, weak nodes at the intersections; aperture quadrate; columella recurved, with 2 plications.

Type locality: *T. variegata*, none given. San Felipe, Baja California, Mexico, is here designated as the type locality; *T. africana*, none; *T. hupei*, none; *T. melia*, "Taboga Is., Panama."

Distribution: From Baja California, Mexico, to the Santa Elena Peninsula of Ecuador and the

---

Plate 39

FIGS. **151a-d,** *T. specillata* Hinds. **151a,** Lectotype British Museum (N.H.) No. 1844.6.7.84.; 39.3 mm. **151b,** Holotype of *T. balaneorum* Dall, National Museum Natural History, Washington No. USNM-110599; 27.0 mm. **151c,** Chamala Bay, W. Mexico; 33.3 mm. **151d,** Off Guaymas, Mexico; 42.4 mm. **152a,b,** *T. variegata* Gray. **152a,** Holotype British Museum (N.H.) No. 1979121; 61.4 mm. **152b,** Holotype of *T. melia* Pilsbry, Academy of Natural Sciences, Philadelphia No. 153870; 34.4 mm. **180a,** *T. hemphilli* Vanatta. Holotype of *T. philippiana* Dall, National Museum Natural History, Washington No. USNM-4943; 23.9 mm (for other illustrations of *T. hemphilli* see Pl. **46,** FIGS. **180b-f**). **153a,b,** *T. carolae* Bratcher. **153a,** Holotype Los Angeles County Museum No. AHF-1178; 52.3 mm. **153b,** Lower whorls. **154a,** *T. plicata* Gray. Holotype British Museum (N.H.) No. 1874.11.10.8.; 38.0 mm.

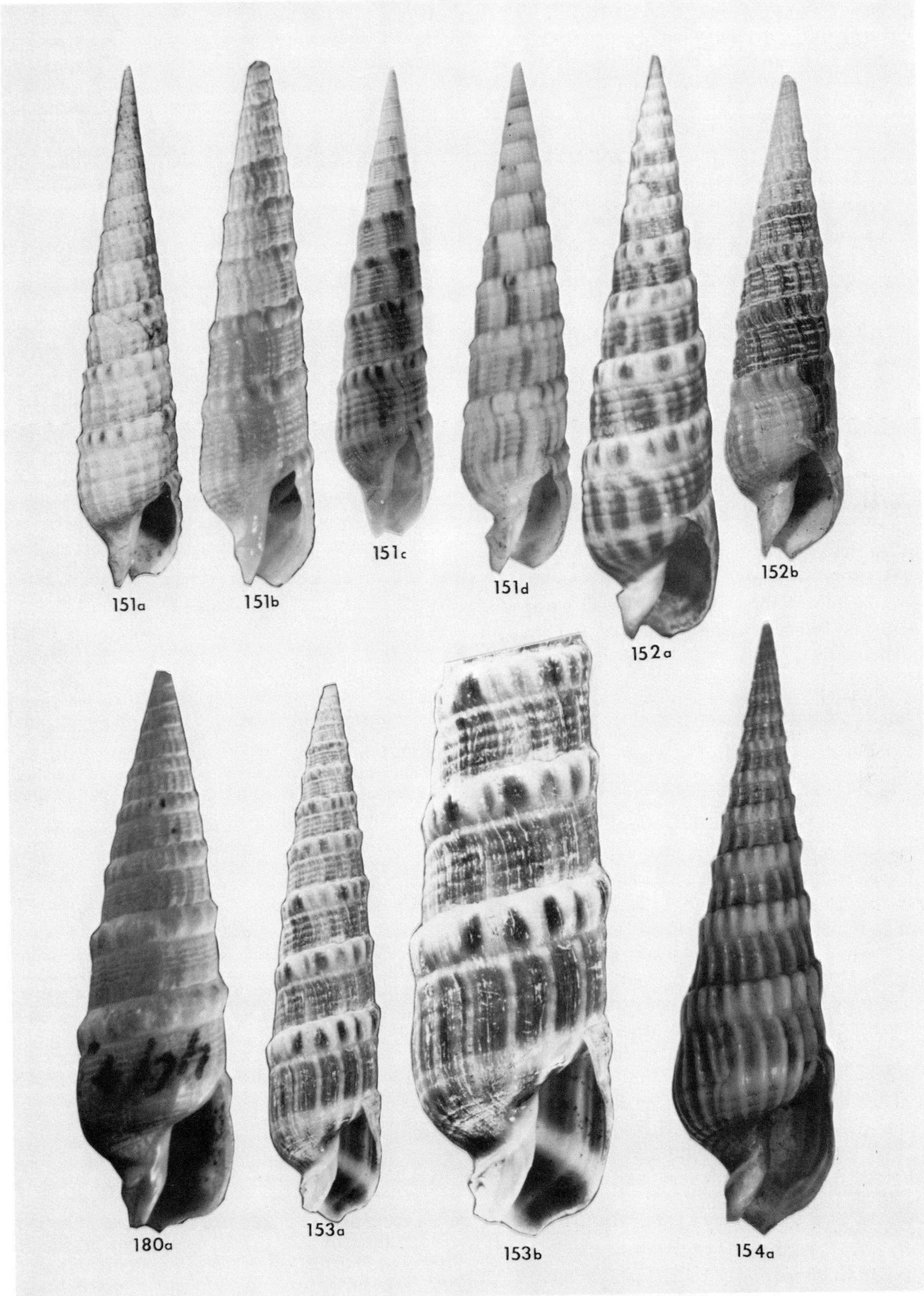

Plate 39. Eastern Pacific *Terebra*.

Galapagos Islands; intertidal to 300 m.; more common in the northern part of the range.

Type: *T. variegata* holotype BM(NH) no. 1979121, 61.4 mm; *T. africana* holotype BM(NH) no. 1872.10.12.12, 51.6 mm (see our pl. 37a, fig. 15e in text); *T. hupei* type probably in MNHNP but not segregated [Dr. Philippe Bouchet *in litt.*]; *T. melia* holotype (pl. 39, fig. 152b) ANSP no. 153870, 34.4 × 9.0 mm.

Discussion: The figure illustrating Lorois' *T. hupei* matches Gray's holotype in both shape and pattern. The wider apical angle, shorter whorls, and broader, flatter subsutural band, almost with a ledge at the posterior, separate this species from *T. intertincta* **(156)** and *T. armillata* **(156)**. *T. intertincta* usually has a row of small nodes at the periphery of the body whorl and may be separated from the two above by its 1½ whorl protoconch. *T. carolae* **(153)** also has a 1½ whorl protoconch, and more closely resembles *T. variegata* but has a somewhat more elongate body whorl and non-planktonic larval development, which the former lacks. *T. plicata* **(154)**, a species found almost exclusively in the Galapagos Islands, is ivory or flesh-colored, except for the first whorls being lavender, and has no blotches or spots.

153. ***Terebra carolae*** Bratcher, 1979

(Pl. 39, figs. 153a-b)

1979 *Terebra carolae* Bratcher, Veliger 22(1):63, figs. 5, 7, 9, 11; 1984 Aubry, Terebridae pl. 4.

Description: Shell to 60 mm; color dull beige-brown with darker brown between nodes of subsutural band and with a light band at periphery of body whorl; outline of whorls flat, protoconch of 1½ mamillate whorls; subsutural band noded, defined by groove cutting through ribs; axial ribs widely spaced, narrow, sometimes obsolete, about 14 on the penultimate whorl; spiral cords, about 4, crossing over ribs, sometimes forming indistinct nodes, becoming obsolete on body whorl; sutures deep; aperture somewhat elongate-quadrate; columella recurved.

Type locality: "Santa Maria Bay, off Hughes Point, Baja California del Sur, Mexico (24° 45′5″N, 112°19′W); 54 m."

Distribution: The west coast of Baja California, Mexico, 15 to 38 m.

Type: Holotype LACM-AHF no. 1178; 52.3 × 11.7 mm.

Discussion: This species is found only on the outer coast of Baja California, Mexico. It is similar in shape to *T. variegata* **(152)** and was considered a form of that species until it was found to have a protoconch of 1½ whorls, denoting direct development. See under *T. variegata* **(152)** to separate it from that and other similar species.

154. ***Terebra plicata*** Gray, 1834

(Pl. 39, fig. 154a;
pl. 40, fig. 154b)

1834 *Terebra plicata* Gray, Proc. Zool. Soc. Lond. pt. 2:61; 1844 Hinds in Sowerby, Thes. Conch. 1:165, pl. 43, fig. 61; 1860 Reeve, Conch. Icon. 12, pl. 17, sp. 76; 1971 Bratcher & Burch, Proc. Calif. Acad. Sci., 4th ser. 37(21):549, figs. 14, 15, 18, 19, 20; 1971 Bratcher & Burch in Keen, Seashells Trop. W. America ed. 2:682, fig. 1556; 1984 Aubry, Terebridae pl. 14.

Description: Shell to 70 mm; color ivory, flesh, or dark brick-red, the red being like a periostracum or a wash of varnish that is not removable with lye or chlorine; first few whorls of teleoconch usually lavender; outline of whorls flat, slightly turreted; protoconch of 3 to 3½ whorls; subsutural band prominent, usually with elongate nodes, occasionally with mere flexuous plications, band defined by subsutural groove; axial ribs straight on early whorls, flexuous later; interspaces wide on some specimens, narrow on others; spiral grooves, usually 3, occasionally only one, usually crossing ribs weakly if at all, sometimes crossing strongly, giving a slightly beaded appearance; grooves clearly punctate on

---

Plate 40

FIG. **154b,** *Terebra plicata* Gray. Conway Bay, Santa Cruz I., Galapagos Is.; 25.6 mm. **155a-c,** *T. crenifera* Deshayes. **155a,** Holotype British Museum (N.H.) No. 197990; 29.8 mm. **155b,** Holotype of *T. ligyrus* Pilsbry & Lowe, Academy Natural Sciences Philadelphia No. 155279; 26.7 mm. **155c,** Paratype of *T. ligyrus* Pilsbry & Lowe, same Institution; 24.5 mm. **156a-c,** *T. armillata* Hinds. **156a,** Lectotype British Museum (N.H.) No. 1979124; 49.2 mm. **156b,** Lectotype of *T. marginata* Deshayes, British Museum (N.H.) No. 197966; 34.0 mm. **156c,** Lectotype of *Myurella albocincta* Carpenter, British Museum (N.H.) No. 197943; 41.0 mm. **157a-e,** *T. intertincta* Hinds. **157a,** Lectotype British Museum (N.H.) No. 1968235; 44.1 mm. **157b,** Paralectotype, same Institution; 40.4 mm. **157c,** Sal-si-puedes I.; 28.5 mm. **157d,** Holotype of *T. malonei* Vanatta, Academy Natural Sciences Philadelphia No. 8873; 15.6 mm. **157e,** Lectotype of *T. rufocinerea* Carpenter, British Museum (N.H.) No. 1979156; 32.0 mm.

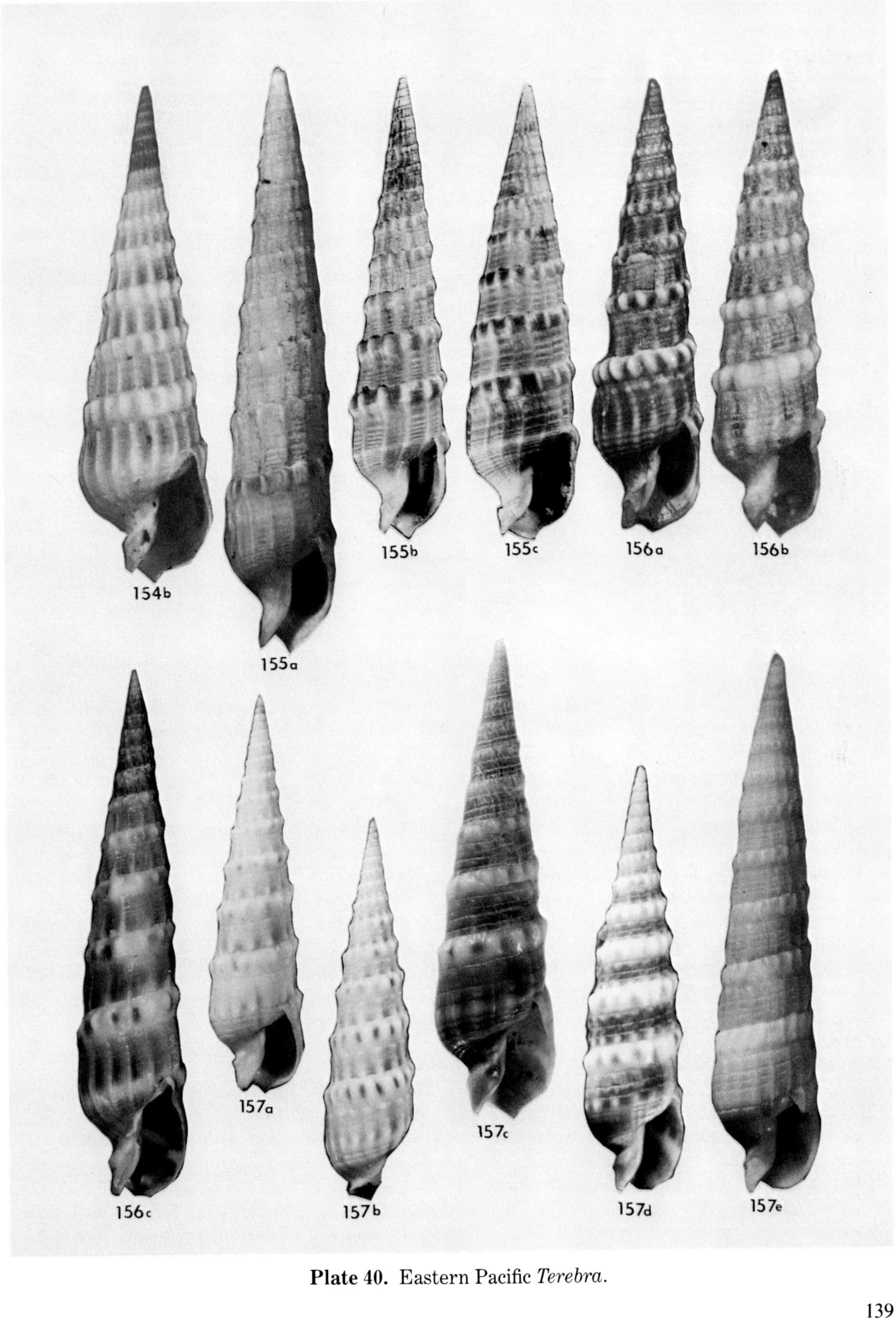

Plate 40. Eastern Pacific *Terebra*.

some specimens, microscopically on others; aperture quadrate; outer lip thin with rib pattern showing through; columella curved.

Type locality: None given. Sombrero Is., Galapagos Is., is here selected as the type locality.

Distribution: Nicaragua to the Galapagos Is.; common in the Galapagos, rare elsewhere; 5 to 128 m.

Type: Holotype BM(NH) no. 1874.11.10.8; 38.0 mm.

Discussion: Specimens with the dark beige to brick-red varnish-like coating have been taken in quantities at various collecting stations from intertidal to deep water and on many types of bottom. Some of these lots have a few normal flesh-colored specimens among them. Refer to *T. variegate* (**152**) for comparison.

155. *Terebra crenifera* Deshayes, 1859

(Pl. 40, figs. 155a-c)

1859 *Terebra crenifera* Deshayes, Proc. Zool. Soc. Lond. pt. 7:298; 1971 Bratcher & Burch in Keen, Seashells Trop. W. America, ed. 2:674, fig. 1531.

1932 *Terebra (Strioterebrum) ligyrus* Pilsbry & Lowe, Proc. Acad. Nat. Sci. Philadelphia 84:40, pl. 1, figs. 10, 11; 1958 Keen, Seashells Trop. W. America ed. 1:492, figs. 968.

Description: Shell to 38 mm; color, from white to tan; outline of whorls slightly concave; protoconch of 3½ conical whorls; subsutural band with elongate nodes, defined by a thin groove, which may be partially or completely obsolete; axial sculpture of ribs from suture to suture beginning at an elongate node on the subsutural band and appearing to hang from that node, 12 to 18 on penultimate whorl; spiral sculpture of 4 to 6 cords, well-developed or obsolete, may be confined to interspaces but usually cross ribs; aperture quadrate; columella recurved.

Type locality: *T. crenifera*, "China," error. Guaymas, Mexico, is here designated as the type locality; *T. ligyrus*, "Guaymas, Mexico."

Distribution: From southern California (where it is rare) to Ecuador; intertidal to 110 m.

Type: *T. crenifera*, holotype BM(NH) no. 197990, 29.8 × 6.0 mm; *T. ligyrus* holotype (pl. 40, fig. 155b) ANSP no. 155279, 26.7 × 5.8 mm.

Discussion: Although this species may vary in number of ribs and may be slender or broad, it may be easily distinguished by the axial ribs that appear to hang from an elongate node below the suture.

156. *Terebra armillata* Hinds, 1844

(Pl. 40, figs. 156a-c)
(Color pl. E, fig. 22)

1844 *Terebra armillata* Hinds, Proc. Zool. Soc. Lond. pt. 11:154; 1844 Hinds in Sowerby, Thes. Conch. 1:173, pl. 43, fig. 49; 1860 Reeve, Conch. Icon. 12: pl. 16, sp. 72a, b; 1966 Keen, Veliger 8(4):273, pl. 47, fig. 18; 1971 Bratcher & Burch, Proc. Calif. Acad. Sci. 4th ser. 37(21):556, fig. 27; 1971 Bratcher & Burch in Keen, Seashells Trop. W. America, ed. 2:672, fig. 1522.

1850 *Terebra albicostata* Adams & Reeve, Zool. Voy. H. M. S. Samarang: p. 30, pl. 10, fig. 21.

1857 *Terebra marginata* Deshayes, Journ. Conchyl. 6:86, pl. 4, fig. 8.

1857 *Myurella albocincta* Carpenter, Cat. Coll. Mazatlan Shells p. 384.

1958 *Terebra (Strioterebrum) albocincta* Carpenter, Keen, Seashells Trop. W. America, ed. 1:490, fig. 956.

1958 *Terebra (Strioterebrum) armillata* Hinds, Keen, Seashells Trop. W. America, ed. 1:490, fig. 956.

1968 *Terebra (Myurella) albocincta* Carpenter, Keen, Veliger 10(4):428, pl. 58, fig. 70.

1971 *Terebra albocincta* Carpenter, Bratcher & Burch in Keen, Seashells Trop. W. America, ed. 2:672, fig. 1520.

Description: Shell to 67 mm; color dark gray or brown with white subsutural band spotted with brown, or monochromatic brown, gray, or rarely creamy peach; outline of whorls almost straight, usually with extremely convex subsutural band; protoconch of 3½ conical whorls; subsutural band with round or elongate nodes which may be large and pearl-like or smaller and less conspicuous; band defined by moderate subsutural groove; axial ribs, 10 to 18 on penultimate whorl, sometimes swollen anteriorly; spiral grooves, about 6, sometimes faintly crossing ribs, breaking them into inconspicuous nodes; aperture moderately elongate; columella recurved; siphonal fasciole striate, with very sharp keel.

Type locality: *T. armillata*, "Between Panama and Bay of Magdalena in Lower California, 5–15 fms, also Galapagos, 10 fms." Mazatlan, Mexico, is here designated as the type locality; *T. albicostata*, "China Sea," error; *T. marginata*, "West Africa," error; *T. albocincta*, "Mazatlan," Mexico.

Distribution: From southern California to Peru and the Galapagos Islands; intertidal to 110 m.

Type: *T. armillata* lectotype BM(NH) no. 1979124, 49.2 mm; *T. albicostata* type lost; *T. marginata* lectotype (pl. 40, fig. 156b) BM(NH) no. 197966, 34.0 × 8.0 mm; *T. albocincta* lectotype here selected (pl. 40, fig. 156c) BM(NH) no. 197943, 41.1 mm.

Discussion: This is an extremely variable species, with one form intergrading into another. Carpenter's *T. albocincta* was separated primarily on color. Those of the type lot which were the slender form were easy to separate from the typical *T. armillata* form, but some of Carpenter's type lot were not so slender and had ribs broken into nodes. Only the color separated them. Had they been light brown or beige with cream-colored ribs and nodes, they would have been recognized immediately as specimens of *T. armillata*. The slender form with ribs swollen anteriorly also occurs in the typical light brown color with cream ribs and nodes. Refer to *T. variegata* (**152**) for comparison.

157. *Terebra intertincta* Hinds, 1844

(Pl. 40, figs. 157a-e; pl. 41, figs. 157f-g)

1844 *Terebra intertincta* Hinds, Proc. Zool. Soc. Lond. pt. 11:155; 1844 Hinds in Sowerby, Thes. Conch. 1:173, pl. 44, fig. 81; 1955 Hertlein & Strong, Bull. American Mus. Nat. Hist. 107(2):212, pl. 1, fig. 12, pl. 3, fig. 9; 1971 Bratcher & Burch in Keen, Seashells Trop. W. America ed. 2:678, fig. 1543; 1984 Aubry, Terebridae pl. 3.

1857 *Myurella hindsii* Carpenter, Cat. Mazatlan Shells p. 385.

1857 *Myurella rufocinerea* Carpenter, Cat. Mazatlan Shells p. 386.

1924 *Terebra malonei* Vanatta, Proc. Acad. Sci. Philadelphia 76:424.

1958 *Terebra (Strioterebrum) intertincta* Hinds, Keen, Seashells Trop. W. America ed. 1:678, fig. 1543.

1958 *Terebra (Strioterebrum) hindsii* (Carpenter), Keen, Seashells Trop. W. America ed. 1:492, fig. 964; 1984 Aubry, Terebridae pl. 8.

1958 *Terebra (Strioterebrum) malonei* Vanatta, Keen, Seashells Trop. W. America ed. 1:492, fig. 970.

1966 *Terebra hindsii* (Carpenter), Morris, Field Guide to Shells p. 200, pl. 61, fig. 20; 1971 Bratcher & Burch in Keen, Seashells Trop. W. America ed. 2:678, fig. 1542.

1968 *Terebra (Myurella) hindsii* (Carpenter), Keen, Veliger 10(4):428, pl. 58, figs. 71, 71b.

1968 *Terebra (Myurella) rufocinerea* (Carpenter), Keen, Veliger 10(4):428, pl. 58, fig. 73.

1971 *Terebra rufocinerea* (Carpenter), Bratcher & Burch in Keen, Seashells Trop. W. America ed. 2:684, fig. 1562; 1984 Aubry, Terebridae pl. 7.

Description: Shell to 61 mm; color cream to pinkish or bluish gray, often with small brown spots between nodes; outline of whorls almost flat to concave; protoconch of 1½ whorls; subsutural band with weak or well-developed nodes, band defined by groove; axial ribs often almost completely obsolete, usually broken into small nodes; spiral grooves or noded cords, 3 or 4; a row of nodes, sometimes obsolete, usually appearing posterior to the suture and at the periphery of the body whorl; aperture quadrate; columella recurved.

Type locality: *T. intertincta*, "Gambia," error. Mazatlan, Mexico, is here designated as the type locality; *M. hindsii*, "Mazatlan," Mexico; *M. rufocinerea*, "Mazatlan," Mexico; *T. malonei*, "Santa Rosalia, opposite San Marcos Is., Lower California," Mexico.

Distribution: Baja California, Mexico, to Ecuador; intertidal to 37 m.

Type: *T. intertincta* lectotype BM(NH) no. 1968235, 44.1 mm; *M. hindsii* lectotype (pl. 41, fig. 157f) BM(NH) no. 1979155, 31.5 mm; *M. rufocinerea* lectotype (pl. 40, fig. 157e) BM(NH) no. 1979156, 32 mm; *T. malonei* (pl. 40, fig. 157d) holotype ANSP no. 8873, 15.6 × 6.3 mm.

Discussion: The intergrades show a gradual transition from one form to another in this species with the extremes of each form being separable. The whorls of those with larger nodes on the band appear concave. *T. variegata* (**152**) may be separated from this species by its wider apical angle, lack of nodes posterior to the suture and at the periphery of the body whorl, and protoconch of 3 instead of the 1½ whorls found on this species.

158. *Terebra tiarella* Deshayes, 1857

(Pl. 41, figs, 158a-c)

1857 *Terebra tiarella* Deshayes, Journ. Conchyl. 6:91, pl. 5, fig. 7.

1958 *Terebra fitchi* S. S. Berry, Leaflets Malac. 1(15):89.

Description: Shell to 34 mm; color cream, reddish brown, or blackish brown; outline of whorls flat except for protruding nodes below suture, weakly turreted; protoconch of 1½ whorls; subsutural band of protruding nodes; subsutural groove absent; sutures deep; axial ribs on early whorls only; spiral grooves fine, evenly spaced; body whorl elongate; aperture elongate; columella curved, with 2 weak plications.

Type locality: *T. tiarella*, "Natal," error. Bahia Santa Maria, Baja California, Mexico, is here designated as the type locality; *T. fitchi*, "Bahia Santa Maria, Isla Magdalena, Baja California."

Distribution: Endemic to the outer coast of Baja California, Mexico.

Type: *T. tiarella* lectotype BM(NH) no. 197974, 31.6 × 7.9 mm; *T. fitchi* holotype Berry Coll., 34.3 × 7.9 mm.

Discussion: This species has an extremely long body whorl with large nodes below the suture, making it easy to distinguish from other Panamic terebrids.

159. *Terebra allyni* Bratcher & R. D. Burch, 1970

(Pl. 41, figs. 159a-b)

1970 *Terebra allyni* Bratcher & Burch, Veliger 12(3):298, pl. 44, figs. 1–41; 1971 Bratcher & Burch in Keen, 2nd ed.: 672, fig. 1521; 1984 Aubry, Terebridae pl. 13.

Description: Shell to 39 mm; color light cream to light gray with or without blotches of rust-brown or chocolate-brown; outline of whorls flat; protoconch of 1½ dome-shaped whorls; subsutural band slightly convex, noded; axial ribs may be strong, weak, or obsolete; spiral cords, up to 5, may be strong, breaking ribs into weak nodes, or may be absent, body whorl with a row of nodes at the periphery in most individuals; aperture semi-elongate; columella straight.
Type locality: "East shore of Marie Madre Island, Tres Marias Group, Mexico (21°35′N, 106° 26′W); 10 to 20 m."
Distribution: From Baja California, Mexico, to Costa Rica; intertidal to 20 m.
Type: Holotype CAS Dept. Geology no. 13278; 25.8 × 6.3 mm.
Discussion: Occasional specimens are more slender than others, and some have a more convex subsutural band because of having larger nodes. This species was named in honor of Allyn G. Smith (1893–1976), a well-known malacologist of California.

160. *Terebra roperi* Pilsbry & Lowe, 1932

(Pl. 41, figs. 160a-c; pl. 42, fig. 160d)

1932 *Terebra (Strioterebrum?) roperi* Pilsbry & Lowe, Proc. Acad. Nat. Sci. Philadelphia 84:41, pl. 1, fig. 14.
1958 *Terebra (Strioterebrum) roperi* Keen, Pilsbry & Lowe, Seashells Trop. W. America ed. 1:494, fig. 974; 1964 Campbell, Veliger 6(3):133, pl. 17, fig. 1.
1964 *Terebra (Strioterebrum) adairensis* Campbell, Veliger 6(3):133, pl. 17, figs. 3–5.
1971 *Terebra roperi* Pilsbry & Lowe, Bratcher & Burch in Keen, Seashells Trop. W. America ed. 2:682, fig. 1561; 1984 Aubry, pl. 8.
1971 *Terebra adairensis* Campbell, Bratcher & Burch in Keen, Seashells Trop. W. America ed. 2:670, fig. 1518; 1984 Aubry, Terebridae pl. 12.

Description: Shell to 29 mm, slender, delicate; color shiny gray to light brown; outline of whorls concave; protoconch of 3½ to 4 whorls; sculpture of 2 rows of nodes, 1 above and 1 below the suture; spiral grooves, sometimes obsolete, vary from none to 4; aperture moderately elongate; columella curved.
Type locality: *T. roperi*, "Mazatlan," Mexico; *T. adairensis*, "Bahia de Adair, approx. 15 mi. N. of Puerto Penasco, Sonora, Mexico."
Distribution: Gulf of California, Mexico, to Ecuador; 4 to 31 m.
Type: *T. roperi* holotype ANSP no. 155287, 18.1 × 3.8 mm; *T. adairensis* holotype (pl. 42, fig. 160d) CAS no. 12539, 29.3 mm.
Discussion: Specimens toward the northern part of the range are usually light gray and tend to have more spiral grooves, as in the holotype of *T. adairensis* Campbell, although specimens with no spiral grooves have been collected at Cholla Bay in the northern part of the Gulf of California. The holotype of *T. roperi* has 2 obsolete spiral grooves on the penultimate whorl and the body whorl, and the paratype has a well marked subsutural groove. This species was named in honor of Edward Warren Roper (1858–98), a newspaper editor on the California coast.

161. *Terebra ninfae* Campbell, 1961

(Pl. 41, figs. 161a-b)

1961 *Terebra (Strioterebrum) ninfae* Campbell, Veliger 4(1):27, pl. 5, figs. 7, 8.
1971 *Terebra ninfae* Campbell, Bratcher & Burch in Keen, Seashells Trop. W. America 2nd ed: 680, fig. 1553; 1984 Aubry, Terebridae pl. 12.

Description: Shell to 9.0 mm; color dark red-brown with a tan peripheral band, the first 2 whorls of teleoconch being transparent white

---

Plate 41
FIGS. **157f,g,** *Terebra intertincta* Hinds. **157f,** Lectotype of *T. hindsii* Carpenter, British Museum (N.H.); 31.5 mm. **157g,** Middle whorls. **158a-c,** *T. tiarella* Deshayes. **158a,** Lectotype British Museum (N.H.) No. 197974; 31.6 mm. **158b,** Paralectotype, same Institution; 31.1 mm. **158c,** Holotype of *T. fitchi* Berry, Berry coll.; 34.3 mm. **159a,b,** *T. allyni* Bratcher & Burch. **159a,** Holotype California Academy Sciences, Dept. Geology No. 13278; 25.8 mm. **159b,** Punta Juluapan, Manzanillo, W. Mexico; 31.7 mm. **160a-c,** *T. roperi* Pilsbry & Lowe. **160a,** Holotype Academy Sciences, Philadelphia No. 155287; 18.1 mm. **160b,** Middle whorls. **160c,** Paratype San Diego Museum Natural History No. 44781; 28.5 mm.

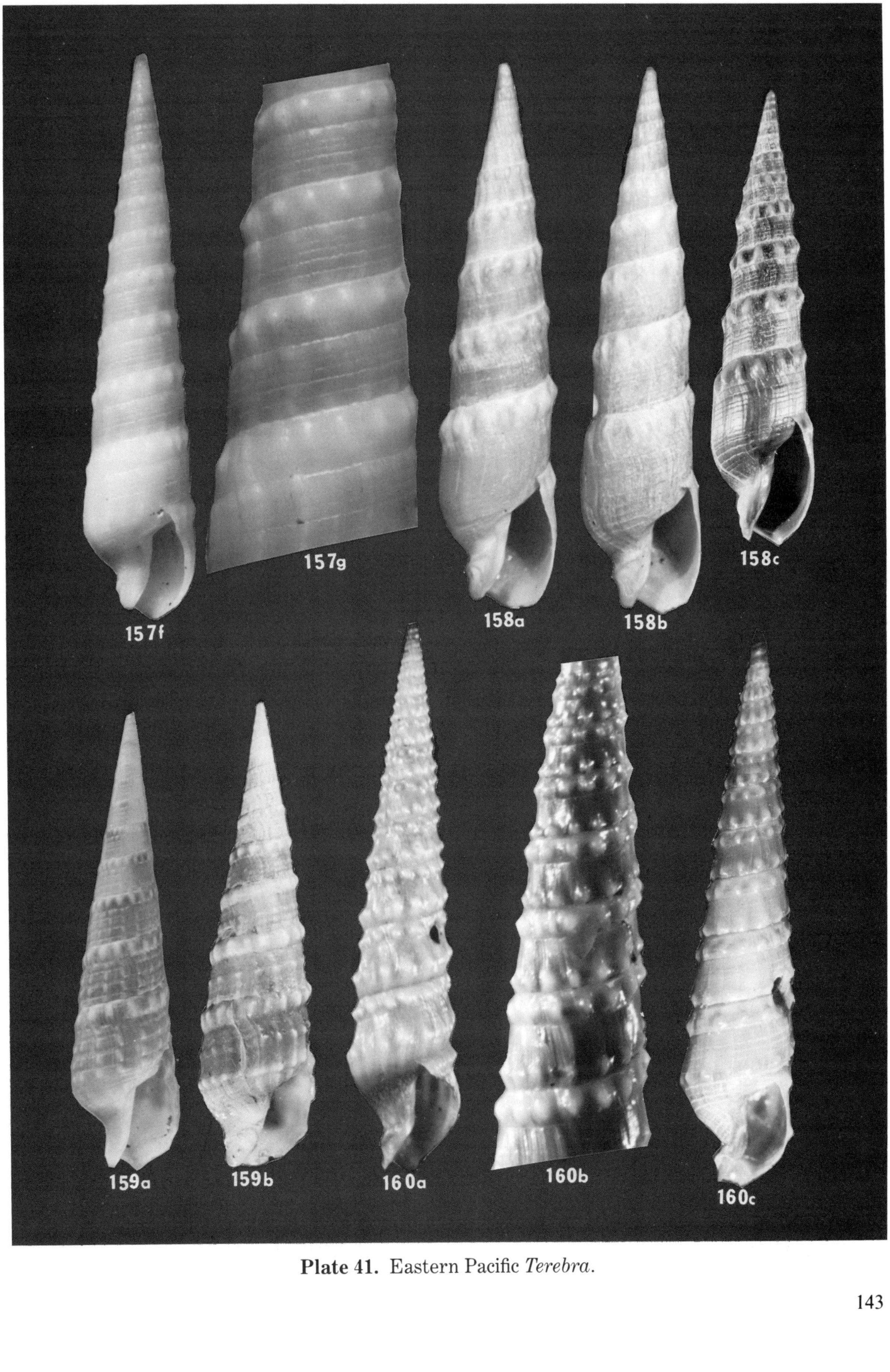

Plate 41. Eastern Pacific *Terebra*.

with a brownish subsutural band; outline of whorls sloping outward from suture to anterior end; protoconch of 3½ whorls; subsutural band narrow, brown, ornamented with light to whitish, sharp, round nodes, subsutural groove absent; axial ribs, about 12, convex, arcuate, swollen at the anterior ends, often appearing to be slightly overhanging the suture; spiral grooves microscopic, over entire teleoconch; aperture elongate; columella recurved.

Type locality: "Between Puerto Madero and 30 mi. N. of the Guatemalan border and Chiapas, Mexico, 14°55′, 92°15′W; 15 to 20 fms."

Distribution: From Manzanillo, Mexico, to Chiapas, Mexico; 27 to 55 m.

Type: Holotype CAS no. 12,353; 6.2 × 2 mm.

Discussion: This is the smallest of the Panamic terebrids. The rather wide-spaced nodes seem to overhang the suture. Refer to *T. bridgesi* **(162)** for comparison. This species was named for Ninfa Mendoza, wife of the captain of the Mexican shrimp boat that first dredged this species in 1959.

162. *Terebra bridgesi* Dall, 1908

(Pl. 42, figs. 162a-c)

1908 *Terebra (Strioterebrum) bridgesi* Dall, Bull. Mus. Comp. Zool. Harvard 43(6):253; 1958 Keen, Seashells Trop. W. America ed. 1:490, fig. 958.

1964 *Terebra (Strioterebrum) dushanae* Campbell, Veliger 6(3):135; 1984 Aubry, Terebridae pl. 9.

1971 *Terebra bridgesi* Dall, Bratcher & Burch in Keen, Seashells Trop. W. America ed. 2:672, fig. 1526; 1984 Aubry, Terebridae pl. 9.

Description: Shell to 11 mm, delicate; color brown, usually with lighter area toward anterior of whorls, darker brown on base of body whorl, axial ribs whitish; outline of whorls slightly convex; protoconch 4 convex whorls; subsutural band of elongate nodes, defined by shallow subsutural groove; axial ribs on remainder of whorl not contiguous with nodes on band; interspaces, including band, with shallow spiral grooves, 2 to 7 on penultimate whorl; aperture moderately quadrate; columella straight.

Type locality: *T. bridgesi*, "Panama"; *T. dushanae*, "Puertecitos, Baja California, Mexico."

Distribution: From the Gulf of California, Mexico, to Ecuador; intertidal to 26 m.

Type: *T. bridgesi*, holotype USNM no. 9404, 10.5 × 2.5 mm; *T. dushanae* holotype (pl. 42, fig. 162c) CAS Geol. type coll. no. 12532, 9.2 × 1.9 mm.

Discussion: This small terebrid is somewhat variable in sculpture. It is a rather fragile shell, but not as fragile as *Terebra ninfae* **(161)**, which may be distinguished from this species by its smaller size and by the swelling of the axial ribs anteriorly into a teardrop shape overhanging the suture. The species was named for the original collector, Robert Bridges (1806–1882), a pharmacist professor who collected on the Pacific coast.

163. *Terebra frigata* Hinds, 1844

(Pl. 42, figs. 143a, b)

1844 *Terebra frigata* Hinds, Proc. Zool. Soc. Lond. pt. 11:162; 1844 Hinds in Sowerby, Thes. Conch. 1:163, pl. 44, fig. 71; 1860 Reeve, Conch. Icon. 12: pl. 24, sp. 132; 1971 Bratcher & Burch, Proc. Calif. Acad. Sci., ser. 4, 37(21):547, figs. 16, 17; 1971 Bratcher & Burch in Keen, Seashells Trop. W. America ed. 2:676, fig. 1537; 1984 Aubry, Terebridae pl. 14.

Description: Shell to 27 mm; color white; outline of whorls slightly concave; protoconch 3½ whorls; subsutural band noded, defined by a row of punctations between ribs; axial ribs weak, almost obsolete in early whorls, becoming well-developed, strong, noded posteriorly and slightly swollen anteriorly in later whorls, 16 to 20 on penultimate whorl; spiral grooves, 6 to 8, fill interspaces; interspaces wider than ribs; aperture quadrate; columella curved.

Type locality: "Galapagos Islands, 6 fms."

Distribution: Galapagos Islands.

Type: Lectotype BM(NH) no. 1968420; 22.9 mm.

Discussion: The difference in sculpture between the early whorls and later ones of this species is

---

Plate 42

FIG. **160d,** *Terebra roperi* Pilsbry & Lowe. Holotype of *T. adairensis* Campbell, California Academy Sciences No. 12539; 29.3 mm. **161a,b,** *T. ninfae* Campbell. **161a,** Holotype California Academy Sciences No. 12353; 6.2 mm. **160b,** Manzanillo, W. Mexico, 15-40 fms.; 7.0 mm. **162a-c,** *T. bridgesi* Dall. **162a,** Holotype National Museum Natural History, Washington No. USNM-9404; 10.5 mm. **162b,** Upper whorls. **162c,** Holotype of *T. dushanae* Campbell, California Academy Sciences, Dept. Geology No. 12532; 9.2 mm. **163a,b,** *T. frigata* Hinds. **163a,** Lectotype British Museum (N.H.) No. 1968420; 22.9 mm. **163b,** Santa Cruz I., Galapagos Is.; 22.5 mm. **164a-d,** *T. jacquelinae* Bratcher & Burch. **164a,** Holotype California Academy Sciences No. 13215; 33.2 mm. **164b,** Middle whorls. **164c,** Juvenile holotype of *T. hertleini* Bratcher & Burch, California Academy Sciences No. 13222; 11.8 mm. **164d,** Upper whorls and protoconch.

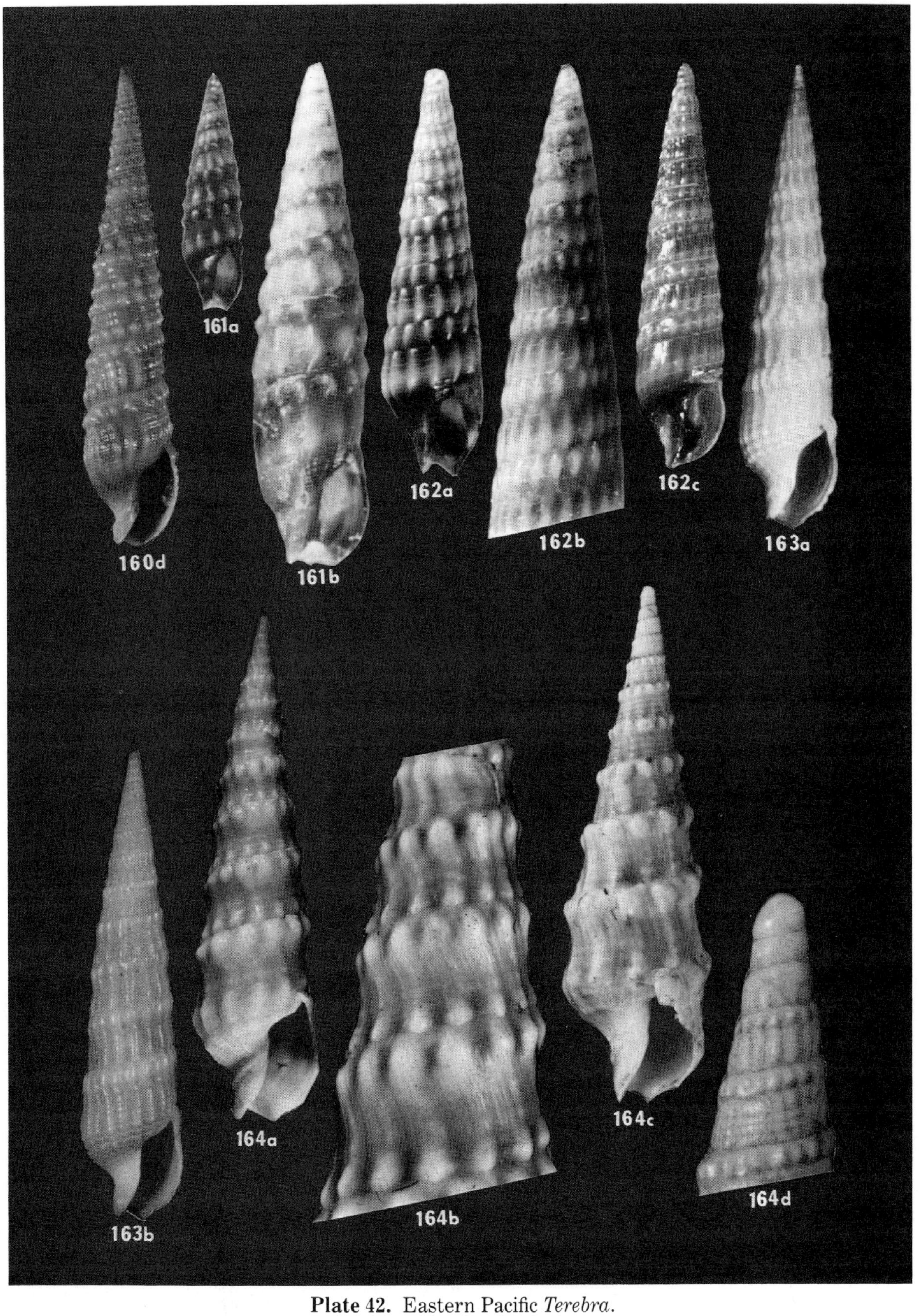

Plate 42. Eastern Pacific *Terebra*.

so great that it would be impossible to recognize the juvenile individuals as the same species without the advantage of seeing a growth series. *T. jacquelinae* **(164)** may be separated from this species by the lack of spiral sculpture, by having a more concave outline, and by the large nodes posterior to the suture and at the periphery of the body whorl, and by its larger size.

164. *Terebra jacquelinae* Bratcher & Burch, 1970

(Pl. 42, figs. 164a-d)

1970 *Terebra jacquelinae* Bratcher & Burch, Contrib. Sci. Los Angeles County Mus. Nat. Hist. 188:2, figs. 3, 4; 1971 Bratcher & Burch in Keen, Seashells Trop. W. America ed. 2:678, fig. 1545; 1984 Aubry, Terebridae pl. 12.

1970 *Terebra hertleini* Bratcher & Burch, Contrib. Sci. Los Angeles County Mus. Nat. Hist. p. 1, figs. 1, 2; 1971 Bratcher & Burch in Keen, Seashells Trop. W. America ed. 2:678, fig. 1541.

Description: Shell to 36 mm; color usually shiny cream, rarely beige, or light brown with cream-colored nodes; outline of whorls concave; protoconch of 1½ somewhat bulbous whorls; subsutural groove absent; axial ribs, with large nodes at both ends, fading in center of whorl after first few whorls, interspaces containing macroscopic axial growth striae; spiral sculpture absent after earliest whorls, which have obsolete grooves that cross ribs; body whorl with about 9 to 18 large nodes at the periphery; aperture moderately elongate.
Type locality: *T. jacquelinae*, "Academy Bay, Santa Cruz Is., Galapagos Is., 0°46′16″S, 90° 9′38″W, at about 20 m"; *T. hertleini*, "Academy Bay, Santa Cruz Is., Galapagos Is."
Distribution: Galapagos Islands; 4 to 40 m.
Type: *T. jacquelinae* holotype CAS 13215, 33.2 × 8.4 mm; *T. hertleini* holotype (pl. 42, fig. 164b) CAS no. 13222, 11.8 × 5.0 mm, juvenile.
Discussion: This species has whorls with an extremely concave outline, and the spiral sculpture becomes obsolete on the later whorls. The holotype of *T. hertleini* is an immature specimen. Refer to *T. frigata* **(163)** for comparison. This species was named for Jacqueline De Roy, who lives in the Galapagos Islands and collected and studied shells.

165. *Terebra concava* (Say, 1826)

(Pl. 43, fig. 165)

1826 *Turritella concava* Say, Journ. Acad. Nat. Sci. Philadelphia 5:207.

1925 *Terebra concava* (Say), Dall, Proc. U. S. Nat. Mus. 66(2554):27, pl. 9, fig. 8; 1954 Abbott, American Seashells ed. 1:266, pl. 261, fig. j.

1974 *Terebra (Strioterebrum) concava* (Say), Abbott, American Seashells ed. 2:259, fig. 2839; 1975 Rios, Brazilian Mar. Moll. Icon. p. 126, pl. 38, fig. 554.

Description: Shell to 25 mm; color white or grayish; outline of whorls concave; protoconch of 2 rounded whorls, broader than following whorl of teleoconch; subsutural band absent; axial sculpture absent; spiral sculpture of 2 rows of elongate nodes, the larger immediately anterior to suture, the smaller posterior with 5 or 6 grooves between; body whorl with a row of nodes at periphery, and a row of larger ones below suture; aperture quadrate; columella recurved.
Type locality: None given. Charleston, South Carolina, is here designated as the type locality.
Distribution: From North Carolina to the Gulf of Mexico and Brazil.
Type: Type lost.
Discussion: Although the type of this species is missing, and it was not figured, the description leaves no doubt as to its identity. It was described as a *Turritella*, but Say placed all terebrids in that genus. *T. vinosa* **(188)** originally described as *T. concava vinosa*, may be separated from this species by its axial ribs and lack of nodes at the periphery of the body whorl.

Plate 43
FIG. **165,** *Terebra concava* (Say). Boynton Beach, Florida; 21.8 mm. **166a-e,** *T. gemmulata* Kiener. **166a,** Lectotype Museum d'Histoire Naturelle, Geneva No. 1152/76; 47.0 mm. **166b,** Middle whorls. **166c,** Damaged holotype of *T. patagonica* d'Orbigny, British Museum (N.H.) No. 1854.12.5.501.; 48.8 mm. **166d,** Holotype of *T. chilensis* Deshayes, Museum National d'Histoire Naturelle, Paris; 40.2 mm. **167a,b,** *T. benthalis* Dall. **167a,** Holotype Museum Comparative Zoology, Harvard No. 7011; 21.0 mm. **167b,** Upper whorls.

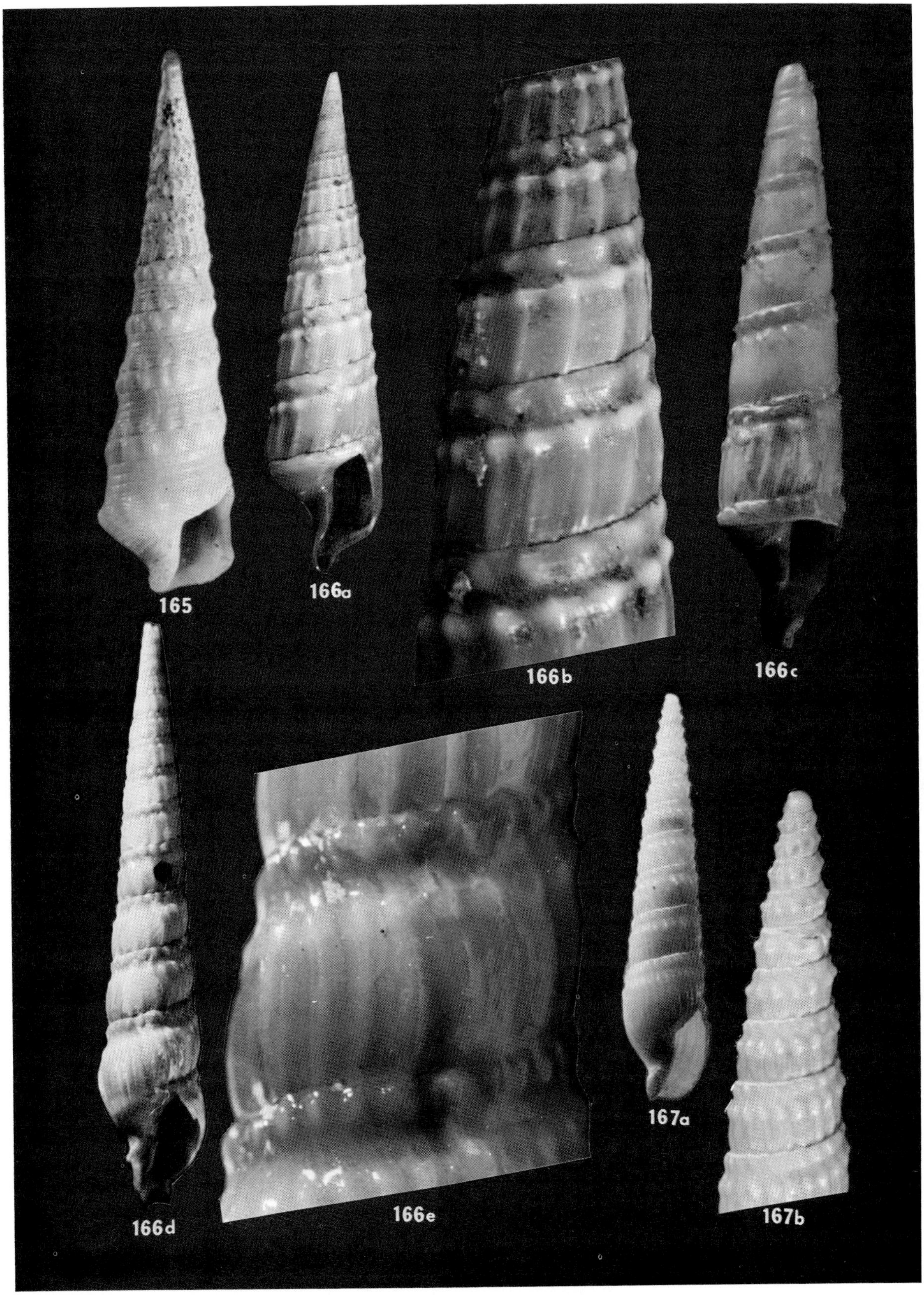

Plate 43. Western Atlantic *Terebra*.

166. *Terebra gemmulata* Kiener, 1839

(Pl. 43, fig. 166a-e)

1839 *Terebra gemmulata* Kiener, Spéc. Gén. Coq Viv. 9:15, pl. 5, figs. 11, 11a. 1860 Reeve, Conch. Icon. 12: pl. 9, sp. 33; 1885 Tryon, Man. Conch. 7:18, pl. 4, figs. 61, 62; 1975 Rios, Brazilian Mar. Moll. Icon. p. 126, pl. 38, fig. 552.

1839 *Terebra patagonica* d'Orbigny, Voy. Ameri. Merid. p. 442, pl. 62, fig. 1.

1859 *Terebra chilensis* Deshayes, Proc. Zool. Soc. Lond. pt. 27:295. 1977 Bratcher, Nautilus 91(2):40, fig. 2.

Description: Shell to 61 mm; color tan with white subsutural nodes and a white band at periphery of body whorl; outline of whorls slightly convex with concave subsutural band; protoconch of 1½ whorls; subsutural band consisting of two rows of nodes divided by a moderately broad channel, the lower row of nodes white; axial ribs extending from lower row of subsutural nodes to following suture, ribs usually widely spaced, occasionally closely set; interspaces sometimes filled with axial striae; spiral sculpture absent; body whorl with ribs ending at periphery; aperture quadrate.
Type locality: *T. gemmulata*, none given. White Bay, Patagonia, is here designated as the type locality; *T. patagonica*, "White Bay, Patagonia"; *T. chilensis*, none given, other than the inference of Chile.
Distribution: From Rio de Janeiro, Brazil, to Chile; 25 to 90 m.
Type: *T. gemmulata* lectotype MHNG no. 1152/76, 47.0 × 4.6 mm; *T. patagonica* holotype BM(NH) no. 1854.12.5.501, 48.8 × 11.4 mm; *T. chilensis* holotype MNHNP, 40.2 × 7.9 mm.
Discussion: This is one of the two species found in frigid water. The other is *T. tristis* (**120**) which ranges into the cold waters of southern New Zealand. In the form Deshayes described as *T. chilensis*, the anterior row of subsutural nodes are almost obsolete.

167. *Terebra benthalis* Dall, 1889

(Pl. 43, figs. 167a,b;
pl. 44, figs. 167c,d)

1889 *Terebra (Acus?) benthalis* Dall, Bull. Mus. Comp. Zool. Harvard 18(2):65, pl. 29, fig. 6.

1889 *Terebra benthalis* Dall, Bull. U.S. Nat. Mus. 37:94, pl. 29, fig. 6.

1890 *Terebra (Acus) benthalis* var. *nodata* Dall, Proc. U. S. Nat. Mus. 12(733):299, pl. 5, fig. 9.

1974 *Terebra (Fusoterebra* Sacco, 1891) *benthalis* Dall, Abbott, American Seashells, ed. 2:261, fig. 2856.

Description: Shell to 27 mm; color yellowish white to pale brown; outline of early whorls elongate, later ones moderately convex; protoconch 1½ whorls; subsutural groove absent on early whorls, shallow groove developing later; axial ribs weak, rising to a sharp node in middle of early whorls, moving to immediately anterior to groove from center of whorl, becoming progressively less sharp, in some individuals fading completely. Microscopic spiral grooves developing; aperture elongate; columella curved.
Type locality: *T. benthalis*, "Off Morro Light, Havana, Cuba, 100 to 400 fms;" *T. benthalis* var *nodata*, "St. 2750, off St. Bartholomew, West Indies, 496 fms., bottom temp. 44° F."
Distribution: From Cuba and the Florida Straits to St. Bartholomew, West Indies; 200 to 992 m.
Type: *T. benthalis* holotype MCZ no. 7011, 21.0 × 5.9 mm; *T. benthalis* var. *nodata* holotype USNM no. 97099, 17.4 × 4.4 mm.
Discussion: There is no other Atlantic species with which this could be confused. Evidently Dall named the var. *nodata* from a very young specimen, as the younger and smaller the specimen, the more it resembles the *nodata* form.

168. *Terebra panamensis* Dall, 1908

(Pl. 44, figs. a-c)

1908 *Terebra (Strioterebrum) panamensis* Dall, Bull. Mus. Comp. Zool. 43(6):250, pl. 5, fig. 10; 1944 M. Smith, Panamic Mar. Shells p. 36, fig. 469; 1958 Keen, Seashells Trop. W. America ed. 1:494, fig. 973.

1971 *Terebra panamensis* Dall, Bratcher & Burch in Keen, Seashells Trop. W. America ed. 2:692, fig. 1555; 1984 Aubry, Terebridae pl. 3.

Plate 44
FIGS. **167c,d,** *Terebra benthalis* Dall. **167c,** Paratype National Museum Natural History, Washington No. USNM-87226; 26.9 mm. **167d,** Holotype of *T. benthalis* var. *nodata* Dall, National Museum Natural History, Washington No. USNM-97099; 17.0 mm. **168a-c,** *T. panamensis* Dall. **168a,** Holotype National Museum Natural History, Washington No. USNM-123084; 22.5 mm. **168b,** Lower and middle whorls of holotype. **168c,** Upper whorls of holotype. **169a-d,** *T. shyana* Bratcher & Burch. **169a,** Holotype Los Angeles County Museum No. 1249; 26.0 mm. **169b,** Upper whorls of holotype. **169c,** Holotype of *T. purdyae* Bratcher & Burch, Los Angeles County Museum No. AHF-1182; 13.9 mm. **169d,** Upper whorls of holotype. **170,** *T. corintoensis* Pilsbry & Lowe. Holotype Academy Natural Sciences, Philadelphia No. 155281; 13.5 mm. **171a,b,** *T. stohleri* Bratcher & Burch. **171a,** Holotype Los Angeles County Museum No. AHF-1180; 21.4 mm. **171b,** Upper whorls of holotype.

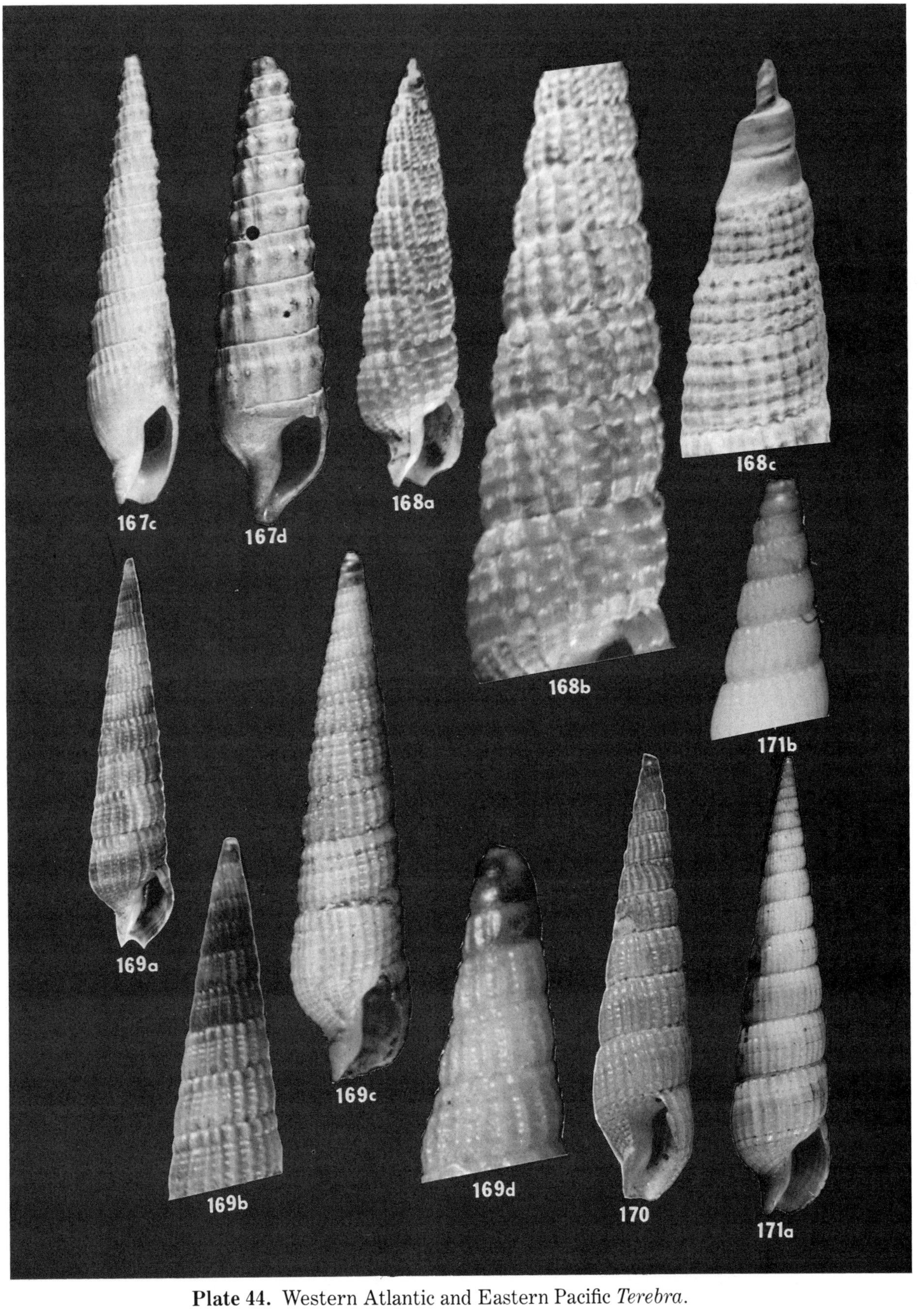

Plate 44. Western Atlantic and Eastern Pacific *Terebra*.

Description: Shell to 49 mm; color yellowish beige; outline of whorls convex; protoconch of 3 whorls, usually eroded; subsutural band inconspicuous, defined by deep groove between axial cords, shallow over cords; sculpture coarsely cancellate, of evenly spaced axial cords crossed by spiral cords of equal strength, forming small round nodes at intersections; aperture quadrate; columella recurved.
Type locality: "Gulf of Panama, 153 fms."
Distribution: From the west coast of Baja California, Mexico, to Peru; 20 to 280 m.
Type: Holotype USNM no. 123084, 22.5 × 6.0 mm.
Discussion: There is an obvious weakness in the protoconch and early whorls of the teleoconch of this species. The outer layer of shell and sculpture peels away, leaving a snaggle-tooth appearance to the apex of the shell of the holotype, paratypes, and numerous other specimens. This species may be separated from *T. shyana* **(169)** by its more convex whorls, coarser sculpture, lack of fulvous blotches and more obese shell.

169. *Terebra shyana* Bratcher & Burch, 1970

(Pl. 44, figs. 169a-d)

1970 *Terebra shyana* Bratcher & Burch, Veliger 12(3):295, pl. 44, figs. 9, 10 (Jan. 1, 1970); 1971 Bratcher & Burch in Keen, Seashells Trop. W. America ed. 1:684, fig. 1564; 1984 Aubry, Terebridae pl. 8.

1970 *Terebra purdyae* Bratcher & Burch, Contrib. Sci. Los Angeles Co. Mus. Nat. Hist. 188:5, figs. 5, 6 (May 4, 1970); Bratcher & Burch in Keen ed. 2:682, fig. 1559.

Description: Shell to 36 mm; color beige with fulvous blotches, sometimes with faintly lighter stripes; outline of whorls almost straight; protoconch of 3 slender conical whorls; subsutural band slightly noded, defined by groove cutting through ribs; sculpture finely cancellate, with about 25 to 37 axial riblets crossed by about 4 or 5 spiral cords, often forming small inconspicuous nodes at intersections, spiral riblets sometimes stronger, often unevenly spaced; aperture quadrate; columella curved.
Type locality: *T. shyana*, "Manzanillo, Colima, Mexico. 19°02′N, 104°21′W, 8.5 to 20 m"; *T. purdyae*, "Santa Maria Is., Galapagos Is."
Distribution: From the outer coast of Baja California, Mexico, to Peru and Galapagos Is.; 3 to 146 m.
Type: *T. shyana* holotype LACM no. 1249, 26.0 × 5.6 mm; *T. purdyae* holotype LACM no. AHF 1182, 13.9 × 3.3 mm, juvenile.
Discussion: This shell has fine cancellate sculpture, although in some individuals the axial sculpture is stronger, causing it to look somewhat more ribbed and less cancellate. It was named for the original collectors, Mr. and Mrs. Carl Shy, of Westminster, California. See *T. panamensis* **(168)** for comparison. *T. purdyae* is a very young individual of this species.

170. *Terebra corintoensis* Pilsbry & Lowe, 1932

(Pl. 44, fig. 170)

1932 *Terebra (Strioterebrum) corintoensis* Pilsbry & Lowe, Proc. Acad. Nat. Sci. Philadelphia 84:42, pl. 1, fig. 5; 1958 Keen, Seashells Trop. W. America ed. 1:491, fig. 960.

1971 *Terebra corintoensis* Pilsbry & Lowe, Bratcher & Burch in Keen, ed. 2:674, fig. 960; 1984 Aubry, Terebridae pl. 3.

Description: Shell to 14 mm; color bluish brown to cinnamon brown with lighter subsutural band; outline of whorls convex; protoconch of 4½ slender whorls; subsutural band defined by shallow groove which cuts through ribs; axial ribs curved, somewhat narrower than interspaces, about 25 on penultimate whorl; spiral cords about 4 to 7; body whorl with groove at periphery; aperture moderately quadrate; columella curved.
Type locality: "Corinto, Nicaragua."
Distribution: From Baja California, Mexico to Ecuador; 13–18 m.
Type: Holotype ANSP no. 155281; 13.5 × 3.2 mm.
Discussion: *T. shyana* **(169)** may be separated from this species by its more cancellate and flatter whorls, its noded conspicuous subsutural band defined by a constricting groove, and by its color which is beige with fulvous blotches while *corintoensis* species has a brown shell.

171. *Terebra stohleri* Bratcher & Burch, 1970

(Pl. 44, figs. 171a,b)

1970 *Terebra stohleri* Bratcher & Burch, Contrib. Sci. Los Angeles County Mus. no. 188:5, figs. 7, 8; 1971 Bratcher & Burch, Proc. Calif. Acad. Sci. 4th ser. 37(21):538, figs. 1, 2; Bratcher & Burch in Keen, Seashells Trop. W. America ed. 2:684, fig. 1567.

Description: Shell to 21 mm; color shiny ivory with slightly darker blotches; whorls convex, slightly shouldered above suture; protoconch multispiral; subsutural band inconspicuous; sutures deeply channeled, constricted; axial ribs slightly curved, about 28 on penultimate whorl, equal to interspaces; spiral grooves, 3 to 5 grooves, aperture elongate; outer lip thin with sculpture pattern showing through; columella straight with no plication.
Type locality: "Braithwaite Bay, Socorro Island, Mexico."
Distribution: From Pulmo Reef, Baja California, Mexico, to the Galapagos Islands; 4 to 42 m.
Type: Holotype LACM no. AHF 1180; 21.4 × 5.1 mm.
Discussion: This shell looks as though it has been carved from shiny ivory. It was named in honor of Dr. Rudolph Stohler, Californian malacologist and for many years editor of *The Veliger.*

172. *Terebra elata* Hinds, 1844

(Pl. 45, figs. a-c)

1844 *Terebra elata* Hinds, Proc. Zool. Soc. Lond. pt. 11:156; 1844 Hinds in Sowerby, Thes. Conch. 1:177, pl. 44, figs. 68, 69; 1966 Keen, Veliger 8(4):273, pl. 47, fig. 14; 1971 Bratcher & Burch in Keen, Seashells Trop. W. America ed. 2:676, fig. 1535 (in part, right fig. only); 1979 Bratcher, Veliger 22(1): fig. 1; 1984 Aubry, Terebridae pl. 9.

1932 *Terebra montijoensis* Pilsbry & Lowe, Proc. Acad. Nat. Sci. Philadelphia 84:42, pl. 1, fig. 1; 1958 Keen, Seashells Trop. W. America, ed. 1, p. 493, fig. 972; 1971 Bratcher & Burch in Keen, Seashells Trop. W. America, ed. 2, p. 680, fig. 1552.

1958 *Terebra (Strioterebrum) elata* Hinds, Keen, Seashells Trop. W. America ed. 1:491, fig. 962.

Description: Shell to 27 mm; color white or pale beige, occasionally with a few yellow blotches; outline of whorls convex; protoconch of 3½ smooth, glassy whorls; subsutural band marked by groove, deeper between ribs; axial ribs narrower than interspaces, 16 to 20 on body whorl; spiral grooves, 4 to 8, sometimes crossing ribs, outer lip thin; aperture moderately elongate; columella curved, with simple basal fold.
Type locality: *T. elata*, "Bay of Montejo, West Coast of America, in 15 fms"; *T. montijoensis*, "Montijo Bay, R. P." [Panama],
Distribution: From southern Mexico to Ecuador; intertidal to 70 m.
Type: *T. elata* lectotype BM(NH) no. 1968240, 26.6 mm; *T. montijoensis* holotype ANSP 155285, 14.8 × 3.3 mm.
Discussion: Most freshly collected specimens of this species have a purplish tinge to the early whorls of the teleoconch. This color fades in time, leaving a brownish tint. Small immature individuals of *T. guayaquilensis* **(173)** resemble this species but have heavier, coarser sculpture. The latter have not been observed to have purplish early whorls, and they become much larger in maturity and have more variable sculpture.

173. *Terebra guayaquilensis* (E. A. Smith, 1880)

(Pl. 45, figs. 173a,b)

1873 *Myurella belcheri* E. A. Smith, Ann. Mag. Nat. Hist. Lond. 11:267; [non *Terebra belcheri* Philippi, 1851]

1880 *Myurella guayaquilensis* E. A. Smith, Proc. Zool. Soc. Lond. p. 481. (nom. nov. pro. *Myurella belcheri* E. A. Smith, 1873)

1932 *Terebra (Strioterebrum) ira* Pilsbry & Lowe, Proc. Acad. Nat. Sci. Philadelphia 84:40, pl. 1, fig. 13.

1979 *Terebra guayaquilensis* (E. A. Smith), Bracher, Veliger 22(1):62, fig. 2.c; 1984 Aubry, Terebridae pl. 13.

Description: Size to 76 mm; color white to beige, often with yellowish blotches or bands; outline of whorls convex, protoconch of 1½ whorls; subsutural band convex, noded, defined by groove between ribs, cutting lightly over ribs; sculpture extremely variable; axial ribs sometimes close-set, usually widely spaced; spiral grooves, as few as 3 per whorl, not overriding ribs; or there may be up to 5 heavy spiral cords, overriding the ribs to form small beads at the intersections; sutures deep, aperture moderately quadrate; columella recurved.
Type locality: *T. guayaquilensis*, "Guayaquil, Ecuador"; *T. belcheri*, same; *T. ira*, "Mazatlan," Mexico.
Distribution: From the outer coast of Baja California, Mexico, to Ecuador; intertidal to 180 m.
Type: *T. guayaquilensis* holotype BM(NH) no. 1844.6.7.77, 38.0 mm; *T. belcheri* same; *T. ira* holotype ANSP no. 155280, 20.3 × 4.6 mm.
Discussion: Although it has a larger, heavier, less delicate shell than that of *T. elata* **(172)**, it was confused with the latter for a time because the early sculpture of both species is extremely similar, except that the early whorls of this species have not been seen with a purple hue. For comparison refer to *T. elata* **(172)**.

174. *Terebra berryi* Campbell, 1961

(Pl. 45, figs. a-c)

1961 *Terebra (Strioterebrum) berryi* Campbell, Veliger 4(1):26, pl. 5, figs. 5–6.

1971 *Terebra berryi* Campbell, Bratcher & Burch, Proc. Cal. Acad. Sci. 4th ser. 37(21):558, figs. 28, 29; 1971 Bratcher & Burch in Keen, Seashells Trop. W. America ed. 2:672, fig. 1524; 1984 Aubry, Terebridae pl. 14.

Description: Shell to 57 mm; color pale gray or beige marbled or mottled with brown; whorls very slightly convex; protoconch of 4, sometimes almost black, whorls; subsutural band convex, defined by impressed groove, which in some specimens becomes a sharply cut groove; axial ribs sharp, flexuous, swelling into nodes on the subsutural band; interspaces much wider than ribs on early whorls, occasionally narrower on later whorls; spiral grooves, in interspaces, about 4, sometimes faintly crossing ribs; aperture elongate; outer lip thin with marbled color showing through; columella curved, slight parietal callus on mature specimens.
Type locality: "Puertocitos, Baja California, Mexico."
Distribution: From Puertocitos, Baja California, Mexico, to Cocos Island, Costa Rica; intertidal to 37 m.
Type: Holotype CAS no. 12352; 32.0 × 7.0 mm.
Discussion: The color varies from those almost entirely lavender-gray to some which are quite brownish, although all have a more or less marbled appearance. This species was named for Dr. S. Stillman Berry (1887–1984), a leading private malacologist of Redlands, California, and editor of *Leaflets in Malacology.*

175. *Terebra churea* Campbell, 1964

(Pl. 45, fig. 175)

1964 *Terebra churea* Campbell, Veliger 6(3):134, pl. 17, figs. 17, 18, 24. 1971 Bratcher & Burch in Keen, Seashells Trop. W. America ed. 2:672, fig. 1528.

Description: Shell to 30 mm; color white to light brown; outline of whorls slightly convex with convex subsutural band; protoconch of $1^1/_2$ rounded whorls; subsutural band with weak nodes or ribs, with several indistinct spiral cords; axial ribs curved, weak, 20 to 24 on penultimate whorl; interspaces wider than ribs; spiral cords weak, usually 6 or 7, sometimes crossing ribs; aperture moderately elongate; columella curved.
Type locality: "Ensenada San Francisco, Guaymas, Sonora, Mexico, intertidal."
Distribution: Gulf of California, Mexico; intertidal.
Type: Holotype SBMNH no. 03461; 17.4 × 3.6mm.
Discussion: This rare species has weak sculpture compared to most of the other Panamic terebrids.

176. *Terebra brandi* Bratcher & Burch, 1970

(Pl. 176, figs. a,b)

1970 *Terebra brandi* Bratcher & Burch, Veliger 12(3):296, pl. 44, figs. 5–6; 1971 Bratcher & Burch in Keen, Seashells Trop. W. America ed. 2:672, fig. 1525; 1984 Aubry, Terebridae pl. 8.

Description: Shell to 24 mm; color bluish gray; outline of whorls slightly convex with moderately convex subsutural band; protoconch of 4 dark, blackish brown, shiny whorls; subsutural band of axial ribs, defined by impressed suture and subsutural groove which crosses ribs after about the 4th whorl of the teleoconch; axial ribs slightly curved, narrower than interspaces, spiral grooves evenly spaced, not crossing ribs, 6 to 9 on penultimate whorl, 2 to 3 on subsutural band; aperture moderately elongate; columella brown, slightly curved, with light parietal callus.
Type locality: "Petatlan Bay, Mexico, 17°31′N, 101°27′W; 10 to 20 m, hard sand and shell bottom."
Distribution: From the Gulf of California, Mexico, to Peru; intertidal to 20 m.
Type: Holotype LACM no. AHF 1252; 16.9 × 3.9 mm.

Plate 45
FIGS. **172a-c,** *Terebra elata* Hinds. **172a,** Lectotype British Museum (N.H.) No. 1968240; 26.6 mm. **172b,** Holotype of *T. montijoensis* Pilsbry & Lowe, Academy Natural Sciences, Philadelphia No. 155285; 14.8 mm. **172c,** Kobbe Beach, Panama; 24.3 mm. **173a,b,** *T. guayaquilensis* (E.A. Smith). **173a,** Holotype British Museum (N.H.) No. 1844.6.7.77.; 38.0 mm. **173b,** Holotype of *T. ira* Pilsbry & Lowe, Academy Natural Sciences, Philadelphia No. 155280; 20.3 mm. **174a-c,** *T. berryi* Campbell. **174a,** Holotype California Academy Sciences No. 12352; 32.0 mm. 174b, Cocos Id.; 16.8 mm. **174c,** Middle whorls. **175,** *T. churea* Campbell. Holotype Santa Barbara Museum Natural History No. 03461; 17.4 mm. **176a,b,** *T. brandi* Bratcher & Burch. **176a,** Holotype Los Angeles County Museum No. AHF-1252; 16.9 mm. **176b,** Upper whorls of holotype.

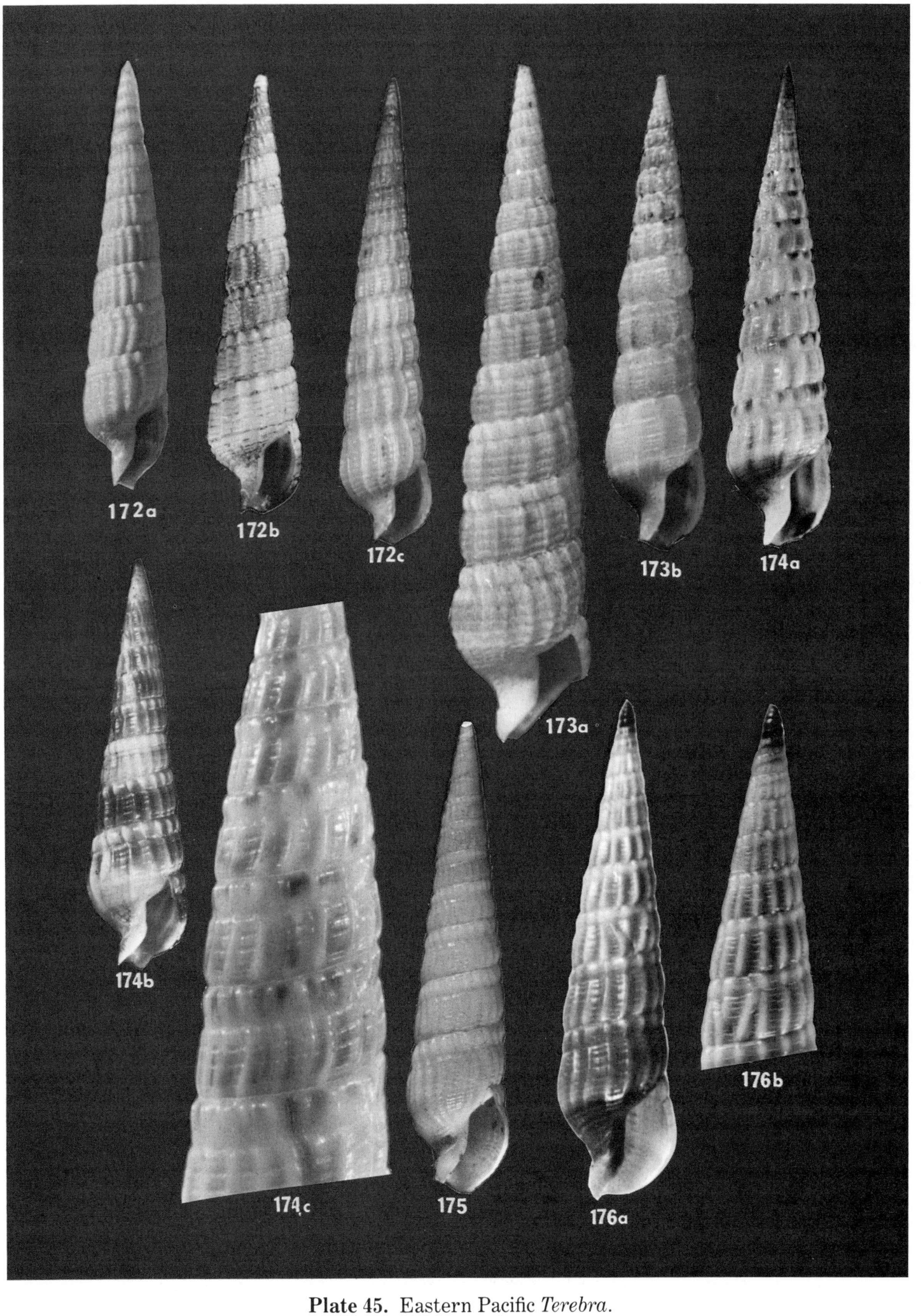

Plate 45. Eastern Pacific *Terebra.*

Discussion: The color of the protoconch varies on some terebrid species, but it appears always to be black in this species as it does on *Hastula bacillus* (**221**). This species was named for Dr. Louis M. Brand, a Texas shell collector.

177. *Terebra sanjuanensis* Pilsbry & Lowe, 1932

(Pl. 46, fig. 177)

1932 *Terebra (Strioterebrum) sanjuanensis* Pilsbry & Lowe, Proc. Acad. Nat. Sci. Philadelphia 84:42, pl. 1, fig. 2; 1958 Keen, Seashells Trop. W. America ed. 1:494, fig. 975.

1971 *Terebra sanjuanensis* Pilsbry & Lowe, Bratcher & Burch in Keen, Seashells Trop. W. America ed. 2:684, fig. 1563; 1984 Aubry, Terebridae pl. 12.

Description: Shell to 26 mm, slender; straw color, sometimes violet-tinted on the base of body whorl; outline of whorls slightly convex; protoconch of $2^1/2$ conical whorls; subsutural band narrow, marked by groove cutting through ribs; axial ribs moderately spaced, narrower than interspaces; spiral sculpture variable, of numerous grooves which faintly cross ribs, few grooves confined to interspaces, or no grooves; aperture elongate; columella recurved.
Type locality: "San Juan, Costa Rica."
Distribution: From Sinaloa, Mexico to Panama; 7 to 27 m.
Type: Holotype ANSP no. 155286; 16 × 2.9 mm.
Discussion: The violet tint on the base of the shell mentioned in the description is seen only on freshly collected specimens. It soon fades to a scarcely noticeable tan.

178. *Terebra polypenus* Pilsbry & Lowe, 1932

(Pl. 46, fig. 178)

1932 *Terebra (Microtrypetes) polypenus* Pilsbry & Lowe, Proc. Acad. Nat. Sci. Philadelphia 84:44, pl. 1, fig. 4; 1958 Keen, Seashells Trop. W. America ed. 1:490, fig. 953.

1971 *Terebra polypenus* Pilsbry & Lowe, Bratcher & Burch in Keen, Seashells Trop. W. America ed. 2:682, fig. 1557.

Description: Shell to 14 mm; color light brown, darker anterior to the periphery of the body whorl; outline of whorls slightly convex; protoconch of 3 to $3^1/2$ whorls; subsutural band ribbed with about 4 weak spiral cords, defined by shallow groove; axial ribs numerous, crowded, weak, unevenly spaced, sinuous, about 33 on penultimate whorl; spiral cords weak, crossing ribs, about 8 on penultimate whorl; aperture elongate; columella curved.
Type locality: "Mazatlan, Mexico, about 20 fms."
Distribution: Mazatlan, Mexico.
Type: ANSP no. 155282; 14.0 × 3.0 mm.
Discussion: This species was described from a single specimen and has not been reported since. It may be a pathological specimen of another species.

179. *Terebra pedroana* Dall, 1908

(Pl. 46, figs. a-c)

1908 *Terebra (Strioterebrum) pedroana* Dall, Bull. Mus. Comp. Zool. Harvard 43(6):251.

1927 *Terebra pedroana* Dall, Oldroyd, Mar. Shells W. Coast N. America 2:59, pl. 6, fig. 6; 1966 Morris Field Guide Shells p. 102, pl. 40, fig. 9; 1969 McLean, Mar. Shells S. California p. 52, pl. 28, fig. 4; 1984 Aubry, Terebridae pl. 9.

Description: Shell to 47 mm; color dull gray to yellowish beige, sometimes blotched with fulvous brown; outline of whorls straight; protoconch of $1^1/2$ whorls; subsutural band with nodes, sometimes completely obsolete, band defined by very weak groove; sculpture extremely variable, some individuals with many weak ribs crossed by weak spiral grooves, others with well developed axial ribs and spiral grooves; aperture quadrate; columella short, recurved.
Type locality: "San Pedro, California."
Distribution: From Santa Monica, California, to southern Baja California, Mexico; subtidal to 260 m.
Type: Holotype USNM no. 118806; 32.0 × 6.5 mm.

Plate 46
FIG. **177,** *Terebra sanjuanensis* Pilsbry & Lowe. Holotype Academy Natural Sciences, Philadelphia No. 155286; 16.0 mm. **178,** *T. polypenus* Pilsbry & Lowe. Holotype Academy Natural Sciences, Philadelphia No. 155282; 14.0 mm. **179a-c,** *T. pedroana* Dall. **179a,** Holotype National Museum Natural History, Washington, No. USNM-118806; 32.0 mm. **179b,** Middle whorls of holotype. **180a-f,** *T. hemphilli* Vanatta. **180a,** Lectotype Academy Natural Sciences, Philadelphia No. 33645; 16.0 mm. **180c, d,** Paralectotypes same Institution No. 354866; 15.3 mm and 14.3 mm respectively. **180e,** Holotype of *T. danai* Berry, Stanford University, Paleontology coll. No. 8655; 31.4 mm. **180f,** San Pedro, California; 20.2 mm.

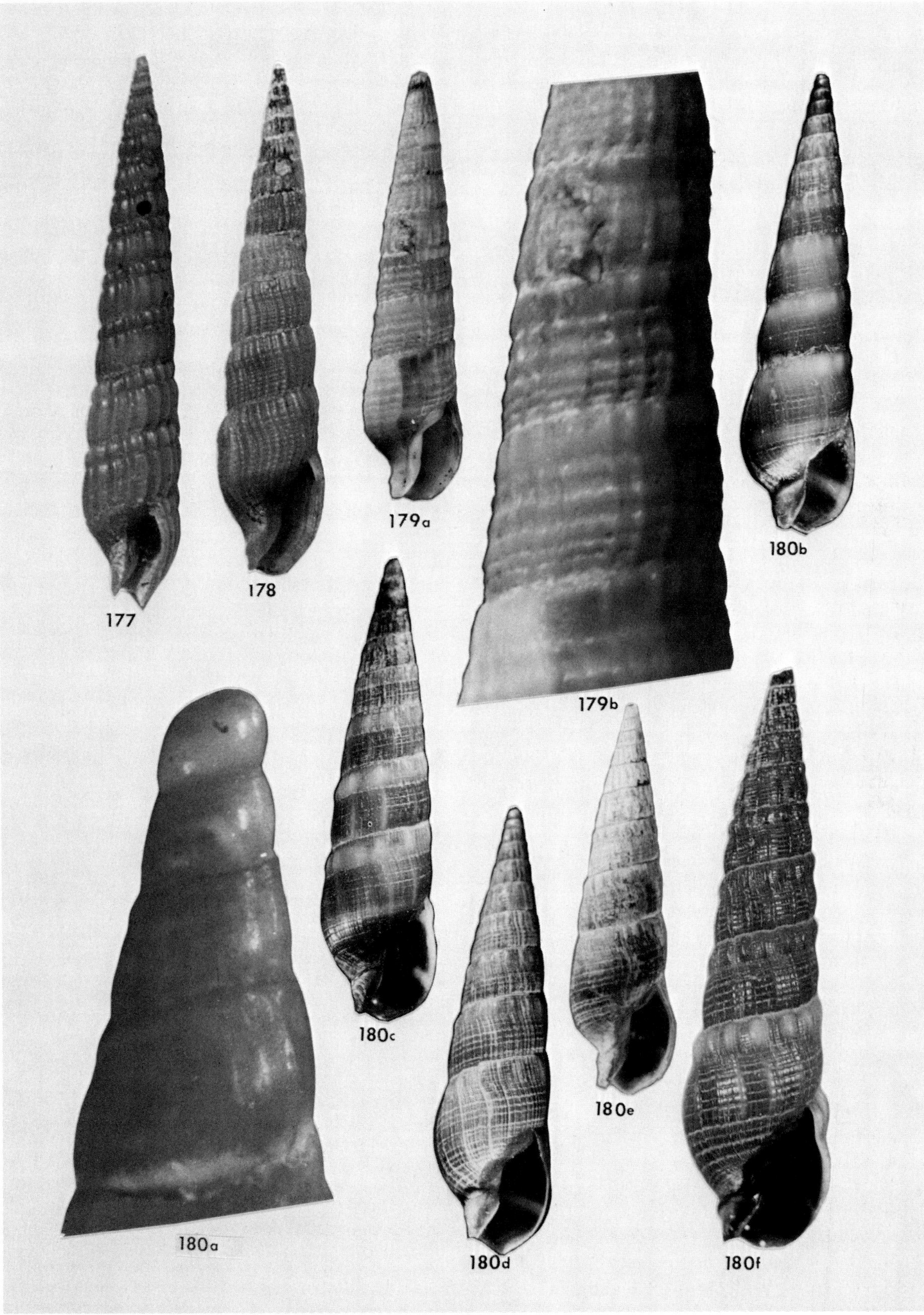

Plate 46. Eastern Pacific *Terebra*.

Discussion: This shell is more elongate and more slender than *T. hemphilli* **(180)** and has a squared-off body whorl and gray to tan color. That of *T. hemphilli* is more inflated and rounded and has darker whorls with a contrasting light subsutural band.

180. *Terebra hemphilli* Vanatta, 1924

(Pl. 46, figs. 180b-f)
(Pl. 39, fig. 180a)

1865 *Myurella simplex* Carpenter, Ann. & Mag. Nat. Hist. 3rd Ser., vol. 15:395 (non *Terebra simplex* Conrad, 1830).
1921 *Strioterebrum pedroanum philippianum* Dall, Bull, U. S. Nat. Mus. no. 112:67. [non *Terebra philippiana* Deshayes, 1859].
1924 *Terebra pedroana hemphilli* Vanatta, Proc. Acad. Nat. Sci. Philadelphia 76:425, fig. 2.
1958 *Terebra (Strioterebrum) danai* Berry, Leaflets Malac. 1(16):96.
1969 *Terebra danai* Berry, McLean, Mar. Shells S. Calif. p. 52, fig. 28.1.
1974 *Terebra (Hastula) danai* Berry, R. T. Abbott, American Seashells ed. 2:261, fig. 2852.
1974 *Terebra hemphilli* Vanatta, Abbott, American Seashells ed. 2:261, fig. 2853.

Description: Shell to 36 mm; color blue-gray or light brown usually with cream subsutural band, sometimes with brown spots on subsutural band and narrow white band at periphery of body whorl and showing through center of aperture; outline of whorls somewhat convex; protoconch of 1½ whorls; subsutural band with obsolete or well-developed nodes, defined by shallow groove; axial ribs narrower than interspaces, may remain sharp and clear-cut in later whorls or may become obsolete; spiral sculpture of 5 to 8 unevenly spaced, sometimes weak, grooves; aperture quadrate; columella curved.
Type locality: *T. hemphilli*, "Scammons Lagoon, Lower California," Mexico; *M. simplex*, "Sta. Barbara," California; *S. pedroanum philippianum*, "San Pedro, California to head of Gulf of California"; *T. danai*, "East San Pedro, California."
Distribution: From Redondo Beach, California, to central Baja California, Mexico; intertidal and subtidal.
Type: *T. hemphilli* holotype ANSP no. 33645, 15.6 × 3.9 mm; *M. simplex* type not found; *S. pedroanum philippianum* holotype USNM no. 4943, 23.9 × 6.9 mm; *danai* holotype Stanford Univ. Paleo coll. no. 8655, 31.4 × 8.6 mm.
Discussion: The shells from the southern end of the range at Scammons Lagoon are somewhat smaller than those in more northern waters but the holotypes show them to be the same species. It was named for Henry Hemphill (1830–1914), an ardent collector of Florida and California shells.

181. *Terebra mariato* Pilsbry & Lowe, 1932

(Pl. 47, figs. 181a,b)

1932 *Terebra (Microtrypetes) mariato* Pilsbry & Lowe, Proc. Acad. Nat. Sci. Philadelphia 84:43, pl. 1, fig. 7; 1958 Keen, Seashells Trop. W. America ed. 1:490, fig. 952.
1944 *Terebra mariato* Pilsbry & Lowe, M. Smith, Panamic Marine Shells p. 36, fig. 459; 1971 Bratcher & Burch in Keen, Seashells Trop. W. America ed. 2:680, fig. 1551.

Description: Shell to 9 mm; color dull brown; outline of whorls convex; subsutural band absent; protoconch unknown; axial ribs almost straight, extending from suture to suture, about 18 on body whorl; interspaces, about equal to ribs, showing faint traces of spiral striae; aperture elongate-quadrate; columella slightly recurved, with distinct parietal callus.
Type locality: "Mariato, Montijo Bay, Panama."
Distribution: Manzanillo, Mexico to Panana; subtidal.
Type: Holotype ANSP no. 155284; 7.3 × 2.3 mm.
Discussion: This small dull brown shell appears not to be found intertidally. Refer to *T. iola* **(182)** for comparison. It was named after the Panamanian village of Mariato.

Plate 47.
FIGS. **181a,b,** *Terebra mariato Pilsbry & Lowe.* **181a,** Holotype Academy Natural Sciences, Philadelphia No. 155284; 7.3 mm. **181b,** Paratype San Diego Museum Natural History No. 44778; 6.2 mm. **182a-c,** *T. iola* Pilsbry & Lowe. **182a,** Holotype Academy Natural Sciences, Philadelphia No. 155289; 14.0 mm. **182b,** Mazatlan, Sinaloa, Mexico: 19.4 mm. **182c,** Guaymas, Sonora, Mexico; 11.5 mm. **183a-f,** *T. larvaeformis* Hinds. **183a,** Lectotype British Museum (N.H.) No. 1968239; 50.7 mm. **183b,** Paralectotype same Institution No. 1968239; 37.3 mm. **183c,** Immature holotype of *T. isopleura* Pilsbry & Lowe, Academy Natural Sciences, Philadelphia No. 155283; 15.0 mm. **183d,** Incomplete holotype of *T. brunneocincta* Pilsbry & Lowe, Academy Natural Sciences, Philadelphia No. 155288; 21.3+ mm. **183e,** Paratype of *T. brunneocincta* Pilsbry & Lowe, San Diego Museum Natural History No. 44775; 30.0 mm. **183f,** 2 mi. W. of Puerto Penasco, Sonora, Mexico; 34.7 mm. **184,** *T. puncturosa* Berry. Holotype San Diego Museum Natural History No. 12954; 30.3 mm.

Plate 47. Eastern Pacific *Terebra*.

182. ***Terebra iola*** Pilsbry & Lowe, 1932

(Pl. 47, figs. 182a-c)

1932 *Terebra (Microtrypetes) iola* Pilsbry & Lowe, Proc. Acad. Nat. Sci. Philadelphia 84:43, pl. 1, fig. 3; 1958 Keen, Seashells Trop. W. America ed. 1:489, fig. 950.

1971 *Terebra iola* Pilsbry & Lowe, Bratcher & Burch in Keen, Seashells Trop. W. America ed. 2:678, fig. 1544; 1984 Aubry, Terebridae pl. 7.

Description: Shell to 19 mm; color uniform light brown; outline of whorls turreted; protoconch of $2^1/_2$ whorls; subsutural groove absent; axial ribs swelling slightly below suture, about 14 per whorl; spiral cords fine, about 15 per whorl, crossing faintly over ribs; aperture elongate; columella curved.
Type locality: "Mazatlan, Mexico; 20 fms."
Distribution: Santa Maria Bay, outer coast of Baja California, Mexico, to Mazatlan, Mexico; intertidal to 55 m.
Type: Holotype ANSP no. 155289; 14.0 × 3.4 mm.
Discussion: This is the type species of the subgenus *Microtrypetes* Pilsbry & Lowe, 1932. They diagnose the genus: "Axial ribs continuous; no subsutural band; one weak fold on the columella. Subgenus of *Terebra*." There is little variation in this species except that some specimens are more slender than others. This species may be distinguished from *T. mariato* (**181**) by its turreted outline, more widely spaced axial ribs, and well-developed spiral sculpture.

183. ***Terebra larvaeformis*** Hinds, 1844

(Pl. 47, figs, 183a-f)
(Color pl. E, fig. 26)

1844 *Terebra larvaeformis* Hinds, Proc. Zool. Soc. Lond. pt. 11:155; 1844 Hinds in Sowerby, Thes. Conch. 1:176, pl. 43, figs. 46, 47; 1860 Reeve, Conch. Icon. 12, pl. 10, sp. 41a, b; 1944 M. Smith, Panamic Mar. Shells p. 36, fig. 462; 1958 Keen, Veliger 8(4):274, pl. 47, fig. 11; Bratcher & Burch in Keen, Seashells Trop. W. America ed. 2:680, fig. 1547; 1984 Aubry, Terebridae pl. 7.

1844 *Terebra varicosa* Hinds, Proc. Zool. Soc. Lond. pt. 11:152; [non *Buccinum varicosum* Gmelin, 1791 = *Terebra*]; 1966 Keen, Veliger 8(4):274, pl. 47, fig. 12, figd. lectotype.

1932 *Terebra (Strioterebrum) isopleura* Pilsbry & Lowe, Proc. Acad. Nat. Sci. Philadelphia 84:41, pl. 1, fig. 6.

1932 *Terebra (Strioterebrum) brunneocincta* Pilsbry & Lowe, Proc. Acad. Nat. Sci. Philadelphia 84:41, pl. 1, fig. 12; 1958 Keen, Seashells Trop. W. America ed. 1:490, fig. 1527.

1958 *Terebra (Strioterebrum) larvaeformis* Hinds, Keen, Seashells Trop. W. America ed. 1:492, fig. 967.

1958 *Terebra (?Microtrypetes) isopleura* Pilsbry & Lowe, Keen, Seashells Trop. W. America ed. 1:489, fig. 951.

1971 *Terebra brunneocincta* Pilsbry & Lowe, Bratcher & Burch in Keen, Seashells Trop. W. America ed. 2:674, fig. 1527.

Description: Shell to 44 mm; color blackish brown, light orange-brown or pale grayish brown with ribs often lighter at the crests and often with a light band at the periphery of the body whorl; outline of whorls convex; protoconch of 3 to $3^1/_2$ whorls; subsutural band defined by spiral grooves between ribs on early whorls, cutting through the ribs later; axial ribs widely spaced, arcuate, 13 to 20 on penultimate whorl; spiral grooves fine, often macroscopic, sometimes completely obsolete; aperture moderately quadrate; columella recurved.
Type locality: *T. larvaeformis*, "St. Elena, Monte Christi, W. Coast of America, 6 to 15 fms," Colombia; *T. varicosa*, "Gulf of Papaagayo, W. Coast of Central America, 23 fms"; *T. isopleura*, "Mazatlan," Mexico; *T. brunneocincta*, "San Juan del Sur, Nicaragua."
Distribution: From the outer coast of Baja California, Mexico, to Ecuador; intertidal to 146 m.
Type: *T. larvaeformis* lectotype BM(NH) no. 1968239, 50.7 mm; *T. varicosa* lectotype BM(NH) no. 1844.6.7.88, 25 mm; *T. isopleura*, holotype ANSP no. 155283, 15.0 × 4.2 mm; *T. brunneocincta* ANSP no. 155288, 21.3 × 7.0 mm, broken specimen.
Discussion: This is another species with shells showing gradual transition in sculpture from one form to another. At first we considered the *T. brunneocincta* form a separate species although the differences were small. Then we examined a number of intergrades which could not be placed in either species with certainty. The almost black color form is more common in the northern end of the range. *T. puncturosa* (**184**), which usually is white, ivory, or pinkish-white, may be distinguished from this species by its flatter outline and deep subsutural punctations between the axial ribs which often coalesce into a subsutural groove after about 15 whorls.

184. ***Terebra puncturosa*** S. S. Berry, 1959

(Pl. 47, fig. 184)

1959 *Terebra (Strioterebrum) puncturosa* S. S. Berry, Leaf. Malac. 1(18):112.

1971 *Terebra puncturosa* Berry, Bratcher & Burch in

Keen, Seashells Trop. W. America ed. 2:682, fig. 1558; 1984 Aubry, Terebridae pl. 7.

Description: Shell to 77 mm; color grayish white to pale ivory; outline of whorls flat; protoconch multispiral; subsutural band slightly noded, defined by deep punctations between ribs, gradually lengthening to become an unbroken groove; axial ribs heavy, arcuate, contiguous with nodes on subsutural band, about 17 on penultimate whorl; spiral sculpture usually absent except for grooves on base of body whorl, occasionally with microscopic threads in early whorls; aperture short, moderately elongate; columella short, recurved.

Type locality: "Bahia Santa Maria, Isla Magdalena, Baja California, Mexico, 15 fms."

Distribution: Bahia Santa Maria, Baja California, Mexico to Peru; 3 to 90 m.

Type: Holotype SDMNH San Diego no. 12954; 30.3 × 7.5 mm.

Discussion: The shell of this species usually has a chalky texture, which Berry mentions in his original description. For comparison see *T. larvaeformis* (**184**).

185 ***Terebra dislocata*** (Say, 1822)

(Pl. 48, figs. 185a-e)

1822 *Cerithium dislocatum* Say, Journ. Acad. Nat. Sci. Philadelphia 1st ser. 2:235; 1955 Abbott, Nautilus, vol. 68, p. 125 (cotype).

1834 *Terebra rudis* Gray, Proc. Zool. Soc. Lond. 2:60.

1839 *Terebra petitii* Kiener, Spéc. Gén. Icon. Coq. Viv. 9:37, pl. 13, fig. 3.

1860 *Terebra dislocata* (Say), Reeve, Conch. Icon. 12: pl. 9, sp. 32; 1954 Abbott, American Seashells ed. 1:265, pl. 26, fig. 1; 1961 Warmke & Abbott, Caribbean Seashells p. 132, pl. 25, fig. d; 1971 Bratcher & Burch in Keen, Seashells Trop. W. America ed. 2:674, fig. 1533; 1974 Abbott, American Seashells ed. 2:259, fig. 2836; 1975 Humfrey, Sea Shells of the West Indies p. 180, pl. 2, fig. 12; 1981 Garcia-Cubas, Moll. Trop Lagoon Syst. p. 74, fig. 84; 1982 Abbott & Dance, Compendium of Seashells p. 276; 1984 Aubry, Terebridae pl. 12.

1873 *Acus dislocatus* (Say), Tryon, American Marine Conchology, pl. 8, fig. 73.

1971 *Terebra (Strioterebrum) dislocata* (Say), Andrews, Sea Shells of the Texas Coast, p. 125, text fig.

1974 *Strioterebrum onslowensis* Petuch, Veliger 17(2):205, figs. 1–4.

1984 *Strioterebrum weisbordi* Gibson-Smith & Gibson-Smith, Tulane Stud. Geo. & Paleo. 18(1&2):61, pl. 1, figs. 1, 2. (Cabo Blanco, Late Pliocene, Venezuela).

Description: Shell to 64 mm; color extremely variable, from creamy, pinkish, or lavender-white to beige, tan or brown, often with white or light subsutural band; outline of whorls slightly convex with whorls short; protoconch of $1^1/_2$ broad whorls; subsutural band convex, ribbed, without spiral sculpture, defined by deep, broad, moderately shallow groove; axial ribs close set or wide spaced, 14 to 32 on penultimate whorl; spiral grooves, or cords, 3 to 7, cutting across ribs, usually forming small, inconspicuous nodes; aperture quadrate; columella recurved.

Type locality: *T. dislocata*, none given. Key West, Florida, is here designated as the type locality; *T. rudis*, none given; *T. petitii*, "Atlantic Ocean, coasts of New York and Philadelphia"; *S. onslowensis*, "Onslow Bay, North Carolina."

Distribution: From Maryland to Brazil and from southern California to west Panama.

Type: *T. dislocata* lectotype ANSP no. 192924 (here selected), 63.6 × 7.9 mm; *T. rudis* holotype BM(NH) no. 1874.11.10.11, 36.8 mm; *T. petitii* holotype MHNG, 30.9 mm; *S. onslowensis* holotype CAS no. 54719, 44.3 × 12 mm; *S. weisbordi* fossil holotype PRI 26324 as illustrated by Weisbord (1962, pl. 40, figs. 12, 13) 38.3 × 9.5 mm, recent paratype USNM no. 255013; 48.3 × 11.0 mm.

Discussion: The animal has a thin yellowish brown operculum. This is one of the most variable of terebrids in both color and sculpture. Specimens with 14 axial ribs and 3 spiral grooves look quite different from those with 32 ribs and 7 grooves and look like different species until one sees the gradual intergrades in both spiral and axial sculpture from a wide range of specimens. The unusually short whorl is a good diagnostic feature. Individuals from the Western Atlantic are impossible to separate from those from the Eastern Pacific.

The holotype of *T. onslowensis*, deposited by Petuch at the California Academy of Sciences, is not the specimen he illustrated with the original description. He did not define the status of the specimen illustrated. His taxon is based on offshore specimens which usually have more numerous ribs.

This species is without radula or poison gland but contains venom in "grooved prickles" (Andrews, 1971). Examination of the stomachs of 35 specimens failed to uncover a single food item. It was suggested that *Balanoglossus* castings may serve as the most concentrated source of organic detritus available on sand flat surfaces and that *T. dislocata* exploits this as a food source (Mollick, 1973).

186. *Terebra brasiliensis* (E. A. Smith, 1873)
(Pl. 48, figs. a-d)

1873 *Abretia brasiliensis* E. A. Smith, Ann. & Mag. Nat. Hist. Lond. ser. 4, 11:271.
1975 *Terebra brasiliensis* (E. A. Smith), Rios, Brazilian Mar. Moll. Icon. p. 126, pl. 38, fig. 551.

Description: Shell to 30 mm; color yellowish beige to light chestnut, with white to lavender-white subsutural band, some specimens having a dark chestnut line at periphery of body whorl; outline of whorls almost straight; protoconch of $1\frac{1}{2}$ dome-shaped whorls; subsutural band with nodes which become elongate in later whorls, subsutural groove absent; remainder of sculpture extremely variable; early sculpture may be of weak axial ribs only, axial ribs with spiral grooves in interspaces, or may be completely smooth; in later whorls axial ribs may become almost completely obsolete; spiral grooves continue and may become stronger, about 9 per whorl; aperture quadrate; columella curved.
Type locality: "Botafogo Bay, Rio de Janeiro; 3 fms., sandy mud."
Distribution: Rio de Janeiro area; 6 to 33 m.
Type: Lectotype BM(NH) no. 1860.5.2.28; 10.3 × 2.9 mm.
Discussion: The sculpture of the shells of this species, like those of *T. frigata* (**163**), differs greatly between juvenile and mature individuals. The three specimens in the type lot are juveniles, the largest being only 10.3 mm. None had developed mature sculpture. Therefore, E. A. Smith's description is of little value for identification. Only a growth series showing the transition from juvenile to mature sculpture confirmed the identity of this rare species, which is unlike any other Atlantic species.

187. *Terebra rushii* (Dall, 1889)
(Pl. 48, figs. 187a,b)

1889 *Acus rushii* Dall, Bull. Mus. Comp. Zoology Harvard 18:64.
1902 *Terebra (Acus) rushii* Dall, Proc. U. S. Nat. Mus. 24:503, pl. 29, fig. 6.

Description: Shell to 17 mm; color shiny white; outline of whorls flat; protoconch of $1\frac{1}{2}$ whorls; subsutural band almost smooth, defined by moderately shallow groove; axial sculpture absent in some individuals, with obsolete ribs on others; spiral sculpture of 3 or 4 inconspicuous cords; aperture quadrate; columella curved.
Type locality: "Off Cape Florida in 8 fms."
Distribution: Off South Florida; to 16 m.
Type: Lectotype USNM no. 82952; 14.4 × 3.2 mm.
Discussion: The larger of the two syntypes is fragmented, the entire aperture missing. The smaller one is juvenile and has an immature aperture. The type specimens do not agree with Dall's subsequent illustration and dimensions, where the larger specimen is shown with a complete aperture, and it may represent a composite drawing of the two specimens. Named in honor of the U. S. Naval physician, William H. Rush, who was an ardent shell collector in the late 19th Century.

188. *Terebra vinosa* (Dall, 1889)
(Pl. 49, figs. 188a,b)

1889 *Acus concavus* var. *vinosus* Dall, Bull. Mus. Comp. Zool. Harvard 18(2):63.
1937 *Terebra concava vinosa* Dall, M. Smith, E. Coast Marine Shells p. 132, pl. 33, fig. 18; 1940 Perry, Bull. Amer. Paleont. 26(95):160, pl. 37, fig. 252.

Description: Shell to 22 mm; color grayish white or ivory-white with tan aperture and columella; outline of whorls slightly convex; protoconch of $1\frac{1}{2}$ whorls; sutures deep; subsutural band with elongate nodes and usually with 3 to 6 spiral cords; axial ribs, straight or sinuous, beginning at nodes on band; spiral cords, 3 to 7, faintly crossing ribs; aperture semi-elongate; columella slightly curved, with parietal callus on mature specimens.

Plate 48.
FIGS. **185a-e,** *Terebra dislocata* (Say). **185a,** Lectotype Academy Natural Sciences, Philadelphia No. 192924; 63.6 mm. **185b,** Holotype of *T. rudis* Gray, British Museum (N.H.) No. 1874.11.10.11.; 36.8 mm. **185c,** Holotype of *T. petitii* Kiener, Museum d'Histoire Naturelle, Geneva; 30.9 mm. **185d,** Holotype of *Strioterebrum onslowensis* Petuch, California Academy Sciences, San Francisco No. 032501; 44.3 mm. **185e,** Panama; 38.9 mm. **186a-d,** *T. brasiliensis* (E.A. Smith). **186a,** Immature lectotype British Museum (N.H.) No. 1860.5.2.28.; 10.3 mm. **186b,** Pio, Brazil (immature specimen); 10.0 mm. **186c,** Pio, Brazil; 29.7 mm. **186d,** Middle whorls. **187a, b,** *T. rushii* (Dall). **187a,** Damaged lectotype National Museum Natural History, Washington No. USNM-82952; 14.4 mm. **187b,** Paralectotype with part of outer lip missing, same Institution No. USNM-82952; 9.7 mm.

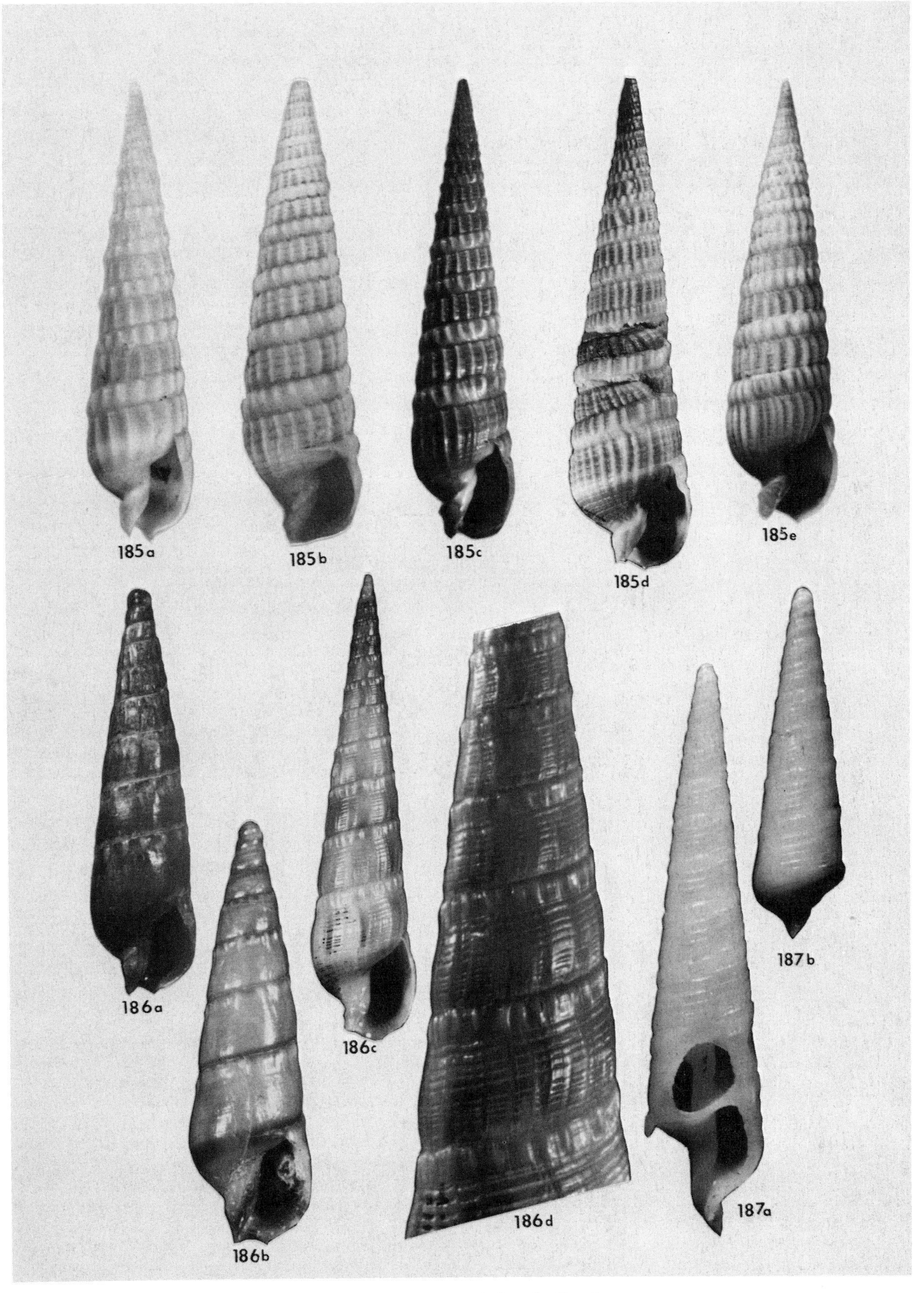

Plate 48. Western Atlantic *Terebra*.

Type locality: "Coast from Hatteras to Florida, 2–30 fms."
Distribution: From North Carolina to northwest Florida; intertidal to 55 m.
Type: Lectotype USNM no. 36076, here designated; 20.2 mm.
Discussion: This species may be separated from the larger *T. concava* **(165)** by its less concave outline; its lack of nodes posterior to the suture of each whorl and at the periphery of the body whorl; and by its ribs which do not become obsolete in the center of the whorls.

189. *Terebra arcas* Abbott, 1954

(Pl. 49, figs. 189a,b)

1954 *Terebra arcas* Abbott, Nautilus 68(2):40, pl. 2, fig. 4; 1974 Abbott, American Seashells ed. 2:259, fig. 2840.

Description: Shell to 27 mm; color white, occasionally with yellowish orange early whorls, and rarely with one wide spiral band of orange; outline of whorls from moderately to extremely convex; protoconch of 1½ whorls; subsutural band defined by groove usually confined to interspaces between ribs, occasionally cutting lightly through ribs, but not indenting outline; axial ribs curved, extending from suture to suture, 12 to 16 per whorl; spiral grooves, 5 to 7 may cut across ribs, but most often are confined to interspaces; aperture quadrate; columella recurved, one basal plication.
Type locality: "East of Arcas Cay, southern Campeche Banks, Mexico, 23 to 24 fms."
Distribution: From the west coast of Florida to Mexico; 4 to 100 m.
Type: Holotype USNM no. 613882; 27.0 × 6.8 mm.
Discussion: The holotype of this species has extremely inflated whorls, but this is a variable feature of the species, which sometimes has only mildly inflated whorls. To distinguish it from *T. glossema* **(190)** refer to the next species.

190. *Terebra glossema* Schwengel, 1942

(Pl. 49, fig. 190)

1940 *Terebra glossema* Schwengel, Nautilus 53(3), pl. 12, figs. 13–14 (figures and name only). [nomen nudum—Art. 13 of ICZN]; 1942 Schwengel, Nautilus, 56(2):65, pl. 6, fig. 3; 1982 Abbott & Dance, Compendium of Seashells p. 276 (paratypes fig'd in color).

1974 *Terebra (Strioterebrum) glossema* Schwengel, Abbott, American Seashells 2nd ed: 260, fig. 2841.

Description: Shell to 28 mm; color white to pale buff with lighter band; outline of whorls convex, slightly turreted; protoconch of 1½ pinkish whorls; subsutural band ribbed, marked by a shallow groove; axial ribs, about 27 on body whorl, crossed by 4 to 6 spiral grooves; aperture quadrate; columella recurved.
Type locality: "Off Pelican Shoals, Florida Keys; 7 to 8 fms."
Distribution: East coast of Florida to Bahama Islands and Cuba.
Type: Holotype ANSP no. 175131; 24.2 × 6.0 mm.
Discussion: *T. arcas* **(189)** may be separated from this species by its whorls which are not turreted and usually are more inflated and by the subsutural band which does not show in outline as it does in this species. See *T. protexta* **(191)** for comparison with that species.

191. *Terebra protexta* (Conrad, 1846)

(Pl. 49, fig. 191)

1846 *Cerithium protextum* Conrad, Proc. Acad. Nat. Sci. Philadelphia 3:26.

1885 *Terebra protexta* (Conrad), Tryon, Man. Conch. 7:25, pl. 6, fig. 98; 1937 M. Smith, E. Coast Mar. Shells p. 132, pl. 51, fig. 134; 1961 Warmke & Abbott, Caribbean Seashells p. 133, pl. 25, fig. C; 1984 Aubry, Terebridae pl. 15.

1940 *Terebra (Strioterebrum) protexta* Conrad, Perry, Bull. Paleont. 26(95):160, fig. 253; 1974 Abbott, American Seashells ed. 2:260, fig. 2842; 1975 Rios, Brazilian Mar. Moll. Icon. p. 127, pl. 38, fig. 5537.

Description: Shell to 25 mm, slender; color brown unless faded to tan or dull white; outline of whorls slightly convex; protoconch of 1½ whorls; subsutural band often scarcely noticeable, marked by a very weak, sometimes obsolete, groove; axial ribs from suture to suture, about 17–24 on penultimate whorl; spiral grooves numerous, fine, sometimes obsolete, (about 8 on

---

Plate 49

FIGS. **188a,b,** *Terebra vinosa* (Dall). **188a,** Gulf Port, Florida; 19.7 mm. **188b,** Early and late whorls of lectotype, National Museum Natural History, Washington No. USNM-36076. **189a,b,** *T. arcas* R.T. Abbott. **189a,** Holotype National Museum Natural History, Washington No. USNM-613882; 27.0 mm. **189b,** Middle whorls of holotype. **190,** *T. glossema* Schwengel. Holotype Academy Natural Sciences Philadelphia No. 175131; 24.2 mm. 191, *T. protexta* (Conrad). Sarasota Bay, Florida; 16.1 mm. **192a,b,** *T. nassula* Dall. **192a,** Holotype Museum Comparative Zoology, Harvard No. 7014; 55.5 mm. **192b,** Middle whorls of holotype.

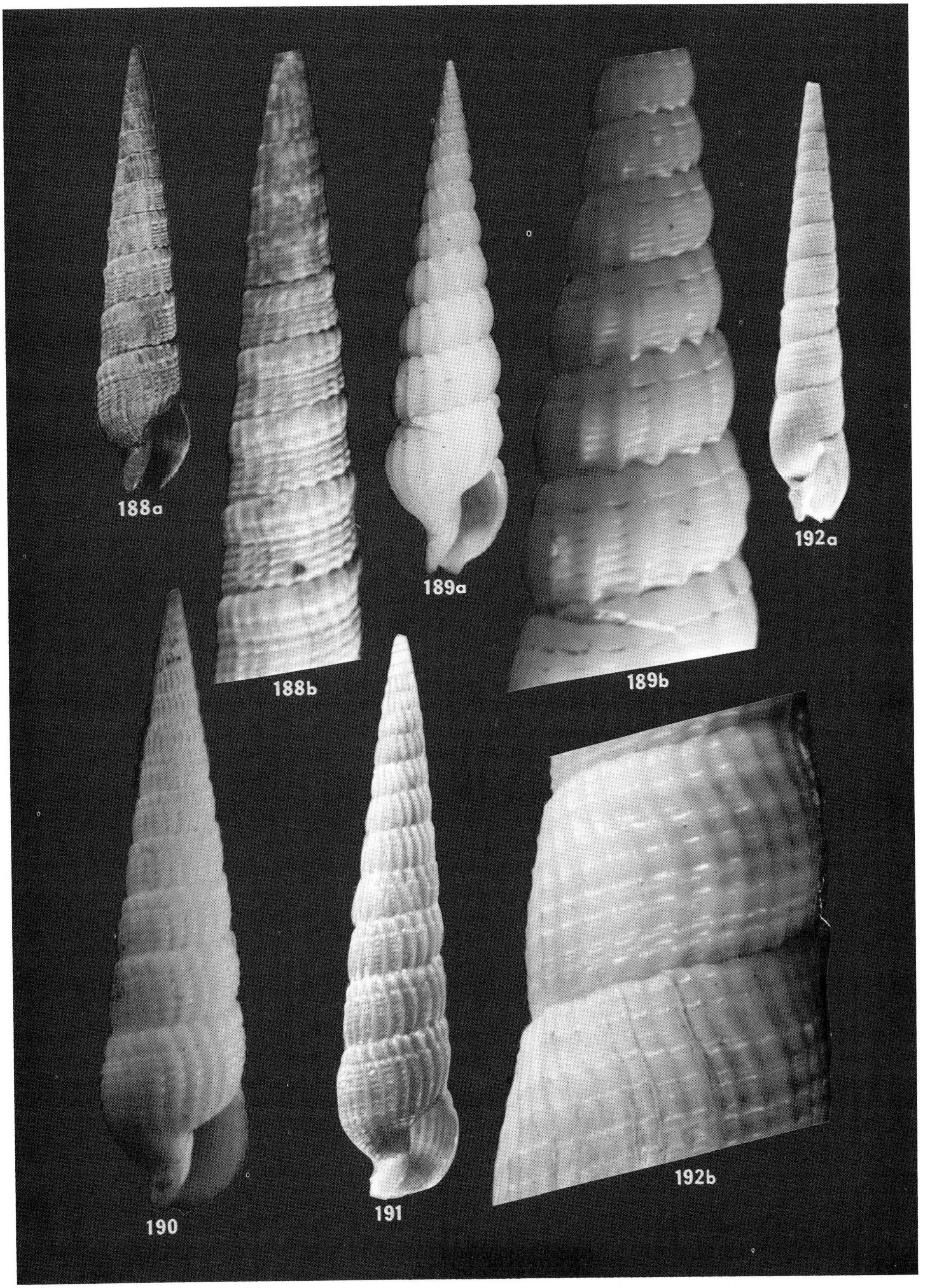

Plate 49. Western Atlantic *Terebra*.

penultimate whorl) crossing ribs lightly if at all; aperture moderately elongate, brown within; columella recurved.
Type locality: "Tampa Bay?, Florida."
Distribution: From North Carolina to Brazil, including Texas, and the Caribbean; intertidal and subtidal.
Type: The type is lost.
Discussion: Although the type is lost, the description fits no other Western Atlantic terebrid. There seems to be some confusion about the date of the description of this species. Tryon (1861) listed it as 1846 and gave the correct journal and volume number. Sherborn also gives 1846. The paper was read on Feb. 24, 1846, but was not published until 1848 in vol. III, which included papers for 1846 and 1847. This species may be separated from *T. arcas* **(189)** and *T. glossema* **(190)** by its more slender shell, brown in color unless faded, and by its more elongate aperture which is brown within.

192. *Terebra nassula* Dall, 1889

(Pl. 49, figs. 192a,b; pl. 50, figs. 192c,d)

1889 *Terebra (Acus?) nassula* Dall, Bull. Mus. Comp. Zool. 18(2):66, pl. 36, fig. 8.
1974 *Terebra nassula* Dall, Abbott, American Seashells ed. 2:261, fig. 2854; 1982 Abbott & Dance, Compendium of Seashells p. 276 (types figured in color).

Description: Shell to 56 mm; color white blending to yellowish on anterior portion of whorls; outline of whorls convex; protoconch of 1½ whorls; subsutural band marked by faint depression between ribs, fading on later whorls to a broad, scarcely indented space; sculpture cancellate, of closely spaced curved axial ribs crossed by spiral cords of equal strength, sometimes forming indistinct nodes at points of intersection; body whorl elongate; aperture elongate; columella straight; siphonal fasciole with coarse rib-like striations and a sharp posterior keel.
Type locality: "North of Yucatan, in the middle of the Gulf of Mexico; 23°32′N, 88°5′W; 95 fms."
Distribution: "From Yucatan, Mexico, to the Lesser Antilles; 190 to 1280 m."
Type: Holotype MCZ no. 7014; 55.5 × 10.0 mm.
Discussion: This cancellate deep-water species is found in few collections. It is a much larger species than *T. alba* **(193)** or *T. doellojuradoi* **(194)**.

193. *Terebra alba* Gray, 1834

(Pl. 50, figs. 193a,b)

1834 *Terebra alba* Gray, Proc. Zool. Soc. Lond. pt. 2:60.

Description: Shell to 15 mm; color white; outline of whorls slightly convex; protoconch of 1½ dome-shaped whorls; subsutural band somewhat narrow with flattish nodes, defined by moderately broad, shallow groove; remainder of whorl cancellate with axial and spiral sculpture of about equal value; aperture moderately quadrate; columella curved.
Type locality: None given. The Gulf of Mexico off Padre Island, Texas, is here designated as the type locality.
Distribution: From Florida and the Gulf of Mexico through the Caribbean Is.
Type: Holotype BM(NH) no. 1979123; 9.3 mm.
Discussion: Although this small white species was described in 1834, the type locality was unknown, and it was only recently recognized as being a Western Atlantic species by comparing the type with some recent Western Atlantic specimens. The cancellate sculpture is much finer than that of *T. doellojuradoi* **(194)**.

194. *Terebra doellojuradoi* Carcelles, 1953

(Pl. 50, figs. 194a,b)

1953 *Terebra (Myrella) doello-juradoi* Carcelles, Com. Zool. Mus. Hist. Nat. Montevideo 4(70):14, pl. 4, fig. 21; 1970 Rios, Coast Brazil Seashells p. 121, fig. upper row right.
1975 *Terebra (Strioterebrum) doello-juradoi* Carcelles, Rios Brazilian Mar. Moll. Icon. p. 126, pl. 38, fig. 555.

Description: Shell to 21 mm; color, beige or yellowish; outline of whorls convex; protoconch

---

Plate 50
FIGS. **192c,d,** *Terebra nassula* Dall. Paralectotypes National Museum Natural History, Washington No. USNM-87225; 38.0 mm and 34.2 mm respectively. **193a,b,** *T. alba* Gray. Holotype British Museum (N.H.) No. 1979123; 9.3 mm. **193b,** Dania, Florida; 10.4 mm. **194a,b,** *T. doellojuradoi* Carcelles. Holotype Museo Ciencias Natural, Argentina No. 23439; 20.5 mm. **195,** *T. juanica* Dall & Simpson. Type-figure from Dall & Simpson, 1901, pl.57, fig.5; 5.2 mm. **196a-c,** *T. evelynae* Clench & Aguayo. **196a,** Holotype Museum Comparative Zoology, Harvard No. 135077; 118.3 mm. **196b,** Middle whorls. **196c,** Lower whorls. 197a,b, *T. senegalensis* Lamarck. **197a,** Lectotype Museum d'Histoire Naturelle, Geneva No. 1152/79; 63.0 mm. **197b,** Lectotype of *T. festiva* Deshayes, British Museum (N.H.) No. 197958; 35.5 mm.

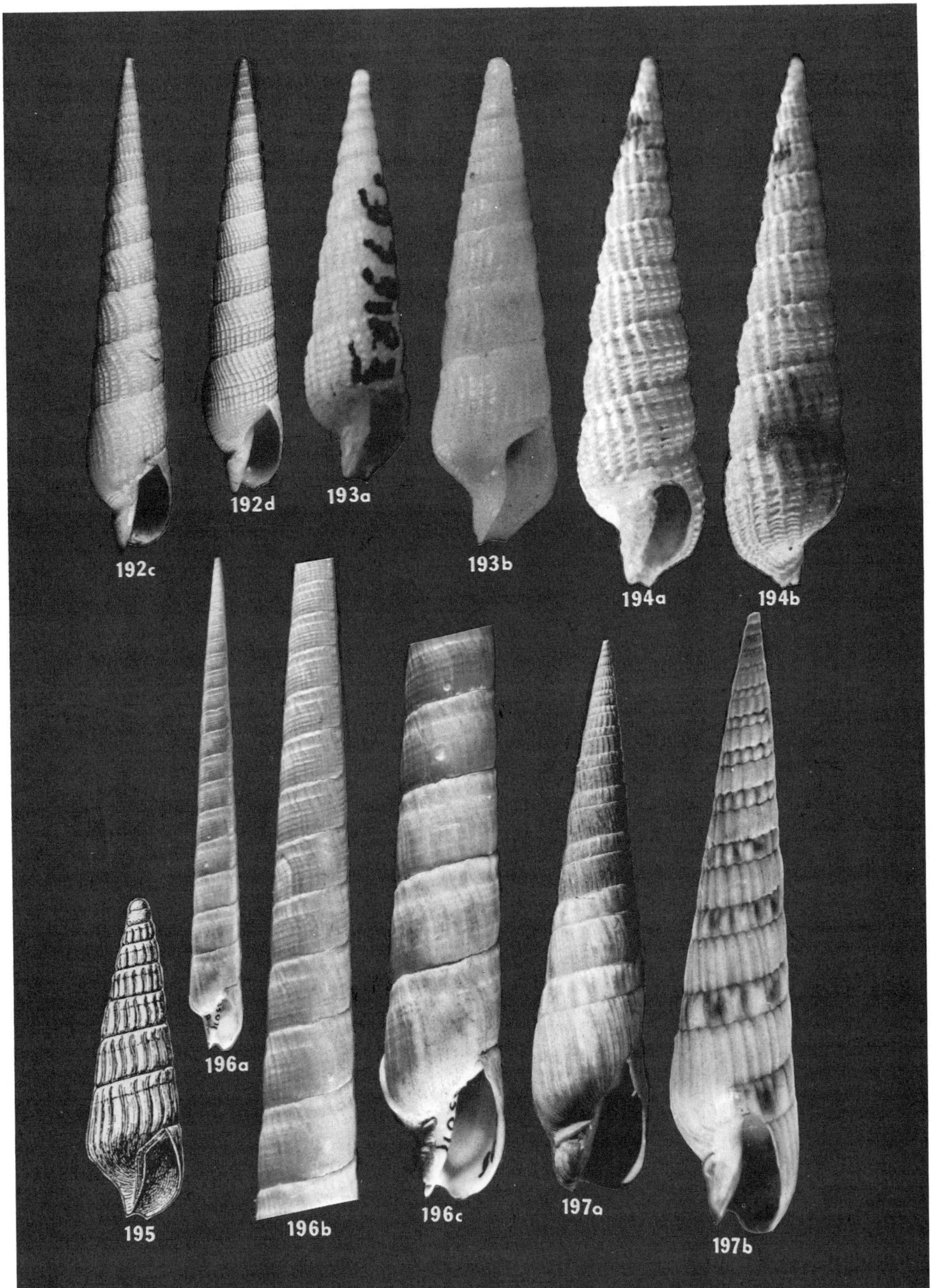

Plate 50. Western Atlantic and West African *Terebra*.

of 1½ whorls; subsutural band sometimes with spiral sculpture between ribs, defined by punctations, square pits, or short grooves between ribs; sculpture cancellate with axial ribs, narrower than interspaces, about 23 on body whorl, crossed by well developed spiral cords, about 10, forming small nodes at intersections and squarish pits between; aperture quadrate; columella recurved.
Type locality: "Uruguay, 35°53′S, 53°54′W, 61.9 m."
Distribution: From the Gulf of Mexico to Argentina; intertidal to 100 m.
Type: Holotype MCNA no. 23439; 20.5 × 5.2 mm.
Discussion: The layer of shell sculpture on the first one or two whorls of the teleoconch tends to peel off leaving a smooth, sometimes shiny, surface and giving the appearance of an extra one or two whorls to the protoconch. This species has rougher and heavier cancellate sculpture than that of *T. alba* or *T. nassula*. It also lacks the sheen and is smaller than the latter. Compare this with *T. riosi* (species **268**) from Brazil.

195. *Terebra juanica* Dall & Simpson, 1901

(Pl. 50, fig. 195)

1901 *Terebra (Acus) juanica* Dall & Simpson, Bull. U. S. Fish comm. 21:382, pl. 57, fig. 5.
1974 *Terebra juanica* Dall & Simpson, Abbott, American Sea Shells ed. 2:261, fig. 2858.

Description: Shell to 15 mm; color shiny reddish brown; outline of whorls slightly convex; protoconch of 1½ whorls; subsutural band marked by spiral grooves between ribs, becoming obsolete on later whorls; axial ribs slightly flexuous, from suture to suture, fading gradually and becoming obsolete on body whorl; aperture quadrate; columella short, curved.
Type locality: "San Juan Harbor, Puerto Rico."
Distribution: Puerto Rico.
Type: Lectotype USNM no. 159690; 5.2 × 1.9 mm.
Discussion: All the specimens in the type lot appear to be immature. All were dead-collected and most are fragmented. The figured type, unfragmented, is only 5.2 × 1.9 mm, making positive identification difficult.

196. *Terebra evelynae* Clench & Aguayo, 1939

(Pl. 50, fig. 196a-c)

1939 *Terebra evelynae* Clench & Aguayo, Mem. Soc. Cubana Hist. Nat. 13(3):196, pl. 29, fig. 1; 1974 Abbott, American Seashells ed. 2:261; 1982 Abbott & Dance, Compendium of Seashells p. 276 (holotype fig'd in color).

Description: Shell to 119 mm, slender; color whitish with indistinct band of straw color, early whorls of teleoconch darker and without straw-colored band; outline of whorls flat; subsutural band slightly concave, bordered by weak spiral cords; axial ribs indistinct, narrow, sinuous, widely spaced, extending from suture to suture, about 11 on penultimate whorl; spiral cords weak, equal in strength to ribs, about 8 on penultimate whorl; body whorl long, with faint white band below periphery; aperture elongate; columella straight.
Type locality: "Off northern Santa Clara Province, Cuba, 22°48′N, 78°48′W; 225 fms."
Distribution: Cuba; 65–421 m.
Type: Holotype MCZ no. 135077; 118.3 × 14.1 mm.
Discussion: This rare species has an extremely long and slender shell, which is quite unlike any other Western Atlantic terebrid.

197. *Terebra senegalensis* Lamarck, 1822

(Pl. 50, figs. 197a,b;
pl. 51, figs. 197c-e)
(Color pl. B, fig. 1)

1757 *Terebra faval* Adanson Histoire Nat. Senegkal p. 54, fig. 5 (non-binomial).
1822 *Terebra senegalensis* Lamarck, Hist. Nat. Anim. S. Vert. 7:287; 1860 Reeve, Conch. Icon. 12: pl. 5, sp. 16; 1942 Fischer-Piette, Journ. Conchyl. 85(2):164, pl. 3, fig. 2; 1950 Nicklés, Moll. Test. Occ. Afrique 2:133, fig. 271; 1983 Bouchet, Boll. Malac. 18(9–12):191, figs. 22–27; 1982 Abbott & Dance, Compendium of Seashells p. 273; 1984 Aubry, Terebridae pl. 5.
1839 *Terebra cingula* Kiener, Spéc. Gén. Icon. Coq. Viv. 9:28, pl. 8, fig. 16.
1839 *Terebra favat* d'Orbigny in Webb & Berthelot, Hist. Nat. Canaries, Zool. 2:92. (error pro *faval*)
1857 *Terebra festiva* Deshayes, Journ. Conchyl. 6:74, pl. 3, fig. 4.
1859 *Terebra speciosa* Deshayes, Proc. Zool. Soc. Lond. pt. 11:279. [non Bean in Thorp, 1844]

Description: Shell to 119 mm; color beige to gray, usually dull, with brown splotches, stripes, and spots; outline of whorls almost straight; protoconch of 1½ whorls, somewhat larger than following whorls; sculpture extremely variable except on early post nuclear whorls; subsutural band flat, defined by narrow shallow groove; axial ribs numerous, fine, evenly spaced on early whorls, becoming more numerous and unevenly spaced or fading out completely except for fine axial striae in later ones; spiral sculpture usually absent, rarely with a few obsolete microscopic spiral grooves; body whorl and aperture elongate; columella straight.

Type locality: *T. senegalensis*, "Senegal," West Africa; *T. cingula*, none; *T. festiva*, "Senegal," *T. speciosa*, none given.

Distribution: West Africa, from the extreme south of Morocco to Angola, so far, not in Cape Verde Islands nor the Gulf of Guinea; subtidal.

Type: *T. senegalensis* lectotype MNHG no. 1152/79, 63.0 × 13.7 mm; *T. cingula* type lost; *T. favat* two syntypes MNHNP -fide Bouchet, 1983; *T. festiva* lectotype BM(NH) no. 197958, 35.5 mm; *T. speciosa* lectotype BM(NH) no. 1979112, 39.8 mm.

Discussion: This species is the most variable of all West African terebrids. Although it is a much larger species, it may be separated from the young of *T. dillwynii* (**198**) by its well-developed, closely set axial ribs, its longer, narrower aperture, and dull beige to gray color blotched with brown. *T. dillwynii* has a shiny yellowish or pinkish shell with dark brown between the elongate nodes on the subsutural band and often immediately anterior to the band, but with no nebulous brown blotches or stripes on the remainder of the whorl, and does not have the closely set axial ribs as in the young of *T. senegalensis*.

198. *Terebra dillwynii* Deshayes, 1859

(Pl. 51, fig. 198)
(Color pl. D, fig. 15)

1857 *Terebra concinna* Deshayes, Journ. Conchyl. 6:98, pl. 3, fig. 10; [non *Buccinum concinnum* Dillwyn, 1817 = *Terebra*]. 1983 Bouchet, Boll. Malac. 18(9–12):193, figs. 28–29.

1859 *Terebra dillwynii* Deshayes, Proc. Zool. Soc. Lond. pt. 27:279; 1860 Reeve, Conch. Icon. 12: pl. 17, sp. 74.

1868 *Terebra divisa* Pease, Amer. Journ. Conch. 4(3); 124. [nom nov. pro *Terebra concinna* Deshayes, 1857].

Description: Shell to 40 mm; color shiny dark yellow to pinkish, with lighter axial ribs, 2 narrow light spiral lines at periphery of body whorl and brown blotches on and anterior to subsutural band; outline of whorls slightly convex; protoconch of 1½ dome-shaped whorls; subsutural band convex with ribs slanting to right, defined by a narrow moderately deep spiral groove; axial ribs, moderately wide-spaced, curving to left; spiral sculpture absent; body whorl rounded, with unevenly spaced weak axial ribs; aperture quadrate; columella recurved.

Type locality: *T. dillwynii*, "Japan," error. N'Gor Bay, Cap Vert Peninsula, Senegal, West Africa, is here designated as the type locality; *T. concinna* Deshayes, Japan, error.

Distribution: Senegal to Guinea, West Africa.

Type: *T. dillwyni* lectotype BM(NH) no. 197991, 38.9 × 9.0 mm; *T. concinna* lectotype MNHNP, 38.0 × 8.7 mm.

Discussion: The color pattern, but not the shape of this shell, resembles that of a mature *Terebra maculata* (**15**), a much larger species which does not occur in West Africa. Deshayes described this species twice, two years apart, and both times gave the type locality as Japan. One can only think that both came from the same lot in Deshayes' collection, a lot with erroneous data. See *T. senegalensis* (**197**) for comparison.

Bouchet, 1983, uses the taxon "*Terebra concinna* Deshayes, 1857" for this species, claiming that no secondary homonymy is involved since *Buccinum concinnum* Dillwyn, 1817 is a *Hastula* and not a *Terebra*. Dr. Bouchet overlooked the emended article 59 (b) (i) of the Code of the ICZN which states that "A junior secondary homonym replaced before 1961 is permanently invalid."

199. *Terebra fatua* Hinds, 1844

(Pl. 51, figs. 199a-c)
(Color pl. E, fig. 14)

1844 *Terebra fatua* Hinds, Proc. Zool. Soc. Lond. pt. 11:150; 1844 Hinds in Sowerby, Thes. Conch. 1:158, pl. 42, fig. 28; 1983 Bouchet, Boll. Malac. 18(9–12):194, figs. 36–37.

1857 *Terebra pura* Deshayes, Proc. Zool. Soc. Lond. pt. 27:82, pl. 5, fig. 8.

Description: Shell to 75 mm; color, all creamy white or with pale orange-brown spots anterior to suture and at periphery of body whorl; outline of whorls straight; subsutural band confined to early whorls on some individuals, weakly indicated throughout or entirely absent in others; axial sculpture of straight ribs resembling those of a *Duplicaria* on early whorls, often fading completely after about 7 whorls of the teleoconch with only axial striae remaining; body whorl exceptionally long; aperture moderately elongate; columella straight.

Type locality: *T. fatua*, "St. Christopher, W. Indies," error. Boavista, Cape Verde Islands, is here designated as the type locality; *T. pura*, "Zanzibar," error.

Distribution: Endemic to the Cape Verde Islands; 2–15 m.

Type: *T. fatua* lectotype BM(NH) no. 1968175, 74.9 mm; *T. pura* lectotype BM(NH) no. 197970, 64.7 mm.

Discussion: This rare species has an extremely long body whorl. It has not been reported from the Western Atlantic subsequent to its original description, and it may be assumed that the species is endemic to the Cape Verde Islands.

200. *Terebra grayi* E. A. Smith, 1877

(Pl. 51, fig. 200a,b)

1834 *Terebra gracilis* Gray. Proc. Zool. Soc. Lond. pt. 2:61 [non Lea, 1833].

1877 *Terebra grayi* E. A. Smith, Ann. Mag. Nat. Hist. ser. 4, 19:227 [nov nom pro *T. gracilis* Gray]; 1983 Bouchet, Boll. Malac. 18(9–12):194, figs. 7, 34.

Description: Shell to 27 mm; color rather shiny gray-white; outline of whorls slightly convex; protoconch of 4 to 4½ whorls; subsutural band convex, with elongate nodes, defined by narrow groove; axial ribs wide-spaced, almost straight; spiral threads fine, numerous; aperture narrowly quadrate; columella curved.

Type locality: The type locality for Gray's *T. gracilis* was given as "Africa." The type locality is here restricted to West Africa.

Distribution: From the Ivory Coast to Angola, West Africa.

Type: Holotype of *T. grayi* (and of *T. gracilis*) BM(NH) no. 1874.10.29.1, 22.0 mm.

Discussion: There has been much confusion about this species, and many specimens of *T. reticulare* **(201)** have been erroneously identified as *T. grayi*, which has a more slender shell with wider-spaced axial ribs, and finer spiral sculpture and some sheen. The type locality was given as S. E. Africa, but both Drs. Gosliner, then of the South African Museum, and Kilburn of the Natal Museum have no examples of this species from South Africa. At one time Bratcher and R. D. Burch believed the species to be from tropical West America, but subsequently decided the specimen in question was an extremely slender, light beige specimen of *T. armillata* Hinds with smaller than usual subsutural nodes. In doing research on West African terebrids, Dr. P. Bouchet examined 17 specimens of this species from various West African localities.

201. *Terebra reticularis* (Pecchioli in Sacco, 1891)

(Pl. 51, fig. a; pl. 52, figs. 201b-e)

1891 *Strioterebrum reticulare* Pecchioli in Sacco, I Molluschi dei Terreini Terziarii, 40, pl. 2, fig. 20; 1983 Bouchet, Boll. Malac. 18(9–12):195, figs. 30–33.

1975 *Terebra pseudoturbonilla* Talavera, Bol. Inst. Espanol Oceanogr. 192:5, pl. 2, fig. 3.

1984 *Terebra reticulare* (Pecchioli in Sacco), Aubry, Terebridae pl. 6.

1984 *Terebra pseudoturbonilla* Talavera, Aubry, Terebridae pl. 3.

Description: Shell to 45 mm, width variable; color beige-gray; protoconch of 3½ to 4 whorls; subsutural band with elongate ribs; band defined by deep, narrow groove developing about the 5th whorl; axial sculpture of narrow, curved ribs which may or may not be widely spaced; spiral sculpture varies from about 6 strong cords to about nine flatter, less obvious ones; aperture semi-elongate; columella recurved.

Type locality: *T. reticularis*, "Pliocene of the Torsero River," Italy; *T. pseudoturbonilla*, "Continental Plateau, Mauritania."

Distribution: From Mauritania to Angola, W. Africa; from 20 to 130 m.

Plate 51

FIGS. **197c-e**, *Terebra senegalensis* Lamarck. **197c**, Lectotype of *T. speciosa* Deshayes, British Museum (N.H.) No. 1979112; 39.8 mm (slender form). **197d**, Lower whorls of paralectotype of *T. speciosa* Deshayes. **197e**, N.Angola, West Africa (smooth form); size not available. **198a,b**, *T. dillwynii* Deshayes. 198a, Lectotype British Museum (N.H.) No. 197991; 38.9 mm. **198b**, Topotype from N'Gor, Cap Vert Peninsula, Senegal, West Africa; 25.0 mm. **199a-c**, *T. fatua* Hinds. 199a, Lectotype British Museum (N.H.) No. 1968175; 74.9 mm. **199b**, Lectotype of *T. pura* Deshayes, British Museum (N.H.) No. 197970; 64.7 mm. **199c**, Boavista Id., Cape Verde Ids.; 38.4 mm. **200a,b**, *T. grayi* E.A. Smith. **200a**, Holotype British Museum (N.H.) No. 1874.10.29.1.; 22.0 mm. **200b**, Abidjan, Ivory Coast, West Africa; 27.0 mm. **201a**, *T. reticularis* (Pecchioli in Sacco). Abidjan, Ivory Coast, West Africa; 34.5 mm.

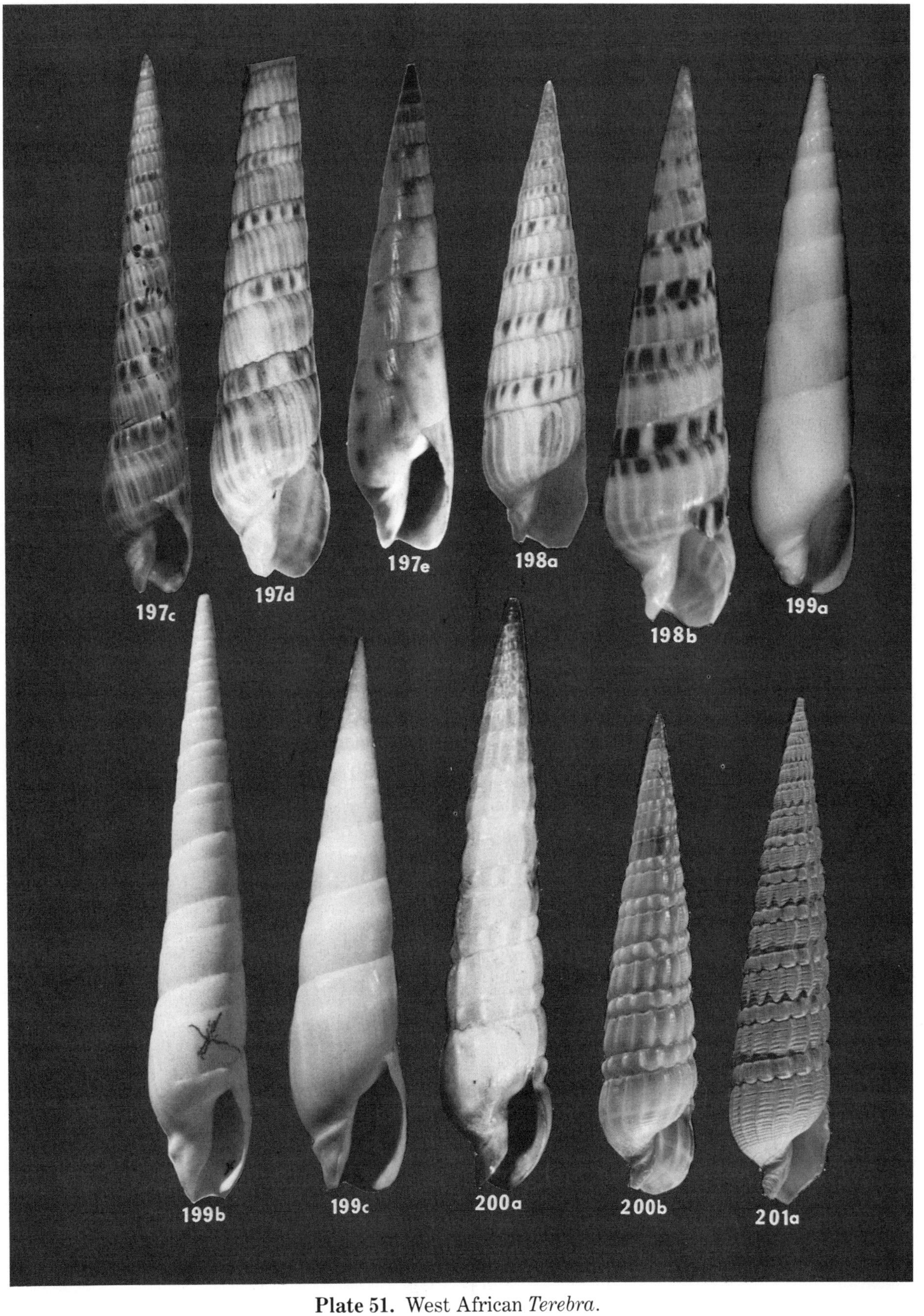

Plate 51. West African *Terebra*.

Type: *T. reticularis* holotype Mus. geol. de Genova (Sacco); *T. pseudoturbonilla* holotype Mus. Insular de Ciencias Nat., Santa Cruz de Tenerife, 5.6 mm, juvenile.
Discussion: *T. pseudoturbonilla,* based on an immature individual with fine sculpture, lies at the end of the variability range. It is indistinguishable from the young of *T. reticularis.* This form and the one with strong axial and spiral sculpture are found coexisting throughout the range.

202. ***Terebra twilae*** Bouchet, 1983

(Pl. 52, fig. 202)

1983 *Terebra twilae* Bouchet, Boll. Malac. 18(9–12):197, figs. 8, 35.

Description: Shell to 22 mm, slender; color creamy white; outline of whorls very slightly convex; protoconch of $1^1/_2$ whorls; subsutural band ribbed with 2 or 3 spiral grooves, defined by deep punctations between ribs; axial ribs, about 10, contiguous with ribs on band; spiral cords, about 6, not crossing ribs; aperture quadrate; columella slightly recurved.
Type locality: "Ghana, West Africa; 05°45′N, 00° 57′E; 17 m; Atlantide Exped. sta. 86."
Distribution: Known only from the holotype.
Type: UZMC, 21.5 × 4.5 mm.
Discussion: This species may be separated from *T. reticulare* **(201)** by having less convex whorls, more numerous axial ribs, and a paucispiral protoconch, indicating direct development. *T. histrio* **(202)** also has a paucispiral protoconch but may be distinguished by its almost obsolete sculpture and brown spots. This species was named after the senior author of this book.

203. ***Terebra histrio*** Deshayes, 1857

(Pl. 52, figs. 203a,b)

1857 *Terebra histrio* Deshayes, Journ. Conchyl. 6:76, pl. 4, fig. 11; 1977 Bratcher, Nautilus 91(2):41; 1983 Bouchet, Boll. Malac. 18(9–12):190, figs. 20–21.

Description: Shell to 48 mm; color creamy beige with large regularly spaced longitudinal brown spots, the subsutural band having tiny reddish brown fringelike lines; outline of whorls almost flat; protoconch of $1^1/_2$ whorls; subsutural band noded on early whorls, nodes giving way to plications in middle whorls, then disappearing in later ones; band defined by a narrow subsutural groove; sculpture of curved axial striae and sometimes very fine spiral striae; body whorl short; aperture quadrate; columella recurved.
Type locality: None given. Senegal, West Africa, is here designated as the type locality.
Distribution: West Africa, from S. Morocco to Senegal, and to the Cape Verde Islands.
Type: Type lost. (Bouchet *pers. comm.*)
Discussion: This species was figured by Deshayes in 1857. Reeve (1860) and Tomlin (1944) both mention that the type is cited from the Deshayes collection but that the specimen in the Cuming collection is merely a form of *T. corrugata* **(144)**. Neither Deshayes figure nor his description is that of *T. corrugata,* which has a much narrower shell, multispiral protoconch, and noded band.

204. ***Terebra brevicula*** Deshayes, 1859

(Pl. 52, figs. 204a-c)

1859 *Terebra brevicula* Deshayes, Proc. Zool. Soc. Lond. pt. 27:296; 1860 Reeve, Conch. Icon. 12: pl. 22, sp. 119; 1952 Cotton, Trans. Royal Soc. S. Australia 75:38.
1912 *Terebra gruveli* Dautzenberg, Ann. Inst. Oceanog. Monaco 5:5, pl. 1, figs. 1, 2; 1983 Bouchet, Boll. Malac. Milano 18(9–12):198, figs. 9, 38–39.

Description: Shell to 26 mm, broad; color gray, with posterior portion of the whorls darker, body whorl having a white stripe at the periphery and dark gray anterior to that, some specimens with a few small brown flecks; outline of whorls convex; protoconch of $1^1/_2$ whorls; subsutural band raised, sometimes with a few obsolete nodes, defined by a weak subsutural indentation; axial sculpture of 10 to 12 widely spaced obsolete

---

Plate 52
FIGS. **201b-e,** *Terebra reticularis* (Pecchioli in Sacco). **201b,** S.E. of Sekondi, Ghana, West Africa, 50 m; 28.0 mm (slender form). **201c,** Pointe Noire, Congo, West Africa; 14.0 mm. **201d,** Ilha de Luanda, Angola, 40-60 m; 14.3 mm (broad form). **201e,** Paratype of *T. pseudoturbinella* Talavera, Museum National d'Histoire Naturelle, Paris; 5.6 mm. 202, *T. twilae* Bouchet. Holotype Zoological Museum, Copenhagen; 21.5 mm. **203a,b,** *T. histrio* Deshayes. **203a,** Type-figure from Deshayes, 1857, pl.4, fig.11; 48.0 mm. **203b,** Near Cape Verde, West Africa; 47.0 mm. Zoological Museum, Copenhagen; dimensions unknown. **204a-c,** *T. brevicula* Deshayes. **204a,b,** Probable syntypes, British Museum (N.H.) No. 197985; 25.6 mm and 21.6 mm respectively. **204c,** Holotype of *T. gruveli* Dautzenberg, Museum National d'Histoire Naturelle, Paris; 15.0 mm. **205,** *T. fernandesi* Bouchet. Holotype Museum National d'Histoire Naturelle, Paris; 12.4 mm.

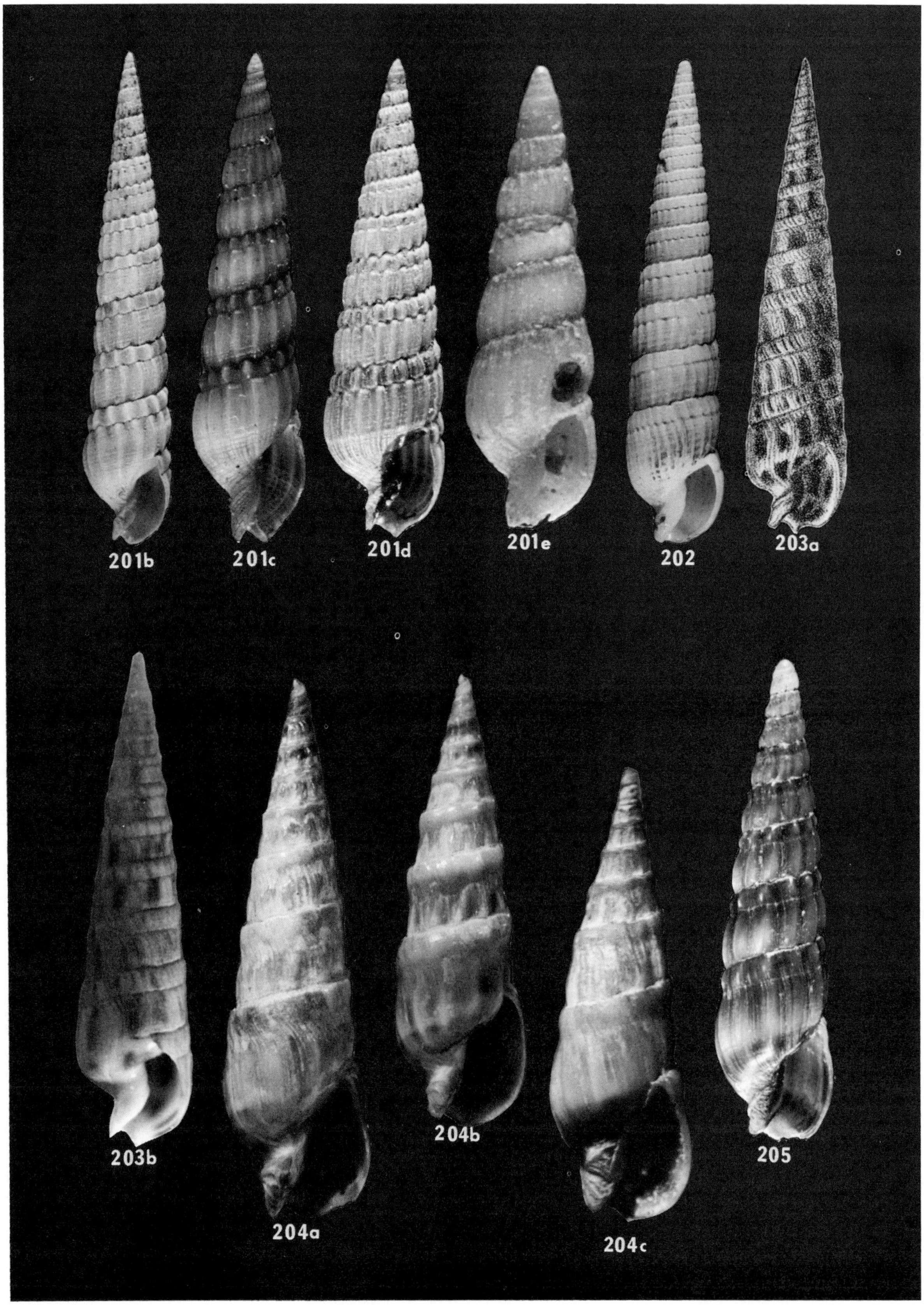

Plate 52. West African *Terebra.*

ribs, sometimes completely faded, and numerous axial striae; spiral sculpture absent; aperture quadrate, with a white and dark gray stripe showing through outer lip, columella curved; operculum yellow, translucent.

Type locality: *T. brevicula*, "Van Diemen's Land" Tasmania, error. Santa Maria Bay, south Angola, West Africa, is here designated as the type locality; *T. gruveli*, Mossamedes, Angola.

Distribution: Angola, West Africa; 1 to 3 m.

Type: *T. brevicula* probable type BM(NH) no. 197985, 21.6 mm; *T. gruveli* holotype MNHNP, 15.0 mm.

Discussion: Even fresh specimens of this species often look as though the sculpture has been worn or eroded. There are about 90 paratypes of *T. gruveli* in the Dautzenberg collection in the Brussels Museum, all looking worn, but specimens in alcohol with the animal and operculum intact also have a worn appearance. Although the type locality of *T. brevicula* was given as Tasmania, no shell resembling it has been collected there. Cotton (1952) stated that the species is not found in Australia, but at the same time he wrongly synonymized it with a Californian species.

205. *Terebra fernandesi* Bouchet, 1983

(Pl. 52, fig. 205)

1983 *Terebra fernandesi* Bouchet, Boll. Malac. 18(9–12):197, figs. 10, 40–4; 1984 Aubry, Terebridae pl. 7.

Description: Shell to 13 mm; color red-violet background with a white band above the suture, the early whorls of the teleoconch usually being white; outline of whorls moderately convex; protoconch of 1½ short mamillate whorls; subsutural band with rounded nodes, defined by narrow, shallow groove; axial ribs wide-spaced, with axial striae in interspaces, ribs becoming almost obsolete on body whorl; spiral sculpture extremely faint or absent; aperture quadrate; columella short, curved.

Type locality: "Barra do Dande, Luanda, Angola, West Africa; 0–1 m."

Distribution: From the estuaries of Zaire and the northern part of Angola, West Africa; intertidal.

Type: Holotype MNHNP, 12.4 × 3.1 mm.

Discussion: This species is replaced to the south by *T. brevicula* (**204**) and may be separated from that species by its more slender and appreciably smaller form, more distinctly marked subsutural band, and lack of flecks of color. The early whorls of the two species are indistinguishable. They are not sympatric. *T. brevicula* is limited to the small bays of the southern provinces of Angola while *T. fernandesi* lives only in estuaries. Their protoconchs indicate that neither have planktonic development, making distribution difficult when the major part of the coasts of Angola consists of uninterrupted cliffs. This species was named for the well-known shell collector, Francisco Fernandes, of Luanda, Angola.

Genus *Hastula* H. & A. Adams, 1853

Type species. *Hastula strigilata* (Linnaeus, 1758).

The shell characteristics of the species we place in this genus are the following: smooth, shiny shells; axial sculpture of crenulations below the suture, or slender ribs which do not swell into nodes; no spiral sculpture, except for a subsutural groove or subsutural punctations.

INDO-PACIFIC *HASTULA*

206. *Hastula strigilata* (Linnaeus, 1758)

(Pl. 53, figs. 206a,b)

1758 *Buccinum strigilatum* Linnaeus, Syst. Nat. ed. 10:741; 1956 Dodge, Bull. Amer. Mus. Nat. Hist. 111(3):226, detailed evaluation.

1817 *Buccinum concinnum* Dillwyn, Desc. Cat. Rec. Shells 2:647, ref. to Gualtieri, pl. 57, fig. 0.

1825 *Buccinum concinnum* Wood, Index Test, p. 113, pl. 24, fig. 141. [non Dillwyn, 1817]

1839 *Terebra striatula* Kiener, Spéc. Gén. Coq. Viv. 9: pl. 9, fig. 18 [non Lamarck, 1822]

1857 *Terebra verreauxi* Deshayes, Journ. Conch. p. 95, pl. 5, fig. 3.

1859 *Terebra argenvillii* Deshayes, Proc. Zool. Soc. Lond. pt. 27:286.

1860 *Terebra strigilata* (Linnaeus), Reeve, Conch. Ico. 12: pl. 18, figs. 85a, b; 1917 Hirase, Terebridae Jap. Emp. p. 28, pl. 5, figs. 67–70; 1984 Aubry, Terebridae pl. 2.

1925 *Terebra strigilata gracilior* Thiele, Wiss. Ergeb. Deut. Tiefsee Exped. Valdivia 17(2):344, pl. 41, fig. 19.

1935 *Hastula (Hastula) strigilata* (Linnaeus). Dautzenberg, Mem. Mus. Roy. D'Hist. Nat. Belgique 2(17):34, pl. 1, fig. 7.

1935 *Hastula verreauxi* (Deshayes), 1935 Dautzenberg, Mem. Royal Inst. Belgique 2(17):37, pl. 1, fig. 9; 1960 Weaver, Hawaiian Mar. Moll. 1(9):34, pl. 9, fig. bottom left; 1961 Oyama & Takemura, Moll. Shells, Resources Exploitation Inst. 5, *Hastula* pl., figs. 1, 2; 1962 Kira, Shells Western Pacific Color, vol. 2, p. 208, pl. 71, fig. 3; 1966 Habe & Kosuge, Shells of World in Color, vol. 2, p. 99, pl. 40, fig. 2; 1975 Okutani, Gakken Illust. Nat. Encyclo. Moll. Japan p. 154, fig. 2.

1966 *Hastula strigilata* (Linnaeus), Cernohorsky & Jennings, Veliger 9(1):63, pl. 7, fig. 50; 1967 Cernohorsky, Mar. Shells Pacific 1:212, pl. 54, fig. 409; 1975 Salvat & Rives, Coq. Polynésie p. 364, fig. 397; 1979 Kay, Hawaiian Mar. Shells p. 390, figs. 126m, n.

1984 *Hastula strigillata* (Linnaeus) Salvat & Rives, Shells of Tahiti p. 138, pl. 36, fig. 7.

Description: Shell to 52 mm; color yellow, gray or brown, with a row of brown spots on a white band anterior to the suture, also a white stripe at periphery of body whorl; outline of whorls flat; protoconch of 3½ to 4 whorls; subsutural band of color only; subsutural groove absent; axial ribs flat, from suture to suture, with extremely narrow groovelike interspaces; no spiral sculpture; aperture elongate; columella curved.

Type locality: "In O. Asiatico;" *B. concinnum* Dill., none; *B. concinnum* Wood, none; *T. striatula*, none; *T. verreauxi*, none; *T. argenvillii*, none; *strigilata gracilior*, "Padang" Sumatra.

Distribution: From the Arabian Sea through the tropical Indo-Pacific to Hawaii and French Polynesia; intertidal to 100 m.

Type: *H. strigilata* probable holotype in the Linnean Society London, 28.1 mm; *B. concinnum* Dillwyn type lost; *B. concinnum* Wood type lost; *T. striatula* lectotype (here designated) MNHNP, 37.9 × 7.8 mm; *T. verreauxi* lectotype MNHNP, 37.9 × 7.8 mm; *T. argenvillii* questionable type MNHNP, 39.9 mm (the measurements are not consistent with those published by Deshayes, 35.0 × 6.5 mm; *T. strigilata gracilior* type not located in MNHUB with other Thiele types.

Discussion: Veligers of *H. strigilata* hatch when one whorl is complete, add two more whorls while planktonic, then settle when 3¼ to 3½ whorls are complete. They are found in the plankton during May and June in Hawaii. (J. B. Taylor, unpubl. Ph.D. thesis). This is the type species of the genus *Hastula*. *H. strigilata* may be separated from *H. matheroniana* **(208)** by its broad, flat ribs with narrow groovelike interspaces. See *T. acumen* **(207)** for comparison with that species.

The diagnosis of *H. strigilata* is doubtful as Linnaeus cited figures of three different species for *B. strigilatum*. The Gualtiere figure (pl. 57, fig. 0) represents the *strigilata* of authors; the Buonanni figure (pl. 3, fig. 110) is that of *D. duplicata* (Linnaeus); and the figure from Rumphius (pl. 30, fig. H) is *T. commaculata* (Gmelin). Dodge (1956) discussed the nomenclatural problems of this species and ended by saying, "In spite of our lack of conviction as to what Linnaeus meant, the name has become so firmly fixed in literature as the *strigilata* of authors that it would be unwise to change it at this late date. The confusion that would result could hardly justify our adherence to the strict letter of any "rule-of-thumb" for the identification of a Linnean species. *Terebra strigilata* is a common and well known species, and there has not been a single voice raised against it in the last hundred years." We are inclined to agree. The name *strigilata* is sometimes misspelled as *strigillata*.

207. *Hastula acumen* (Deshayes, 1859)

(Pl. 53, figs. 207a-d)

1859 *Terebra acumen* Deshayes, Proc. Zool. Soc. Lond. pt. 27:187.

Description: Shell to 36 mm; color variable: white, gray, beige, or light brown with or without a light band anterior to the suture and/or brown subsutural spots on band; outline of whorls slightly convex; protoconch 1½ dome-shaped whorls; subsutural groove or punctations absent; axial ribs broad, flat, extending from suture to suture with extremely narrow interspaces, rarely ribs fade out and are discernible on only part of shell; aperture elongate; columella curved with inner edge brown.

Type locality: None given. Alor Is., Indonesia, is here designated as the type locality.

Distribution: From Indonesia to the Philippines and Hawaii, subtidal to 116 m.

Type: Lectotype MNHNP, 21.3 × 3.0 mm.

Discussion: This species has been mentioned in molluscan literature only once since it was described in 1859, when Reeve synonymized it with *H. strigilata* **(206)** in 1860. Only the discovery of the protoconch denoting direct development resurrected it from obscurity. The shell of *H. acumen* is so similar in shape, color, and sculpture to that of *H. strigilata* that a definitive allocation to either species may be difficult at times if the protoconch is missing. *H. strigilata* has a protoconch of 3½ to 4 whorls while *H. acumen* has only 1½ embryonic whorls. The aperture of *H. acumen* is narrower, the inside edge of the columella is stained brown, and the subsutural spots may be missing or irregularly placed; but the only dependable diagnostic feature is the protoconch, although the maximum length of *H. acumen* we have examined is only 34.2 mm, that of *H. strigilata* 52.0 mm.

208. *Hastula matheroniana* (Deshayes, 1859)

(Pl. 53, figs. 208a-d)

1859 *Terebra matheroniana* Deshayes, Proc. Zool. Soc. Lond. pt. 27:287; 1984 Aubry, Terebridae pl. 1.
1859 *Terebra modesta* Deshayes, Proc. Zool. Soc. Lond. pt. 27:288.
1869 *Terebra lauta* Pease, American Journ. Conch. 5(1):66.
1925 *Terebra strigilata summatrana* Thiele, Wiss. Ergeb. Deut. Tief. Exped. "Valdiva" 17:344, pl. 29, fig. 20.
1960 *Hastula lauta* (Pease), Weaver, Hawaiian Mar. Moll. 1(8):30, pl. 8, fig. top right; 1966 Cernohorsky & Jennings, Veliger 9(1):60, pl. 7, fig. 51; 1967 Cernohorsky, Mar. Shells Pacific 1:210, pl. 54, fig. 404; 1984 Aubry, Terebridae pl. 6.
1976 *Hastula matheroniana* (Deshayes) Cernohorsky & Bratcher, Rec. Auckland Inst. & Mus. 13:139, figs. 24–27; 1977 Bratcher, Nautilus 91(2):41, fig. 6; 1978 Cernohorsky, Tropical Mar. Shells p. 149, pl. 53, fig. 7; 1978 Salisbury, Hawaiian Shell News 26(8):8, text fig.; 1979 Kay, Hawaiian Mar. Shells p. 388, fig. 126L; 1984 Salvat & Rives, Shells of Tahiti p. 138, lower fig.

Description: Shell to 33 mm; color variable, shiny cream, gray, beige, or purplish brown, ornamented with brown and white spots below suture and a narrow white line at periphery of body whorl; outline of whorls flat; protoconch of 2½ to 3 conical whorls; subsutural band indicated only by punctations between ribs; axial ribs sharp, straight, from suture to suture, equal to interspaces; spiral sculpture absent; aperture elongate; columella straight, with moderate parietal callus.
Type locality: *H. matheroniana*, "Tahiti"; *T. modesta*, "The mouth of the Indus" Pakistan; *T. lauta*, "Oahu," Hawaii; *T. strigilata summatrana*, "Padang," Sumatra.
Distribution: Tropical Indo-Pacific; intertidal to 100 m.
Type: *H. matheroniana* lectotype MNHNP, 18.0 × 3.0 mm; *T. modesta* lectotype BM(NH) no. 1979104, 21.6 × 4.0 mm; *T. lauta* type lost, Deshayes' published measurements 26 × 6 mm; *T. strigilata sumatrana* paratypes BM(NH) no. 1948.12.10.3.4, 18.7 and 12.8 mm.
Discussion: *Hastula parva* **(209)** may be separated from this species by *parva*'s weak or obsolete axial ribs and its protoconch of 1½ whorls. Refer to *H. strigilata* **(206)** for comparison with that species. Deshayes named this species after the French paleontologist, Philippe Matheron.

209. *Hastula parva* (Baird, 1873)

(Pl. 53, figs. 209a,b)

1873 *Terebra (Leiostoma) parva* Baird in Brenchley, Cruise *H. M. S. Curacao*, p. 435, pl. 37, figs. 5, 6.

Description: Shell to 24 mm; color shiny opaque white to bluish gray with evenly spaced yellow-brown subsutural spots; outline of whorls flat, slightly turreted; protoconch of 1½ yellow-brown whorls; sculpture almost smooth except for very slight crenulations immediately below suture; aperture moderately elongate; columella straight.
Type locality: "New Caledonia."
Distribution: Tropical Indo-Pacific; intertidal to 7 m.
Type: Lectotype BM(NH) no. 1979153; 22.8 mm.
Discussion: *H. rufopunctata* **(210)** may be separated from this species by its stripe at the periphery of the body whorl, and its translucent early whorls of the teleoconch with a stripe visible within.

Plate 53
FIGS. **206a,b,** *Hastula strigilata* (Linnaeus). **206a,** Probable holotype, Linnean Society, London, 28.1 mm. **206b,** Questionable type of *Terebra argenvillii* Deshayes, Museum National d'Histoire Naturelle, Paris; 39.9 mm. **207a-d,** *H. acumen* (Deshayes). **207a,** Lectotype Museum National d'Histoire Naturelle, Paris; 21.3 mm. **207b,** Keehi lagoon, Oahu, Hawaii, 46-107 m; 20.4 mm. **207c,** Laminusa, Siasi, Sulu, Philippines; 32.0 mm. **207d,** Upper whorls. **208a-d,** *H. matheroniana* (Deshayes). **208a,** Lectotype Museum National d'Histoire Naturelle, Paris; 18.0 mm. **208b,** Lectotype of *T. modesta* Deshayes, British Museum (N.H.) No. 1979104; 21.6 mm. **208c,** Paratype of *T. strigilata sumatrana* Thiele, British Museum (N.H.) No. 1948.12.10.3.4.; 18.7 mm. **208d,** Efate Id., Vanuatu; 27.0 mm. **209a,b,** *H. parva* (Baird in Brenchley). **209a,** Paralectotype British Museum (N.H.); 18.4 mm. **209b,** Upper whorls of paralectotype.

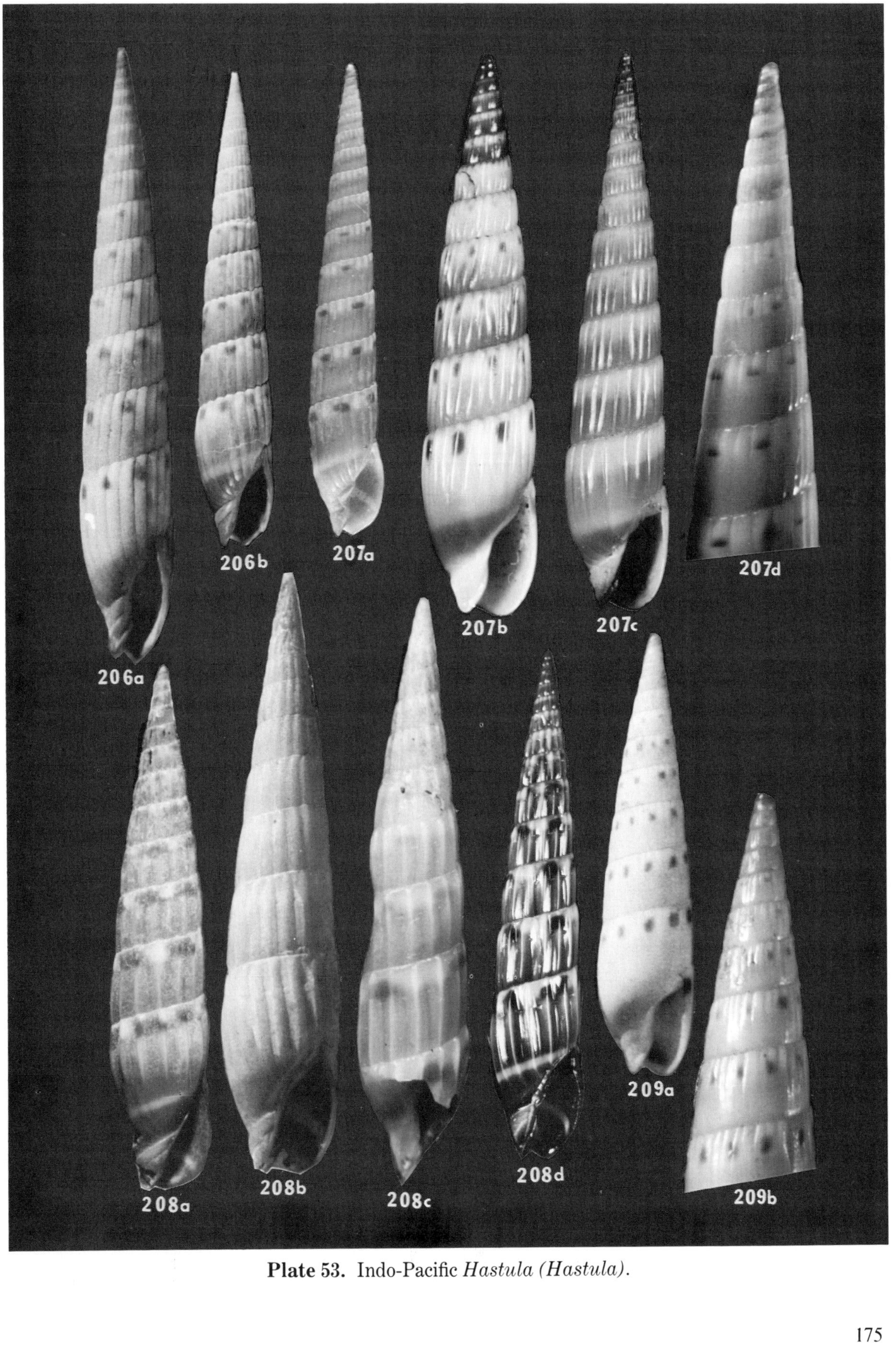

Plate 53. Indo-Pacific *Hastula (Hastula)*.

210. ***Hastula rufopunctata*** **(E. A. Smith, 1877)**

(Pl. 54, figs. 210a-e)
(Color pl. D, figs. 8–12)

1877 *Terebra (Hastula) rufopunctata* E. A. Smith, Ann. & Mag. Nat. Hist. ser. 4, 19:229; 1892 Sowerby, Man. Shells S. Africa p. 12.
1901 *Terebra (Abretia) diversa* E. A. Smith, Journ. Conch. 10:115, pl. 1, fig. 6; 1973 Kensley, Seashells S. Africa p. 214, fig. 868.
1961 *Hastula diversa* (Smith), Oyama & Takemura, Moll. Shells, Resources Exploitation Inst. 5: *Hastula* fig. 3.
1971 *Hastula rufopunctata* (E. A. Smith), Wilson & Gillett, Australian Sea Shells: 158, pl. 106, figs. 11, 11a,b.
1984 *Terebra rufopunctata* E. A. Smith, Aubry, Terebridae pl. 3.
1984 *Terebra diversa* (E. A. Smith), Aubry, Terebridae pl. 12.

Description: Shell to 36 mm; color variable, from white or beige to gray or dark brown with a white subsutural and peripheral stripe and often with reddish brown punctations on subsutural stripe; outline of whorls straight; protoconch of 1½ whorls; first whorls of teleoconch faintly translucent with a very narrow darker stripe showing through; subsutural groove absent; axial plications narrow, usually confined to subsutural area; spiral sculpture absent; aperture moderately elongate; columella recurved.
Type locality: *T. rufopunctata*, none. Natal, S. Africa is here designated as the type locality; *T. diversa*, "Umzinto, Natal, S. Africa."
Distribution: From South Africa to Japan and Western Australia; intertidal to 10 m.
Type: *H. rufopunctata* lectotype BM(NH) no. 1874.10.1.3, 21.6 mm; *T. diversa* lectotype BM(NH) no. 1899.4.14.118, 22.5 mm.
Discussion: E. A. Smith (1901) mentions spiral striations in *H. rufopunctata*, which are not at all evident in the types. Neither the type of *H. rufopunctata* nor *H. diversa* are striate, and spiral striae are unknown in the *Hastula* group. *H. parva* **(209)** may be separated from this species by its more turreted outline and by the lack of the somewhat translucent early whorls of this species.

Olive Schoenberg reported finding at least 25 live *H. rufopunctata* at extreme low tides on the beaches of Broome, Western Australia, encased in lumps of dried sand (Hawaiian Shell News 32(4):11). She speculated that they may not be able to dig deeply enough to protect themselves from desiccation or predation and have devised this way to escape desiccation during low tides.

211. ***Hastula lanceata*** **(Linnaeus, 1767)**

(Pl. 54, figs. 211a,b)

1767 *Buccinum lanceatum* Linnaeus, Syst. Nat. ed. 12:1206; 1956 Dodge, Bull. Amer. Mus. Nat. Hist. III (3):231.
1860 *Terebra lanceata* Reeve, Conch. Icon. 12, pl. 9, sp. 36.
1921 *Terebra lanceata oahuensis* Pilsbry, Proc. Acad. Nat. Sci. Philadelphia 72:307, pl. 12, fig. 7.
1966 *Hastula lanceata* (Linnaeus), Cernohorsky & Jennings, Veliger 9(1):60, pl. 7, fig. 53; 1967 Cernohorsky, Mar. Shells Pacific 1:210, pl. 54, fig. 403; 1972 Hinton, Shells New Guinea & Cent. Indo-Pacific p. 46, pl. 23, figs. 35, 36; 1975 Salvat & Rives, Coq. Polynèsie p. 363, fig. 395; 1979 Kay, Hawaiian Mar. Moll. p. 388, fig. 125F; 1982 Abbott & Dance, Compendium of Seashells p. 274; 1984 Trondle, Xenophora 15:14, fig. 35; 1984 Aubry, Terebridae pl. 4; 1984 Salvat & Rives, Shells of Tahiti p. 138, pl. 36, fig. 8.
1975 *Acuminia lanceata* Okutani, Gakken Illust. Nat. Encyclo. Moll. Japan p. 155, fig. 6.
1984 *Terebra oahuensis* Pilsbry, Aubry, Terebridae pl. 5.

Description: Shell to 61 mm; color shiny white with thin wavy reddish brown axial lines; outline of whorls mildly convex, slightly turreted; protoconch of 4 whorls; subsutural band absent; axial ribs well-developed, absent, or mere crenulations anterior to suture; spiral sculpture absent; aperture elongate; columella slightly curved.
Type locality: *H. lanceata*, "India"; *T. lanceata oahuensis*, "off Honolulu, Hawaii, 6–8 fms."
Distribution: Throughout the tropical Indo-Pacific; intertidal to 120 m.
Type: *H. lanceata* holotype Linnean Soc. London, 37.2 × 7.0 mm; *T. lanceata oahuensis* holotype ANSP no. 117,036, 38.6 × 6.6 mm.
Discussion: This species is the type-species of the genus, *Acuminia* Dall. It was diagnosed as: "Sulcus wholly absent. Adult slender, smooth."

Plate 54
FIGS. **210a-e,** *Hastula rufopunctata* (E.A. Smith). **210a,** Lectotype British Museum (N.H.) No. 1874.10.1.3.; 21.6 mm. **210b,** Middle whorls of lectotype. **210c,** Lectotype of *T. diversa* E.A. Smith, British Museum (N.H.) No. 1899.4.14.118.; 22.5 mm. **210d,** Middle whorls of lectotype. **210e,** Transkei coast; 26.0 mm. **211a,b,** *H. lanceata* (Linnaeus). **211a,** Holotype Linnean Society, London; 37.2 mm. **211b,** Holotype of *T. lanceata oahuensis* Pilsbry, Academy Natural Sciences, Philadelphia No. 117036; 38.6 mm. **212a-e,** *H. penicillata* (Hinds). **212a,** Lectotype British Museum (N.H.) No. 1969244; 35.0 mm. **212b,** Lectotype of *T. venosa* Hinds, British Museum (N.H.) No. 1968245; 35.0 mm. **212c,** Paralectotype of *T. venosa* Hinds, same Institution; 34.9 mm. **212d,** Holotype of *T. crossii* Deshayes, Museum National d'Histoire Naturelle, Paris; 24.0 mm. **212e,** Holotype of *H. betsyae* R. D. Burch, Museum Comparative Zoology, Harvard No. 251238; 34.0 mm.

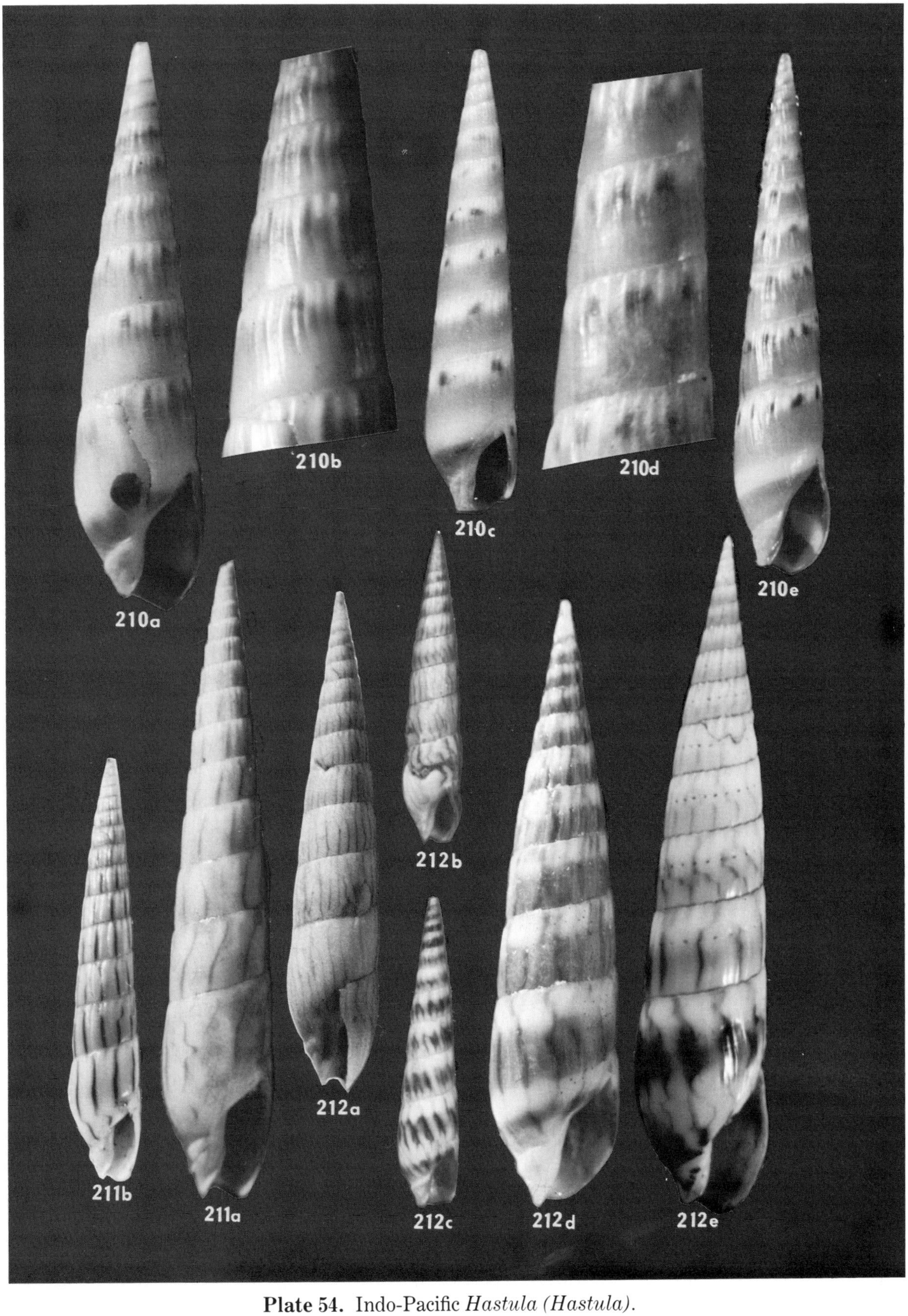

Plate 54. Indo-Pacific *Hastula (Hastula)*.

The species may be separated from *H. penicillata* **(212)** by its more turreted outline, narrower aperture, and color pattern of clean lines without streaks, blotches or bands of color.

212. *Hastula penicillata* (Hinds, 1844)

(Pl. 54, figs. 212a-e)

1844 *Terebra penicillata* Hinds, Proc. Zool. Soc. Lond. pt. 11:157; 1844 Hinds in Sowerby, Thes. Conch. 1:179, pl. 44, fig. 91; 1917 Hirase, Terebridae Jap. Emp. p. 27, pl. 2, fig. 3; 1958 Tinker, Pacific Sea Shells ed. 2:198, figs. upper row center.

1844 *T. venosa* Hinds, Proc. Zool. Soc. Lond. pt. 11:157; 1844 Hinds in Sowerby, Thes. Conch. 1:179, pl. 44, fig. 90; 1860 Reeve, Conch. Icon. 12, pl. 20, sp. 95a,b; 1984 Aubry, Terebridae pl. 4.

1859 *T. crossii* Deshayes, Proc. Zool. Soc. Lond. pt. 27:289; 1977 Bratcher, Nautilus 91(2):40, fig. 4.

1961 *Acuminia penicillata* (Hinds), Oyama & Takemura, Moll. Shells, Resources Exploitation Inst. 5, *Acuminia* fig. 5.

1964 *Hastula (Hastula) penicillata* (Hinds), R. D. Burch, Veliger 6(4):213.

1965 *H. (Punctoterebra) betsyae* R. D. Burch, Veliger 7(4):243, pl. 31, fig. 2.

1966 *H. penicillata* (Hinds), Cernohorsky & Jennings, Veliger 9(1):61, pl. 7, figs. 54, 54a-c, text. fig. 13, radula and operculum; 1967 Cernohorsky, Mar. Shells Pacific 1:212, pl. 54, fig. 406; 1977 Salvat & Rives, Coq. Polynèsie p. 386, fig. 396; 1979 Kay, Hawaiian Mar Shells p. 389, fig. 126G; 1982 Abbott & Dance, Compendium of Seashells p. 274; 1984 Trondle, Xenophora 15:14, fig. 37; 1984 Aubry, Terebridae pl. 6.

1980 *Acuminia venosa* (Hinds), Rehder, Smithsonian Contrib. Zool. 289:92.

Description: Shell to 40 mm; color white to ivory with wavy brown sometimes branching axial lines, some specimens blotched with brown; outline of whorls flat; protoconch of 2½ to 3½ whorls; subsutural punctations on some specimens; axial ribs sharp, straight on early whorls, usually becoming obsolete and fading completely later; spiral sculpture absent; aperture elongate; columella straight.

Type locality: *H. penicillata*, "Seychelles Is."; *T. venosa*, none; *T. crossii*, "Indian Ocean"; *H. betsyae*, "off Honokohua, Maui, Hawaii."

Distribution: From the Seychelles Is. to Easter Is. and Hawaii; intertidal to 114 m.

Type: *H. penicillata* lectotype BM(NH) no. 1968244, 40.0 mm; *T. venosa* lectotype BM(NH) no. 1968245, 35.0 mm; *T. crossii* holotype MNHNP, 24.0 × 5.5 mm; *H. betsyae* holotype MCZ no. 251238, 34.0 × 7.0 mm. The latter was named for Betsy Clarke Harrison of Hawaii.

Discussion: The radular ribbon is translucent white, 1.7 mm long and 0.012 mm wide in an animal with a shell 39 mm in length; fully-formed rows numbered 37, and teeth in the first three front rows were shorter than others. Radular teeth consist of a pair of slender and curved teeth which lack accessory cusps or hooks. (Cernohorsky and Jennings 1966).

The form with a spiral row of subsutural punctations, described as *T. crossii* and *H. betsyae* is most commonly found in the Hawaiian Islands but is also found to be sympatric with the smooth form elsewhere. The color form with extremely dark blotches, sometimes covering the greater part of the shell was described both as *T. venosa* Hinds and *T. crossii* Deshayes. Refer to *T. lanceata* **(211)** for comparison.

213. *Hastula brazieri* (Angas, 1871)

(Pl. 55, figs. 213a,b)

1871 *Terebra (Hastula) brazieri* Angas, Proc. Zool. Soc. Lond. p. 16, pl. 1, fig. 15.

1923 *Terebra brazieri* Angas, May, Index Tasmanian Shells p. 73, pl. 34, fig. 10.

1932 *Acuminia brazieri* (Angas), Cotton & Godfrey, S. Austral. Nat. 13(2):65, pl. 3, fig. 6; 1952 Cotton, Trans. Roy. Soc. S. Australia 75:41.

1971 *Hastula brazieri* (Angas), Wilson & Gillett, Austral. Shells p. 158, pl. 106, fig. 15; Hinton, Guide Austral. Shells pl. 59, fig. 3.

Description: Shell to 35 mm; color pale straw with brownish orange flammules and occasionally with brown sutural spotting; outline of whorls almost flat; protoconch 4 to 4½ purplish brown whorls; subsutural band absent; axial ribs distant or obsolete, weakly noded posteriorly; sculpture well marked on rare individuals; spiral sculpture microscopic; aperture moderately elongate; columella curved.

Type locality: "Brisbane Water, New South Wales," Australia.

Plate 55

FIGS. **213a, b,** *Hastula brazieri* (Angas). **213a,** Lectotype British Museum (N.H.) No. 1871.7.5.15.; 27.8 mm. **213b,** Paralectotype Australian Museum, Sydney No. C-11615; 28.0 mm. **214a-g,** *H. albula* (Menke). **214a,** Type-figure of *Terebra casta* Hinds, from Sowerby, 1844, pl.44, fig.84, **214b,** Type-figure of *T. mera* Hinds, from Sowerby, 1844, pl.45, fig.108. **214c,** Lectotype of *T. incolor* Deshayes, British Museum (N.H.) No. 197996; 33.4 mm. **214d,** Lectotype of *T. bipartita* Deshayes, British Museum (N.H.) No. 197981; 22.3 mm. **214e,** Holotype of *T. medipacifica* Pilsbry, Academy Natural Sciences, Philadelphia No. 117039; 18.8 mm. **214f,** Holotype of *T. medipacifica melior* Pilsbry, Academy Natural Sciences, Philadelphia No. 117021; 28.0 mm. **214g,** Upper whorls. **215a, b,** *H. nana* (Deshayes). **215a,** Holotype British Museum (N.H.) No. 1979105; 9.9 mm. **215b,** Persian Gulf; 18.1 mm. **216a, b,** *H. solida* (Deshayes). **216a,** Holotype British Museum (N.H.) No. 197972; 29.9 mm. **216b,** Okinawa, Ryukyu Ids.; 28.7 mm.

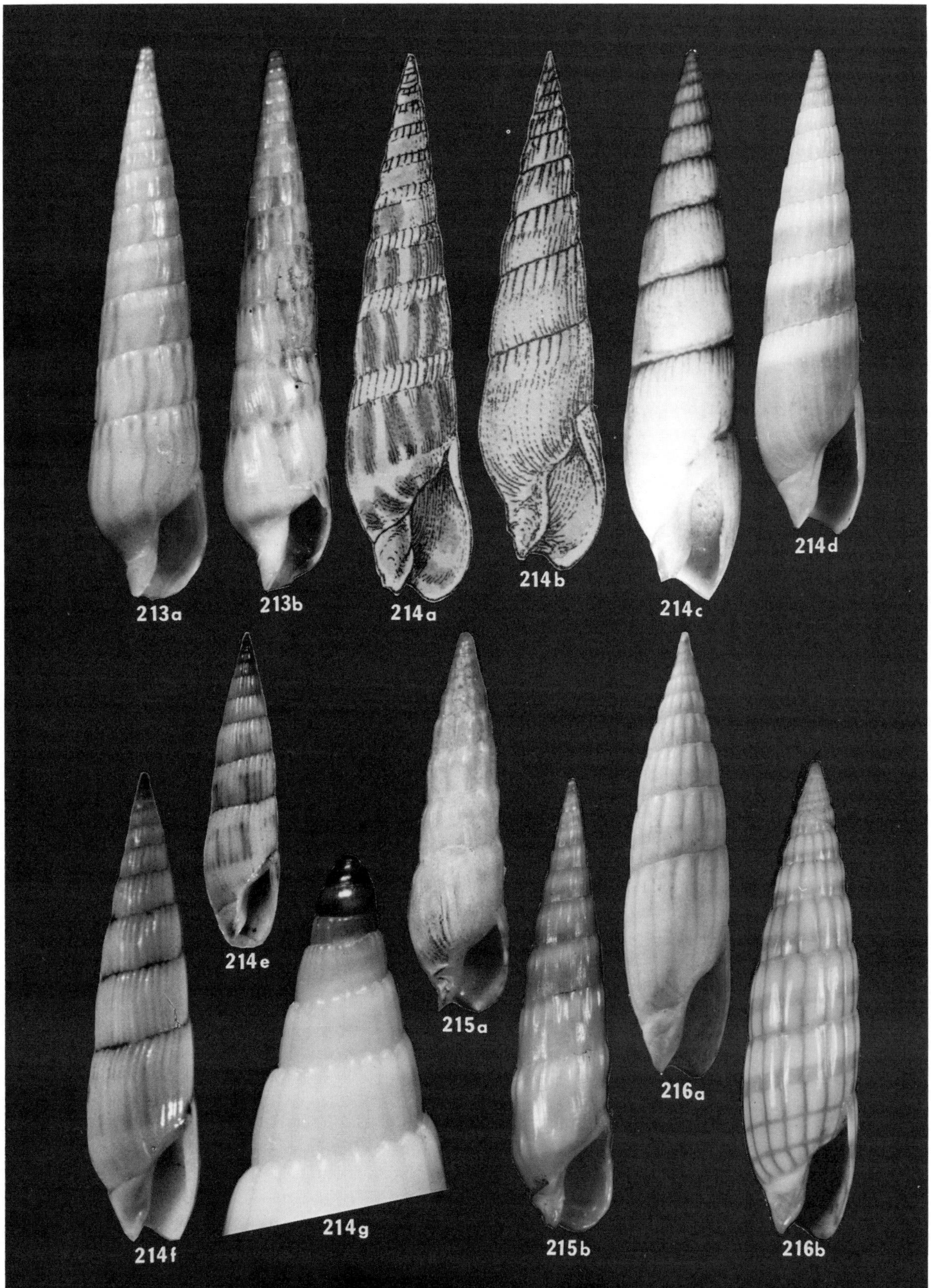

Plate 55. Indo-Pacific *Hastula (Hastula).*

Distribution: New South Wales to South Australia and Tasmania; subtidal to 48 m.
Type: Lectotype BM(NH) no. 1871.7.5.15; 27.8 × 6.0 mm.
Discussion: The pale straw color with brownish flammules of this uncommon species is quite unlike any other in Australia, and it is not likely to be confused with any other. It was named for John Brazier (1842–1930), a leading Australian malacologist and ardent shell collector.

214. ***Hastula albula*** (Menke, 1843)

(Pl. 55, figs. 214a-g)

1843 *Terebra albula* Menke, Moll. Novae-Hollandiae p. 30; 1844 Hinds in Sowerby, Thes. Conch. 1:182, pl. 45, fig. 126.
1844 *Terebra casta* Hinds, Proc. Zool. Soc. Lond. pt. 11:156; 1844 Hinds in Sowerby, Thes. Conch. 1:165, pl. 44, fig. 84.
1844 *Terebra mera* Hinds, Proc. Zool. Soc. Lond. pt. 11:158; 1844 Hinds in Sowerby, Thes. Conch. 1:184, pl. 45, figs. 108, 114; 1984 Aubry, Terebridae pl. 5.
1859 *Terebra incolor* Deshayes, Proc. Zool. Soc. Lond. pt. 27:283.
1859 *Terebra bipartita* Deshayes, Proc. Zool. Soc. Lond. pt. 27:284.
1903 *?Terebra (Hastula) casta* var. *natalensis* E. A. Smith, Proc. Malac. Lond. 5:360.
1921 *Terebra medipacifica* Pilsbry, Proc. Acad. Nat. Sci. Philadelphia 72:308, pl. 12, figs. 8–10.
1921 *Terebra medipacifica melior* Pilsbry, ibid. 72:308.
1965 *Hastula albula* (Menke), R. D. Burch, Veliger 7(4):252; 1966 Cernohorsky & Jennings, Veliger 9(1):59, pl. 7, figs. 56, 56a; 1971 Bratcher & Burch, Proc. California Acad. Sci. 37(21):562, figs. 32, 33; 1971 Bratcher & Burch in Keen ed. 2:686, fig. 1572; 1979 Kay, Hawaiian Mar. Shells p. 385, fig. 126d; 1984 Trondle, Xenophora 15:14, fig. 33.

Description: Shell to 40 mm; color white to pale brown, often with darker bands, streaks, or splotches and frequently with a white subsutural band; outline of whorls slightly convex; parietal callus 3½ to 4½ glassy white, pink, or purple whorls; subsutural groove absent; axial ribs, about equal to interspaces, extending from ¼ to the entire length of the whorl; suture appearing crenulated because of the formation of ribs; aperture elongate; columella curved.
Type locality: *H. albula*, "Western shores of New Holland" Western Australia; *T. casta*, "Ilo-Ilo Is., Panay, Philippines"; *T. mera*, "Straits of Malacca"; *T. incolor*, "Philippine Islands"; *T. bipartita*, "Hawaiian Is."; *T. casta* var. *natalensis*, "Umkomaas, Natal, S. Africa"; *T. medipacifica*, "off Honolulu, Hawaii"; *T. medipacifica melior*, "Kaneohe Bay, Oahu, Hawaii."
Distribution: From the Red Sea throughout the tropical Indo-Pacific to the Revillagigedo Islands, Mexico; intertidal to 116.
Type: *H. albula* type was lost or destroyed in World War II; *T. casta* lectotype BM(NH) no. 1968242, 29.0 mm; *T. mera* type lost (K. Way, *in litt*); *T. incolor* lectotype BM(NH) no. 197996, 33.4 × 7.8 mm; *T. bipartita* lectotype BM(NH) no. 197981, 22.3 × 5.0 mm; *T. casta* var. *natalensis* type lost; *T. medipacifica* holotype ANSP 117039, 18.8 mm; *T. medipacifica melior* holotype ANSP 117021, 28.0 mm.
Discussion: Veligers of *H. albula* vie with those of *Terenolla pygmaea* **(262)** and *H. lanceata* **(211)** for the title of the largest veligers. Veligers of *H. albula* hatch when 1½–2 whorls are complete, add 1½ more whorls while in the plankton stage, then settle when they reach lengths from 1.1 to 1.4 mm when five whorls are complete (J. B. Taylor, unpubl. Ph.D. thesis). Unlike most Indo-Pacific species which are found rarely on the West American off-shore islands, this species is the most common terebrid found on the Revillagigedo islands off Mexico. *T. casta* (now *T. albula*) is the type species of the genus *Hastulina* Oyama, 1961.

215. ***Hastula nana*** (Deshayes, 1859)

(Pl. 55, figs. 215a,b)

1859 *Terebra nana* Deshayes, Proc. Zool. Soc. Lond. pt. 27:291; 1860 Reeve, Conch. Icon. 12, pl. 26, sp. 138; 1885 Tryon, Man. Conch. 7:36, pl. 11, fig. 3; 1917 Melvill & Standen, Journ. Conch. 15:210.

Description: Shell small, slender, to 10 mm; color shiny white or cream, possibly with small fulvous dots; outline of whorls turreted; protoconch of 1½ whorls; subsutural band absent; axial ribs narrow on early whorls, reaching almost to following suture; becoming more rounded and fading at about ¼ the distance from suture on later whorls; microscopic "pin prick" sculpture visible under magnification; body whorl extremely elongate with obsolete elongate nodes sometimes occurring at periphery; aperture moderately quadrate, with outer lip extending beyond the siphonal canal; columella curved.
Type locality: "Mouth of the Indus," Pakistan.
Distribution: Persian Gulf to the Arabian Sea; subtidal to 6 m.

Type: Holotype BM(NH) no. 1979105; 9.9 × 2.5 mm.

Discussion: Both Deshayes and Reeve mention that there are fulvous spots on the shell, but they must have faded in one and a quarter centuries. The remainder of the original description fits the holotype. Fresh specimens have an appreciably different appearance from the worn and faded holotype.

216. *Hastula solida* (Deshayes, 1857)

(Pl. 55, figs. 216a,b)
(Color pl. E, fig. 9)

1791 *Buccinum aciculatum* Gmelin, Syst. Nat. ed. 13:3505 (ref. Chemnitz, vol. 4, pl. 155, fig. 1457) *[nomen oblitum]*.
1857 *Terebra solida* Deshayes, Journ. Conchyl. 6:78, pl. 3, fig. 11; 1917 Hirase, Terebridae Jap. Emp. p. 30, pl. 5, figs. 73, 74.
1912 *Terebra clarkei* M. Smith, Nautilus 26(6):75, pl. 4, fig. 18.
1962 *Hastulina solida* (Deshayes), Kira, Shells W. Pacific Col. 1:208, pl. 71, figs. 7a,b; 1875 Okutani, Gakken Illust. Nat. Encyclo. Moll. Japan p. 154, fig. 17.
1964 *Terebra (Hastula) solida* Deshayes, Shikama, Sel. Shells World Col. 1:125, 211/1.
1966 *Hastula solida* (Deshayes), Cernohorsky & Jennings, Veliger 9(1):62; pl. 7, fig. 55; 1967 Cernohorsky, Mar. Shells Pacific 1:212, pl. 54, fig. 408; 1978 Hinton, Guide Austral. Shells pl. 56, fig. 17; 1978 Hinton, Guide Shells Papua New Guinea, pl. 47, fig. 14; 1979 Kay, Hawaiian Mar. Shells p. 389, fig. 126c; 1984 Trondle, Xenophora 15:15, fig. 40; 1984 Aubry, Terebridae pl. 6; 1984 Salvat & Rives, Shells of Tahiti p. 138, pl. 36, fig. 5.

Description: Shell to 32 mm; color shiny ivory with flesh or gray interspaces between ribs and a stripe of the same color at or near the center of whorl, 3 on body whorl, dividing it into rectangles; outline of whorls convex; protoconch of 4½ conical whorls; subsutural band absent; axial ribs broad, convex, with narrow interspaces; spiral sculpture absent; body whorl elongate; aperture elongate; columella curved.

Type locality: *H. solida*, "Japan"; *T. clarkei*, none given.

Distribution: Throughout the tropical Indo-Pacific; intertidal to 120 m.

Type: *H. solida* holotype BM(NH) no. 197972, 29.9 × 8.0 mm; *T. clarkei* type not found, but possibly syntypes are in the Royal Scottish Museum.

Discussion: The animal of this species is rose-pink. Although Maxwell Smith's type of *T. clarkei* could not be found, his figure leaves no doubt that it is conspecific with Deshayes' *T. solida*. Maxwell Smith's synonym was named for Mr. Eagle Clarke, then Curator of Mollusca at the Royal Scottish Museum in Edinburgh.

217. *Hastula philippiana* (Deshayes, 1859)

(Pl. 56, figs. 217a-c)

1859 *Terebra philippiana* Deshayes, Proc. Zool. Soc. Lond. pt. 27:289; 1860 Reeve, Conch. Icon. 12, pl. 27, sp. 153; 1984 Trondle, Xenophora 15:15, fig. 38.
1965 *Hastula (Hastula) tiedemani* R. D. Burch, Veliger 7(4):241, pl. 31, fig. 6; 1979 Kay, Hawaiian Mar. Shells p. 390, fig. 126h.

Description: Shell to 11 mm, slender; color shiny white or brown, usually a combination of the two in broad bands; occasionally with a dark blotch on the dorsum of the body whorl; outline of whorls slightly convex; protoconch of 3½ conical whorls; subsutural band or groove absent; axial ribs from suture to suture, about equal to interspaces, often fading at the middle of the dorsum; spiral sculpture absent; outer lip usually translucent and without ribs, often colorless; aperture moderately quadrate; columella short, slightly curved; operculum translucent brown.

Type locality: *T. philippiana*, "Marquesas Islands?"; *H. tiedemani*, "Maalaea Bay, off Kihei Bay, Maui, Hawaii."

Distribution: French Polynesia and Hawaii to Guam.

Type: *H. philippiana* lectotype BM(NH) no. 1979101, 9.3 × 2.6 mm; *H. tiedemani* holotype MCZ no. 251237, 6.0 × 1.5 mm. It was named after the shell collector, Alan Tiedeman of Hawaii.

Discussion: This is the smallest species in the genus, usually being less than ⅓ of an inch in length. It somewhat resembles *H. albula* **(214)** and has a similar protoconch, but measures much shorter than that species with shells having the same number of whorls; it is more slender, and has a translucent outer lip. It may be separated from *H. celidonota* **(218)** by its lack of a subsutural band and groove. Deshayes named this species in honor of his German malacological friend, Rudolf Amandus Philippi (1808–1904).

218. *Hastula celidonota* (Melvill & Sykes, 1898)

(Pl. 56, figs. 218a,b)
(Color pl. D, fig. 6)

1898 *Terebra celidonota* Melvill & Sykes, Proc. Malac. Soc. Lond. 3(1):42, pl. 3, fig. 2.

Description: Shell to 15 mm; color shiny white to flesh-white, with protoconch and early whorls of teleoconch rosy brown, conspicuous brown stain on dorsum of body whorl; outline of whorls slightly convex; protoconch of 3½ amber-brown conical whorls; subsutural band marked by punctations between ribs; axial ribs extending from suture to suture, about equal to interspaces, 19 or 20 on penultimate whorl; spiral sculpture absent; body whorl with ribs reaching the periphery except on the dorsum which is smooth with a rosy brown blotch; aperture moderately quadrate; columella curved.
Type locality: "Andaman Is."
Distribution: From the Andaman Islands to the Marquesas Islands.
Type: Lectotype BM(NH) no. 1898.4.30.5; 15.0 × 3.0 mm.
Discussion: Occasionally the shell surface has a tendency to peel off somewhat at the area of the brown spot on the dorsum and appear thinner there. This is the case with a paralectotype in the National Museum of Wales no. 199.158.56, and it has been observed in other specimens.

219. *Hastula apicitincta* (G. B. Sowerby, 1900)

(Pl. 56, fig. 219a,b)

1900 *Terebra apicitincta* G. B. Sowerby, Proc. Malac. Soc. Lond. 4(1):5, pl. 1, fig. 14; 1973 Kensley, Seashells S. Africa p. 214, fig. 861.

Description: Shell to 15 mm; color shiny white with some whorls streaked with amber; outline of whorls straight; protoconch of 1½ amber whorls; subsutural band and groove absent; axial sculpture of fine riblet-striae; spiral sculpture absent; aperture moderately quadrate; columella slightly curved.
Type locality: "The Kowie, South Africa."
Distribution: South Africa.
Type: Lectotype BM(NH) no. 1900.5.22.42; 12.6 mm.
Discussion: All material examined was in a worn condition. This species differs from *H. albula* **(214)** by being more slender and flat-sided and by having an amber protoconch of 1½ whorls, while the latter has a protoconch of 3½ to 4½ whorls.

220. *Hastula filmerae* (Sowerby, 1906)

(Pl. 56, figs. 220a,b)

1906 *Terebra filmerae* Sowerby, Proc. Mal. Soc. Lond. 7(1):37, text fig.

Description: Shell to 22 mm; color cream with lighter subsutural band; outline of whorls slanting inward in both directions from shoulder; protoconch of 2 whorls; subsutural band slanting outward from suture, band ribbed; subsutural groove absent, but obsolete nodes at rib endings mark band; remainder of whorl almost smooth; spiral sculpture absent; columella straight. The aperture on the holotype is broken.
Type locality: "Pondoland, South Africa."
Distribution: Known only from the type locality.
Type: Holotype Natal Museum no. B-1806; 22.3 × 5.3 mm.
Discussion: The strange inward-slanting subsutural band makes this species easily distinguishable. It has not been reported subsequent to the original description and may be a pathological individual. It was named after a Mrs. Filmer, a shell collector during the 1890s in Pondoland, South Africa.

Plate 56
FIGS. **217a-c,** *Hastula philippiana* (Deshayes). **217a,** Lectotype British Museum (N.H.) No. 1979101; 9.3 mm. **217b,** Type-figure of *H. tiedemani* R. D. Burch, from Burch, 1965, pl.31, fig.6; 6.0 mm. **217,** Tahiti, Society Ids.; 9.0 mm. **218a, b,** *H. celidonota* (Melvill & Sykes). **218a,** Lectotype British Museum (N.H.) No. 1898.4.30.5.; 15.0 mm. **218b,** Paralectotype National Museum Wales, Cardiff No. 199.158.56.; 14.0 mm. **219a, b,** *H. apicitincta* (Sowerby). **219a,** Lectotype British Museum (N.H.) No. 1900.5.22.42.; 12.6 mm. **219b,** Paralectotype, same Institution; 12.6 mm. **220 a, b,** *H. filmerae* (Sowerby). **220a,** Holotype Natal Museum No. B-1806; 22.3 mm. **220b,** Lower whorls of holotype. **221a-c,** *H. hastata* (Gmelin). **221a,** Illustrated lectotype from Chemnitz, 1780, pl.154, fig.1454. **221b,** Virgin Gorda, Virgin Ids.; 20.2 mm. **221c,** Lectotype of *T. crassula* Deshayes, Museum National d'Histoire Naturelle, Paris; 23.6 mm.

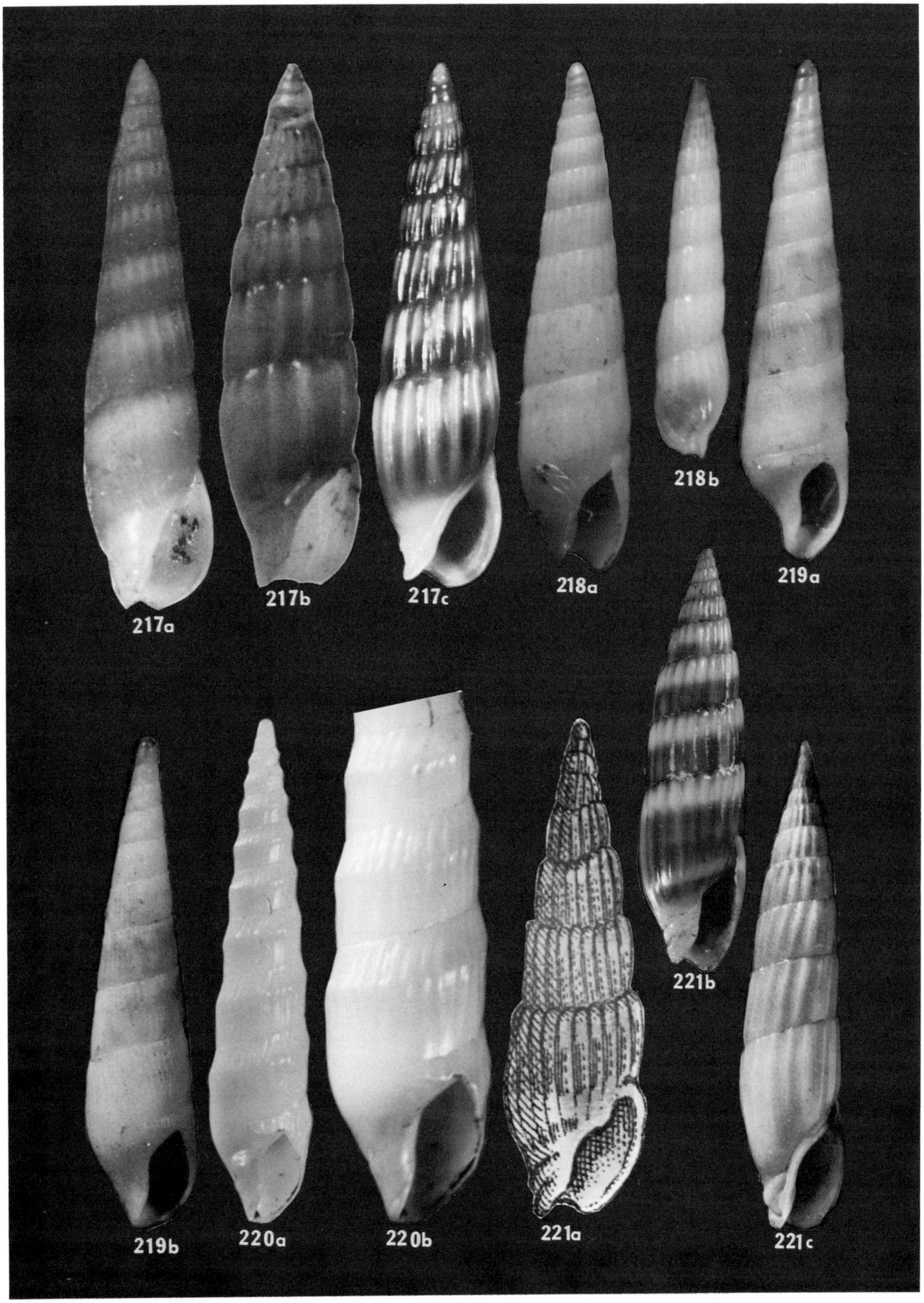

Plate 56. Indo-Pacific and Western Atlantic *Hastula (Hastula)*.

EASTERN PACIFIC AND ATLANTIC *HASTULA*

221. ***Hastula hastata*** (Gmelin, 1791)

(Pl. 56, figs. 221a-c)

1791 *Buccinum hastatum* Gmelin, Syst. Nat. ed. 13:3502, (ref. Chemnitz, 1780, vol. 4, pl. 154, figs. 1453, 1454.

1828 *Terebra costata* Menke, Syn. Meth. Moll. p. 51 [non Borson, 1825]. (nom. nov. pro *Buccinum hastatum* Gmelin)

1859 *Terebra crassula* Deshayes, Proc. Zool. Soc. Lond. pt. 27:282.

1860 *Terebra hastata* (Gmelin), Reeve, Conch. Icon. 12, pl. 18, sp. 81; 1975 Humpfrey, Sea Shells of the West Indies p. 180, pl. 20, fig. 9; 1984 Abbott, Collectible Florida Shells p. 34, figs. 4, 5; 1984 Aubry, Terebridae pl. 6.

1937 *Hastula hastata* (Gmelin), M. Smith, E. Coast Mar. Shells p. 133, pl. 5, fig. 2; 1975 Rios, Brazilian Mar. Moll. Icon. p. 127, pl. 38, fig. 559; 1982 Abbott & Dance, Compendium of Seashells p. 276.

1954 *Terebra (Hastula) hastata* (Gmelin), Abbott, American Seashells ed. 1:266, pl. 26, fig. h; 1961 Abbott & Warmke, Caribbean Seashells p. 132, pl. 25, fig. a; 1974 Abbott, American Seashells ed. 2:260, fig. 2856.

Description: Shell to 32 mm, somewhat obese; color shiny white to ivory, often with beige to brown bands, sometimes interrupted; outline of whorls almost straight; protoconch of $3\frac{1}{2}$ whorls; subsutural band and groove absent; axial ribs well developed, evenly spaced, about 22 on penultimate whorl, sometimes fading anteriorly, ribs about equal to interspaces; suture crenulate; spiral sculpture absent; aperture moderately elongate; columella curved.
Type locality: *H. hastata*, none given. Virgin Gorda, British Virgin Islands, is here selected as the type locality; *T. crassula*, none.
Distribution: From Bermuda through the West Indies to Brazil.
Type: *H. hastata*, the specimen illustrated by Chemnitz (1780), vol. 4 on plate 154, fig. 1454, is here designated as the illustrated lectotype; *T. crassula* lectotype MNHNP, 23.6 × 5.2 mm.
Discussion: Although this species is heavier and more obese, it resembles the Indo-Pacific species *H. albula* (**214**). There is no other Atlantic *Hastula* with which this species could be confused.

222. ***Hastula knockeri*** (E. A. Smith, 1872)

(Pl. 57, figs. 222a,b)

1872 *Terebra knockeri* E. A. Smith, Proc. Zool. Soc. Lond. for 1871: 730, pl. 75, fig. 7.

1983 *Hastula knockeri* (E. A. Smith), Bouchet, Boll. Malac. 18(9–12):201, figs. 13, 47.

Description: Shell to 28 mm; color shiny pale tan, usually with a white band below but not contiguous with suture, and a narrow white band at periphery of body whorl; outline of whorls slightly convex; protoconch $3\frac{1}{2}$ conical amber whorls with narrow brown stripe at suture; subsutural band absent; axial ribs narrow, widely spaced, from suture to suture, often weak, more closely spaced in early whorls; spiral sculpture absent; aperture moderately elongate; columella curved.
Type locality: "Whydah, West Africa."
Distribution: Ivory Coast, West Africa.
Type: Lectotype BM(NH) no. 1870.1.3.47; 18.9 mm.
Discussion: Although this uncommon West African species has no subsutural band, the narrow white band a short distance below the suture gives the appearance of a band. Refer to *H. lepida* (**223**) for comparison. Named after H. H. Knocker who collected in West Africa (Dahomey) in the 1860s.

223. ***Hastula lepida*** (Hinds, 1844)

(Pl. 57, figs. 223a,b)

1844 *Terebra lepida* Hinds, Proc. Zool. Soc. Lond. pt. 11:158; 1844 Hinds in Sowerby, Thes. Conch. 1:82, pl. 45, fig. 102; 1860 Reeve, Conch. Icon. 12: pl. 20, sp. 96a-b; 1952 Knudsen, Vid. Medd. Dansk Nat. Foren. 114:180.

Plate 57

FIGS. **222a, b,** *Hastula knockeri* (E. A. Smith). **222a,** Lectotype British Museum (N.H.) No. 1870.1.3.47.; 18.9 mm. **222b,** Abidjan, Ivory Coast, West Africa; 18.8 mm. **223 a, b,** *H. lepida* (Hinds). **223a,** Paralectotype British Museum (N.H.) No. 1968249; 24.5 mm. **223b,** Middle whorls of lectotype. **224,** *H. leloeuffi* Bouchet. Holotype Museum d'Histoire Naturelle, Paris; 31.0 mm. **225a-c,** *H. cuspidata* (Hinds). **225a,** Lectotype British Museum (N.H.) No. 1968247; 29.4 mm. **225b,** Lectotype of *T. traillii* Deshayes, British Museum (N.H.) No.1979114; 23.8 mm. **225c,** Mouth of Benga river, Luanda, West Africa; 19.8 mm. **226a, b,** *Hastula (Impages) hectica* (Linnaeus). **226a,** Probable type Linnean Society, London; 51.8 mm. **226b,** Holotype of *T. castanea* Kiener, Museum d'Histoire Naturelle, Geneva No. 1152/72; 55.0 mm. **227a-c,** *H. (I.) stylata* (Hinds). **227a,** Lectotype British Museum (N.H.) No. 1968230; 48.8 mm. **227b,** Andaman Ids.; 40.2 mm. **227c,** Tahiti, Society Ids.; 35.1 mm.

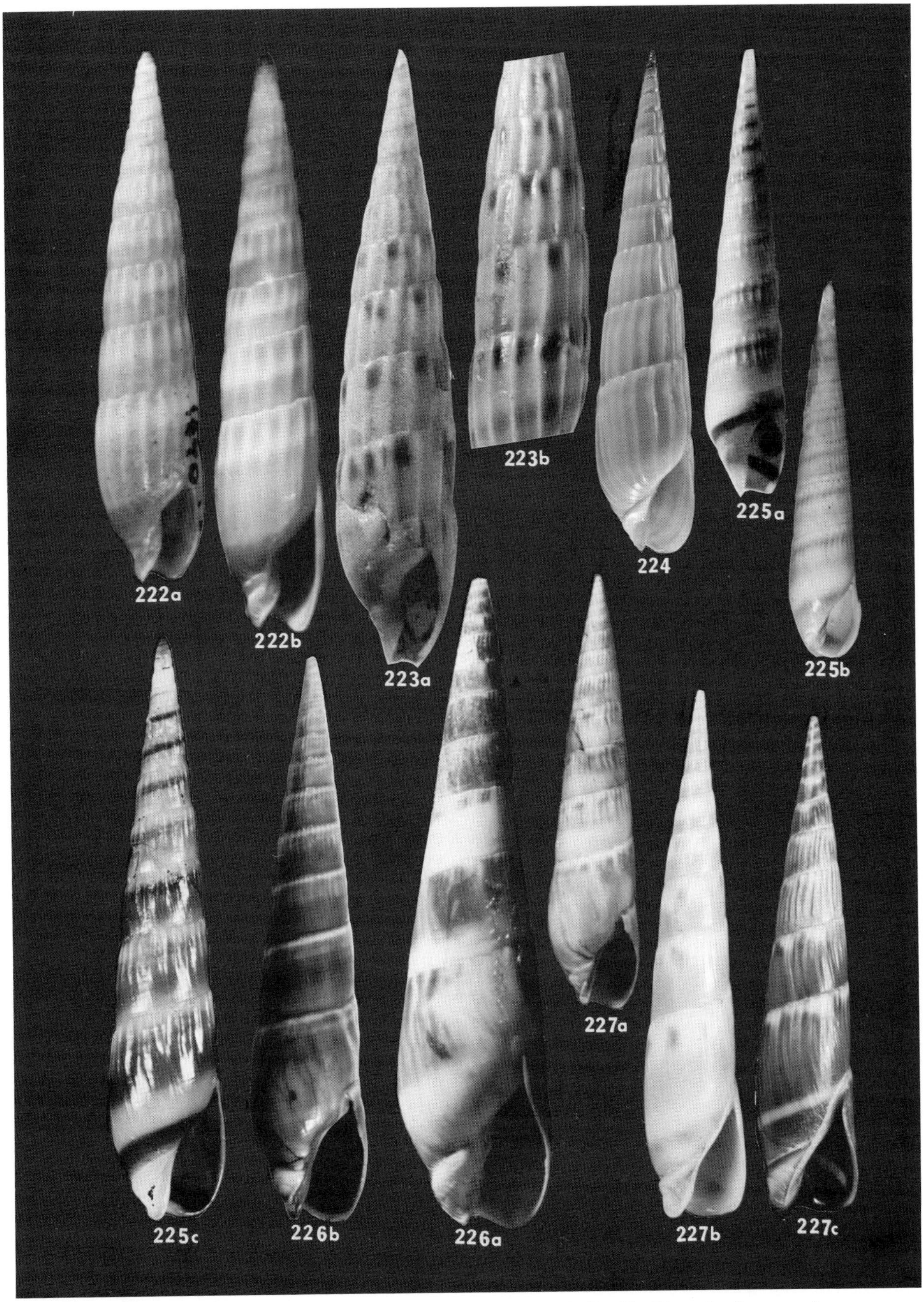

Plate 57. West African *Hastula (Hastula)* and Indo-Pacific *Hastula (Impages)*.

1981 *Hastula lepida* (Hinds), Bouchet, Malacologia 21(1–2):366, fig. 1/14–16; fig. 5/2–4; 1983 Bouchet, Boll. Malac. 18(9–12):199, figs. 15, 42–45; 1984 Aubry, Terebridae pl. 2.

Description: Shell to 30 mm; color cream to light brown, with dark brown spots below suture; outline of whorls flat; protoconch of 4½ conical whorls; subsutural band or groove absent; axial sculpture of narrow, sharp ribs from suture to suture, about 14 to 16 on penultimate whorl; interspaces smooth, wider than ribs; aperture elongate; columella straight.
Type locality: "Guinea," West Africa.
Distribution: West Africa, from the Canary and Cape Verde Islands to Senegal and Angola; 5 to 55 m.
Type: Lectotype BM(NH) no. 1968249; 23.0 mm.
Discussion: Some of the shells of this species are much more slender than others. *Hastula knockeri* **(222)** may be separated by its salmon-orange coloration, its lack of subsutural spots, and its less inflated protoconch. *Hastula matheroniana* **(208)** is from a different faunal province and differs by having subsutural punctations.

224. *Hastula leloeuffi* Bouchet, 1983

(Pl. 57, fig. 224)

1983 *Hastula leloeuffi* Bouchet, Bolletino Malacologico 18(9–12):200, figs. 11, 46; 1984 Aubry, Terebridae pl. 14.

Description: Shell to 31 mm; color rosy-beige with a white spiral band visible only in the interior of the aperture; outline of whorls straight; protoconch of 4 slender brown conical whorls; subsutural band absent; suture moderately shallow; axial ribs about equal to interspaces, very slightly flexuous on later whorls, about 17 on the penultimate whorl; interspaces appear smooth, but microscopic spiral lines are present; aperture elongate; columella straight, edged in orange-brown.
Type locality: "The Ivory Coast, West Africa, 20 m."
Distribution: The Ivory Coast and central part of the coast of Angola, West Africa.
Type: Holotype MNHN Paris; 31.0 × 7.5 mm.
Discussion: This species may be separated from *H. lepida* **(223)** by a less convex protoconch, more shallow suture, and lack of subsutural spots. *H. matheroniana* **(208)** is from a different faunal province and differs by having subsutural punctations. It was named for Dr. Pierre Le-Loeuff who collected shells in the Ivory Coast.

225. *Hastula cuspidata* (Hinds, 1844)

(Pl. 57, figs. 225a-c)
(Color pl. D, fig. 4)

1844 *Terebra cuspidata* Hinds, Proc. Zool. Soc. Lond. pt. 11:157.
1859 *Terebra traillii* Deshayes, Proc. Zool. Soc. Lond. pt. 27:285; 1860 Reeve, Conch. Icon. 12, pl. 26, sp. 142.
1983 *Hastula cuspidata* (Hinds), Bouchet, Boll. Malac. 18(9–12):202, figs. 16, 48–50; 1984 Aubry, Terebridae pl. 2.

Description: Shell to 30 mm; color yellowish, shiny, with narrow grayish brown subsutural stripe, white at periphery of body whorl with dark brown stripe just anterior to periphery, dark brown continuing on some individuals to anterior of shell; early whorls of teleoconch translucent with narrow dark stripe showing through; outline of whorls flat; protoconch of 3½ translucent whorls; subsutural groove absent, subsutural band in color only; axial plications narrow, extending about ⅓ length of whorl; remainder of whorl smooth; aperture quadrate with brown and white stripes showing through outer lip; columella slightly curved.
Type locality: *H. cuspidata*, "Cape Coast Africa," error? *H. trailii*, "Vasigapatam, Indian Ocean." Vizagapatam, central east India.
Distribution: West Africa.
Type: *H. cuspidata* Hinds lectotype BM(NH) no. 1968247, 29.4 mm; *H. trailii* lectotype BM(NH) no. 1979114, 23.8 × 4.5 mm.
Discussion: The existence of *H. cuspidata* in Angola, West Africa, is definitely confirmed through recent collections of Francisco Fernandes, living in Angola. The occurrence of this species in India and South Africa is unconfirmed. It may be separated from *H. cinerea* **(226)** by its lack of tiny pinpricks in the sculpture and by its more colorful striping. *Buccinum cuspidatum* Gmelin, 1791, may be a *Terebra* or *Hastula*, but is considered a *nomen dubium* by Bouchet, 1982, p. 202.

Subgenus *IMPAGES* E. A. Smith, 1873

Type species, *Terebra caerulescens* Lamarck, 1822 = *Hastula hectica* (Linnaeus, 1758).

The shell characters of the species we place in this subgenus fit quite well in the genus *Hastula*, but the shell may be less shiny and smooth than that of a typical *Hastula*. In addition it has an almost triangular aperture which flares broadly at the anterior and narrows to a point at the posterior, with the point being somewhat drawn out. (See text fig. 2 of *Impages* aperture). It has an almost straight columella.

Until it was discovered that *Hastula (Impages) stylata* (Hinds) had neither radula nor poison sac, we believed that this was a full genus based on both shell and animal morphology, and that all species with this shell morphology possessed a typical toxoglossate radula with harpoon-shaped teeth and a poison gland. We had planned to use the genera, *Terebra, Hastula, Duplicaria,* and *Terenolla,* merely as a convenient tool for grouping shells, to give greater ease in locating the proper identification. We also planned to treat *Impages* as a full genus, but with the discovery that at least one of the species in that group lacks radula and poison gland, we now must consider *Impages* to be a poor grouping, and use it also for expediency only in a subgeneric sense.

INDO-PACIFIC *HASTULA*
subgenus *(IMPAGES)*

226. ***Hastula hectica*** (Linnaeus, 1758)

(Pl. 57, figs. 226a,b)
(Color pl. C, figs. 2,3)

1758 *Buccinum hecticum* Linnaeus, Syst. Nat. ed. 10:741; 1956 Dodge, Bull. Americ. Mus. Nat. Hist. 111(3):222.

1791 *Buccinum niveum* Gmelin, Syst. Nat. ed. 13:3504. [non Gmelin, 1791, p. 3491 nec p. 3495].

1791 *Buccinum edentulum* Gmelin, Syst. Nat. ed. 13:3505. (ref. D'Argenville, pl. 11, fig. 0).

1817 *Buccinum bifasciatum* Dillwyn, Descr. cat. rec. shells 2:65. (ref. Buonanni, 3, fig. 109; Petiver, pl. 53, fig. 5; and Klein, pl. 7, fig. 117).

1822 *Terebra caerulescens* Lamarck, Hist. Nat. Anim. S. Vert. 7:288; 1839 Kiener, Spéc. Gén. Icon. Coq. Viv. 9:17, pl. 6, figs. 12a-b; 1844 Hinds in Sowerby, Thes. Conch. 1:159, pl. 41, figs. 5, 6, 15; 1860 Reeve, Conch. Icon. 12, pl. 7, sp. 26; 1984 Aubry, Terebridae pl. 7.

1828 *Buccinum terebrale* Menke, Synopsis meth. moll. ed. 1:84.

1830 ?*Terebra caerulescens* var. *otaitensis* Lesson, Voy. Coquille 2(1):407.

1839 *Terebra castanea* Kiener, Spéc. Gén. Icon. Coq. Viv. 9:19, pl. 7, fig. 14.

1843 *Bullia fuscus* Gray in Dieffenbach, Travels in New Zealand, p. 235. (ref. Martyn, Univ. Conch. 4: pl. 121, fig. 3 on left) [non *Terebra fusca* Perry, 1811].

1844 *Terebra nimbosa* Hinds, Proc. Zool. Soc. Lond. pt. 11:151; 1844 Hinds in Sowerby, Thes. Conch. 1:159, pl. 42, fig. 21.

1880 *Terebra flammulata* von Martens, Beit. Meeresf. Mauritius & Seychellen p. 230, pl. 20, fig. 5.

1935 *Terebra (Impages) hectica* Dautzenberg, Mem. Mus. Royal Hist. Nat. Belgique 2(17):38, pl. 2, fig. 3.

1935 *Terebra (Impages) hectica* var. *alba* Dautzenberg, Mem. Roy. Hist. Nat. Belgique 2(17):42, pl. 2, fig. 4 [non Gray, 1834].

1935 *Terebra (Impages) hectica* var. *fusca* "Martyn" Dautzenberg, Mem. Mus. Roy. His. Nat. Belgique 2(17):42, pl. 2, fig. 2 [non Perry, 1811].

1935 *Terebra (Impages) hectica* var. *nimbosa* Dautzenberg, Mem. Roy. Hist. Nat. Belgique 2(17):00, pl. 2, fig. 5.

1961 *Impages hectica* (Linnaeus) Oyama & Takemura, Moll. Shells, Resources Exploitation Inst. 5, *Impages* figs. 5–7; 1966 Cernohorsky & Jennings, Veliger 9(1):64, pl. 6, figs. 47, 47a; 1967 Cernohorsky, Mar. Shells Pacific 1:209, pl. 53, fig. 399; 1972 Hinton, Shells New Guinea & Cent. Indo-Pacific p. 46, pl. 23, fig. 37; 1975 Salvat & Rives, Coq. Polynesie p. 365, fig. 400; 1975 Okutani, Gakken Illust. Nat. Encyclo. Moll. Japan p. 154, fig. 3; 1982 Abbott & Dance, Compendium of Seashells p. 273; 1984 Trondle, Xenophora 15:15, fig. 43; 1984 Aubry, Terebridae pl. 10.

1979 *Hastula hectica* (Linnaeus), Kay, Hawaiian Mar. Shells p. 387, figs. 6b, 126e; 1984 Salvat & Rives, Shells of Tahiti p. 138, pl. 36, fig. 6.

Description: Shell to 78 mm; color white, cream, gray or occasionally yellow background, with markings, or interrupted stripes or dots of black, brown or tan, sometimes a few completely dark whorls; outline of whorls flat; subsutural band absent; axial ribs on early whorls, becoming curved axial striae; spiral sculpture absent; aperture very narrow at posterior end, flaring anteriorly, as in all species of the *Impages* group; columella straight; siphonal fasciole broad.

Type locality: *H. hectica*, "African Ocean"; *T. caerulescens*, "New Holland," W. Australia; *T. caerulescens* var. *otaitensis*, "Tahiti"; *T. castanea*, "Ile de France" = Reunion Is.; *T. nimbosa*,

none; *T. flammulata*, "Mauritius & Seychelles."
Distribution: Tropical Indo-Pacific; intertidal to 4 m.
Type: *H. (I). hectica* probable type, Linnean Society, London, 51.8 × 12.3 mm; *T. caerulescens* holotype MHNG no. 1101/24/2, 56.3 × 13.2 mm; *T. caerulescens* var. *otaitensis* type lost; *T. castanea* holotype MHNG no. 1152/72, 55.0 × 7.3 mm; *T. nimbosa* lectotype BM(NH) no. 1968226, 58.8 mm; *T. flammulata* syntypes MNHU.
Discussion: This species is usually found on open surf-swept beaches, appearing briefly, usually at the turn of the tide, as the wave recedes, then disappearing in the sand. The color is extremely variable. Three yellow specimens are in the museum at Leiden, Holland, and a golden orange one is in the Trondle collection formerly in Tahiti, now in France.

227. *Hastula stylata* (Hinds, 1844)

(Pl. 57, fig. 227a-c)

1844 *Terebra stylata* Hinds, Proc. Zool. Soc. Lond. pt. 11:152; 1944 Hinds in Sowerby, Thes. Conch. 1:161, pl. 44, fig. 79.
1966 *Hastula stylata* (Hinds), Cernohorsky & Jennings, Veliger 9(1):63, pl. 7, fig. 49H; 1967, Cernohorsky, Mar. Shells Pacific 1:214, pl. 54, fig. 410; 1975 Salvat & Rives, Coq. Polynesie p. 364, fig. 398; 1982 Abbott & Dance, Compendium of Seashells p. 2274; 1984 Aubry, Terebridae pl. 10; 1984 Salvat & Rives, Shells of Tahiti p. 138, pl. 36, fig. 9.

Description: Shell to 49 mm; color from beige to blackish brown, usually with an uneven white band below the suture, dark splotches on or below the white band, and a white stripe at the periphery of the body whorl; outline of whorls flat; protoconch of 3 conical whorls; subsutural band and groove absent; axial ribs weak, slightly curved, usually fading before reaching the following suture; aperture almost triangular, with outer lip flaring anteriorly, coming to a point at the posterior extremity; columella faintly curved, with a parietal callus, the outer edge protruding into a keel.
Type locality: "Japan and the Philippine Is." Cagayan, Mindanao Is., Philippines is here designated as the type locality.
Distribution: From Japan through the tropical Pacific to Easter Is. Not found in the Hawaiian Is.
Type: Lectotype BM(NH) no. 1968230; 48.8 mm.
Discussion: This is the type species of the genus *Egentelaria* Rehder, 1980, which he diagnosed as, "Shell moderately long, subulate, with straight sides and flattened whorls; whorls with more or less distant axial ribs that may become evanescent towards the base of the whorls; with a pale band below the suture and the periphery of the body whorl." It is here considered a subjective synonym of *Impages*. There is a strong similarity between this species and *H. inconstans* **(228)** which occurs in the same range of colors. *H. inconstans* probably replaces *H. stylata* in the Hawaiian Islands. There are differences between the species. The axial ribs of *H. inconstans* reach from suture to suture, and the parietal callus does not develop so early nor protrude as a keel.

228. *Hastula inconstans* (Hinds, 1844)

(Pl. 58, figs. 228a-c)
(Color pl. A; D, fig. 4)

1844 *Terebra inconstans* Hinds, Proc. Zool. Soc. Lond. pt. 11:156; 1844 Hinds in Sowerby, Thes. Conch. 1:179, pl. 44, fig. 83; 1952 Tinker, Pacific Sea Shells 2:192, figs. upper row and center pair 2nd row; 1984 Aubry, Terebridae pl. 2.
1844 *Terebra cinerea* Hinds in Sowerby, Thes. Conch. 1: pl. 45, fig. 130 (non *Buccinum cinereum* Born, 1778).
1860 *Terebra aciculina* (pars), Reeve, Conch. Icon. 12: pl. 23, fig. 121B only [non Lamarck, 1822].
1877 *Terebra (Hastula) confusa* E. A. Smith, Ann. Mag. Nat. Hist. ser. 4, 19:229 [nom. nov. pro *T. cinerea* Hinds in Sowerby, 1844].
1966 *Impages inconstans* (Hinds), Habe & Kosuge, Shells World Color col. 2:98, pl. 39, fig. 2; 1975 Okutani, Gakken Illust. Nat. Encyclo. Moll. Japan p. 154, fig. 5.0
1978 *Hastula inconstans* (Hinds), Cernohorsky, Trop. Pacific Marine Shells p. 149, pl. 53, fig. 6; 1978 Salisbury, Hawaiian Shell News 26(8):9, text fig. p. 10; 1979 Kay, Hawaiian Mar. Shells p. 387, fig. 126b; 1982 Abbott & Dance, Compendium of Seashells p. 275.

Description: Shell to 35 mm; color variable, shiny, pale blue-gray, beige, or brown, striped with light cream, dark brown, or both, sometimes axially streaked with brown; outline of whorls straight; protoconch of 3½ conical whorls; subsutural band and groove absent; axial ribs strong, straight, evenly spaced, from suture to

Plate 58
FIGS. **228a-c,** *Hastula (Impages) inconstans* (Hinds). **228a,** Lectotype British Museum (N.H.) No. 1968243; 33.2 mm. **228b,** Paralectotype same Institution; 32.6 mm. **228c,** Barber's Pt., Oahu, Hawaii, 18 m; 22.0 mm. **229a-c,** *H. (I.) anomala* (Gray). **229a,** Lectotype British Museum (N.H.) No. 1979118; 42.6 mm. **229b,** Middle whorls of lectotype. **229c,** Rabaul, Papua New Guinea; 43.6 mm. **230a-c,** *H. (I.) cernohorskyi* R. D. Burch. **230a,** Holotype Museum Comparative Zoology, Harvard No. 251245; 54.0 mm. **230b,** Lower whorls of holotype. **230c,** Middle whorls of holotype.

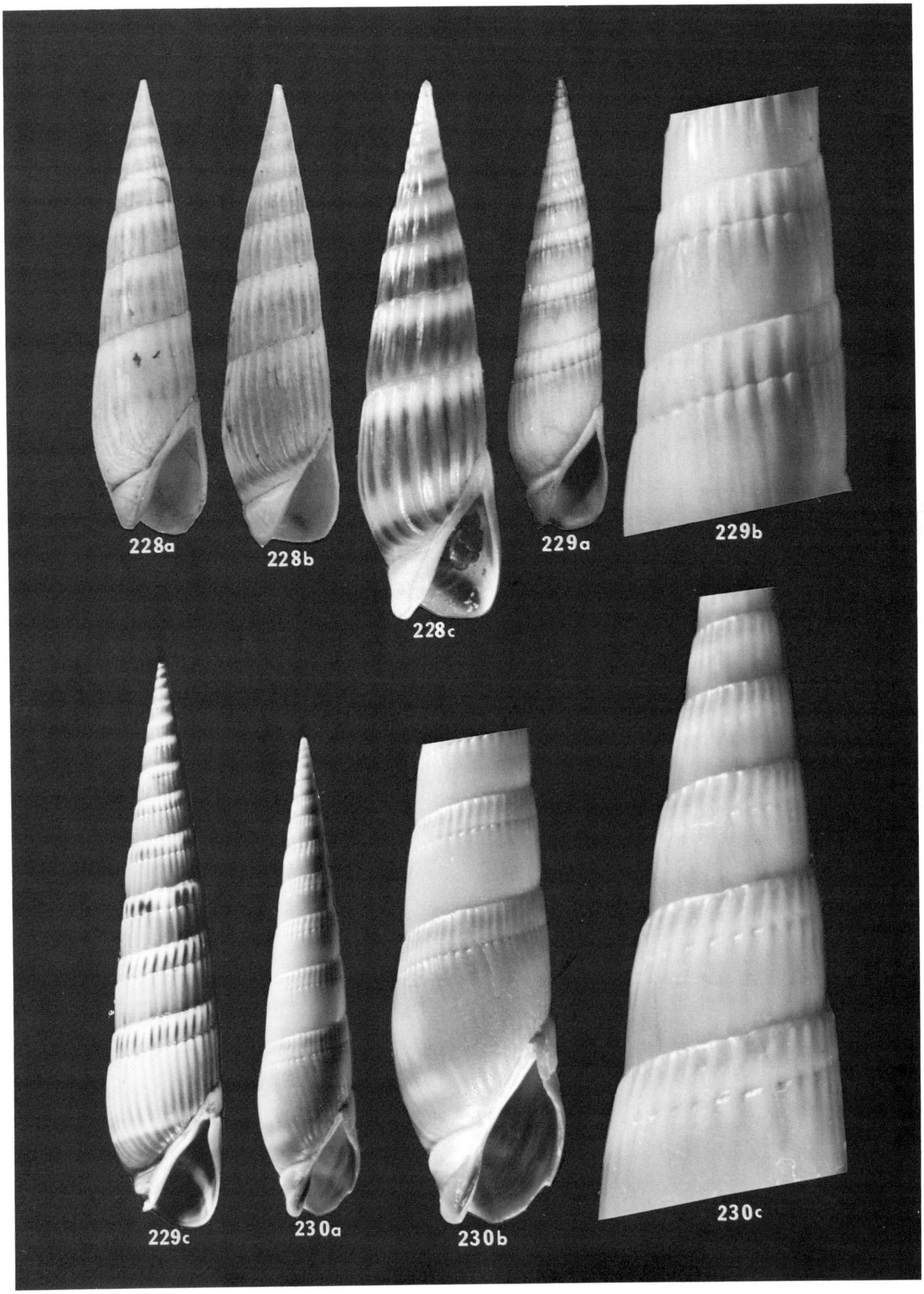

Plate 58. Indo-Pacific *Hastula (Impages).*

suture, about 22 to 27 on penultimate whorl; spiral sculpture absent; body whorl elongate; outer lip almost triangular, flared at anterior, pointed at posterior; columella straight, with ridged parietal callus on mature specimens.
Type locality: "Hawaiian Is."
Distribution: Hawaiian Is.
Type: Lectotype BM(NH) no. 1968243; 33.2 mm.
Discussion: This species has planktonic development, the veligers hatching when one whorl is complete. They add two and a half more whorls before settling. (J. B. Taylor, unpubl. Ph.D. thesis). *H. inconstans* is prey specific, using its radula to inject toxin into the polychaete, *Dispio magna*, on fine sand beaches in the surf. (B. A. Miller, 1970). See *H. stylata* **(227)** for comparison.

229. *Hastula anomala* (Gray, 1834)

(Pl. 58, figs. 229a-c)
(Color pl. D, fig. 7)

1834 *Terebra anomala* Gray, Proc. Zool. Soc. Lond. pt. 2:62; 1844 Hinds in Sowerby, Thes. Conch. 1:180, pl. 44, fig. 97; 1917 Hirase, Terebridae Jap. Emp. p. 33, pl. 2, figs. 6–7; 1984 Aubry, Terebridae pl. 4.
1961 *Noditerebra (Diplomeriza?) anomala* (Gray), Oyama, Venus: Jap. Journ. Malac. 21(2):182.

Description: Shell to 45 mm; color variable, usually of diffused bands of white, gray, and brown fading into one another, occasionally of yellowish gray with some maroon-brown staining between ribs; outline of whorls very slightly convex; protoconch 3½ conical whorls; subsutural band defined by a narrow groove; axial ribs close-set, well-developed, 24 to 30 on the penultimate whorl, ribs fading shortly below subsutural groove on some individuals, continuing to the following suture on others; body whorl with white peripheral stripe; aperture almost triangular with apex at posterior; columella almost straight with heavy parietal callus on mature specimens.
Type locality: None given. Singapore, Malaysia, is here designated as the type locality.
Distribution: From Madagascar to the Philippine Islands; 8 to 14 m.
Type: Lectotype BM(NH) no. 1979118; 42.6 mm.
Discussion: Two of the three specimens in the type lot of this species have continuous ribs on the whorls, while in one specimen the ribs fade out centrally. This species is easily separated from *H. stylata* **(227)** which has no subsutural groove and from *H. cernohorskyi* **(230)** which has a subsutural groove consisting of punctations between the ribs. All three have apertures typical of the subgenus *Impages*. *H. stylata* has finer ribs which may extend to the suture or half way. Those of *H. anomala* are coarser. *H. cernohorskyi* has more evenly spaced ribs which fade at the end of the punctate subsutural groove.

230. *Hastula cernohorskyi* R. D. Burch, 1965

(Pl. 58, figs. 230a-c)

1965 *Hastula (Punctoterebra) cernohorskyi* R. D. Burch, Veliger 7(14):244, pl. 31, fig. 3.

Description: Shell to 58 mm; color light brown or olive green, somewhat lighter anterior to the suture; subsutural band defined by interstician punctations which coalesce to form a weak shallow groove on later whorls; axial sculpture of low, rounded, close-set ribs from suture to suture on early whorls, fading anteriorly until they occupy only the subsutural band; spiral sculpture absent; aperture almost triangular, with flaring outer lip at anterior end, becoming a narrow point at posterior; columella recurved, brown on inner edge, with light parietal callus.
Type locality: "Natadola, Fiji Islands; 18° 6′40″S, 177°19′30″E."
Distribution: Fiji Is. to the Philippines.
Type: Holotype MCZ no. 251245; 54.0 × 7.0 mm.
Discussion: The uniform color and the punctate subsutural groove of this species make it easy to distinguish from *H. anomala* **(229)**. See that species for comparison. This species as named for the original collector and junior author, Walter Cernohorsky, now curator at the Auckland Institute and Museum, New Zealand.

231. *Hastula continua* (Deshayes, 1859)

(Pl. 59, figs. 231a,b)

1859 *Terebra continua* Deshayes, Proc. Zool. Soc. Lond. pt. 27:286; 1860 Reeve, Conch. Icon. 12: pl. 23, sp. 122.

Description: Shell to 33 mm; color tannish cream; outline of whorls straight; protoconch multispiral; subsutural band absent; subsutural constriction weak; axial ribs straight, fading toward anterior of whorl on early whorls, becoming somewhat sinuous from suture to suture on later whorls; spiral sculpture absent; aperture

elongate; columella curved, with weak parietal callus.
Type locality: None given. (Japan is on the lectotype label.)
Distribution: Japan?
Type: Lectotype BM(NH) no. 197898; 32.5 × 7.0 mm.
Discussion: This species is known to us only by the lectotype in the British Museum and two paralectotypes in the Paris Museum. The locality of "Japan" also requires confirmation.

232. *Hastula bacillus* (Deshayes, 1859)

(Pl. 59, figs. 232a-d)
(Color pl. D, figs. 1,2)

1859 *Terebra bacillus* Deshayes, Proc. Zool. Soc. Lond. pt. 27:285; 1860 Reeve, Conch. Icon. 12: pl. 25, sp. 134, figs. 134 a-c; 1984 Aubry, Terebridae pl. 10.
1859 *Terebra apicina* Deshayes, Proc. Zool. Soc. Lond. 27:284.
1859 *Terebra lactea* Deshayes, Proc. Zool. Soc. Lond. pt. 27:285.
1961 *Impages bacillus* (Deshayes), Oyama & Takemura, Moll. Shells, Resources Exploitation Inst. Tokyo 5: *Impages* figs. 8, 9.
1964 *Hastula (Impages) bacillus* (Deshayes), R. D. Burch, Veliger 6(4):212.

Description: Shell to 27 mm; color white with black protoconch, all black, or any combination of white, gray, and black, rarely with inconspicuous spots anterior to the suture; outline of whorls almost straight; protoconch $3^{1/2}$ black conical whorls; a narrow, flat, inconspicuous calloused band occurring posterior to the shallow suture; subsutural groove or band absent; fine crenules extending about $^{2}/_{3}$ of whorl from suture; body whorl smooth except for occasional growth striae and crenules below the suture; aperture quadrate; columella straight, with brown inner edge; siphonal fasciole almost smooth.
Type locality: *H. bacillus*, "Hawaii," error. Singapore, Malaysia, is here designated as the type locality; *T. apicina*, "Singapore"; *T. lactea*, "Hawaii," error.
Distribution: Japan to the Philippines and Papua New Guinea; intertidal.
Type: *H. bacillus* lectotype BM(NH) no. 197982, 22.9 × 5.0 mm; *T. apicina* lectotype BM(NH) no. 197981, 22.7 × 5.0 mm; *T. lactea* lectotype BM(NH) no. 1979100, 19.6 × 4.0 mm.
Discussion: This species seems always to have a blue-black protoconch. The animal is dark gray. (pers. comm. O. K. McCausland, Indonesia). Although it is from a different faunal province, *H. bacillus* has been confused with *H. aciculina* of West Africa (**236**). Refer to that species for comparison.

EASTERN PACIFIC AND ATLANTIC
HASTULA (subgenus) *IMPAGES*

233. *Hastula cinerea* (Born, 1778)

(Pl. 59, figs. 233a-e;
pl. 60, figs. 233f-i)

1778 *Buccinum cinereum* Born, Index rer. Nat. Mus. Caes. Vind. p. 262; 1780 Born, Test. Mus. Caes. Vind. p. 267, pl. 10, figs. 11, 12.
1884 *Terebra luctuosa* Hinds, Proc. Zool. Soc. Lond. pt. 11:157; 1844 Hinds in Sowerby, Thes. Conch. 1:181, pl. 45, fig. 121.
1844 *Terebra laurina* Hinds, Proc. Zool. Soc. Lond. pt. 11:152; 1844 Hinds in Sowerby, Thes. Conch. 1:161, pl. 42, fig. 27.
1850 *Terebra jamaicensis* C. B. Adams, Contrib. Conch. 4:58; 1857 Deshayes, Journ. Conch. 6:80, pl. 5, fig. 9; 1950 Clench & Turner, Occ. Pap., Mus. Comp. Zool. 1(15):297, pl. 39, fig. 22, invalid neoholotype selection; 1968 Morrison, Texas Conch. 4(9):70, lectotype selected, USNM no. 6597.
1860 *Terebra cinerea* (Born), Reeve, Conch. Icon. 12, pl. 9, sp. 35; 1954 Abbott, American Seashells ed. 1:266, pl. 26, fig. g; 1961 Warmke & Abbott, Caribbean Seashells p. 133, pl. 25; 1961 Abbott, Seashells North America p. 164, fig. 3.
1860 *Terebra acuminata* "Gray" in Reeve, Conch. Icon. 12: pl. 26, fig. 143 [non Borson, 1820].
1891 *Terebra exacuminata* Sacco, Moll. Piem. Liguria. pt. 10:18 [nom. nov. pro *Terebra acuminata* Reeve, 1860.
1937 *Terebra (Hastula) cinerea* (Born), M. Smith, E. Coast Mar. Shells p. 133, pl. 51, fig. 2; 1964 Shikama, Sel. Shells World Col. 1:124, pl. 101, fig. 9; 1974 Abbott, American Seashells ed. 2:260, fig. 2847.
1958 *Hastula luctuosa* (Hinds), Keen, Seashells Trop. W. America ed. 1:494, fig. 981; 1968 Morrison, Texas Conch. 4(9):67, fig. L, radula; 1971 Keen, Seashells Trop. W. America ed. 2:686, fig. 1573.
1967 *Hastula cinerea* (Born), Morrison, Texas Conchologist 4(9):70 (radula); 1969 Rudman, Veliger 12(1):59, anatomy; 1975 Rios, Brazilian Mar. Moll. Icon. p. 127, pl. 38, fig. 558; Abbott & Dance, Compendium of Seashells p. 276; 1984 Aubry, Terebridae pl. 10.

Description: Shell to 68 mm; color cream, gray or dark brown, occasionally with a light band and/or dark spots or streaks below the suture; outline of whorls straight; protoconch of 4 to 5 whorls; subsutural band absent; axial sculpture of low riblets extending from suture down 1/3 or 1/2 of whorl; spiral sculpture of rows of microscopic punctations; aperture almost triangular with anterior somewhat flaring and posterior pointed; columella short, straight.

Type locality: *H. cinerea*, none given. Jamaica is here designated as the type locality; *T. luctuosa*, "Gulf of Nicoya, Puerto Portrero, W. coast of America"; *T. laurina*, "Western Africa"; *T. jamaicensis*, "Jamaica"; *T. exacuminata*, none given.

Distribution: West Africa and from Florida and Texas to Brazil; also from Tenacatita Bay, Mexico, to Ecuador; intertidal to 27 m.

Type: *H. cinerea* holotype NHMV, 40.4 × 9.8 mm; *T. luctuosa* lectotype BM(NH) no. 1968246, 38.4 mm; *T. laurina* lectotype BM(NH) no. 1968229, 67.6 mm; *T. jamaicensis* lectotype USNM no. 6597, 30.6 × 7.0 mm; *T. exacuminata* holotype BM(NH) no. 1979117, 29.5 mm.

Discussion: *H. cinerea* paralyzes with its poison and then swallows whole small polychaete worms of the families *Siphonidae* and *Opheliidae* (Marcus & Marcus, 1960).

Morrison (1967) argues that in *H. luctuosa* Hinds, described from the eastern Pacific, the microscopic surface pattern of fine punctations is slightly coarser and the radular teeth slightly longer and narrower and that both dark and light color phases (which he observed living together in Panama Bay) are darker than in the Atlantic species of *H. cinerea*. We find those differences fall within the variability range of this species and that specimens from Brazil may differ equally as much from those living in more northern areas. Both Atlantic and Pacific populations have slender protoconchs of about 4 whorls, which are seldom found intact. Marcus & Marcus figure the protoconch and state that frequently it is broken off, even in living snails. If the protoconch is worn down by sand after breaking, it may look as though the specimen has a protoconch with only 1 1/2 whorls. We believe the two taxa to be conspecific. The long protoconch denotes a long planktonic existence which probably accounts for its wide dispersal. See *H. salleana* **(234)** for comparison.

234 ***Hastula salleana*** (Deshayes, 1859)

(Pl. 60, figs. 234a,b)

1859 *Terebra sallaeana* Deshayes, Proc. Zool. Soc. Lond. pt. 17:287; 1984 Aubry, Terebridae pl. 13.

1968 *Hastula salleana* (Deshayes), Morrison, Texas Conchologist, 4(9):68, fig. S, radula and microsculpture; 1975 Rios, Brazilian Mar. Moll. Icon. p. 127, pl. 28, fig. 560; 1971 Andrews, Sea Shells of the Texas Coast p. 125, text fig.

1974 *Terebra (Hastula) salleana* Deshayes, Abbott, American Seashells ed. 2:260, fig. 2848.

Description: Shell to 32 mm; color bluish gray to brownish gray, usually with a light stripe at periphery of body whorl, followed by a darker area; outline of whorls slightly concave; protoconch of 3 1/2 brown conical whorls; subsutural groove absent; axial ribs evenly spaced, well formed, occupying about 1/3 of whorl; spiral rows of moderately wide spaced microscopic punctations throughout shell; aperture quadrate; columella straight.

Type locality: "Mexico."

Distribution: From North Carolina and the Gulf of Mexico to Brazil; mostly in surf zone.

Type: Lectotype BM(NH) no. 197911; 25.3 × 5.0 mm.

Discussion: Deshayes' original spelling of the epithet was *sallaeana*, but since he named it for A. Sallé (1820–1896), the name was correctly emended to *salleana*. This species has heavier, coarser subsutural riblets, more sparse and wide spaced microscopic punctations, and is generally a smaller species, with much smaller radular teeth than *H. cinerea* **(233)**. *H. maryleeae* **(235)** differs from both *H. cinerea* and *H. salleana* by the absence of spiral punctations.

Plate 59

FIGS. **231a, b,** *Hastula (Impages) continua* (Deshayes). **231a,** Paralectotype Museum National d'Histoire Naturelle, Paris; 30.4 mm. **231b,** Middle whorls of paralectotype. **232a-d,** *H. (I.) bacillus* (Deshayes). **232a,** Lectotype British Museum (N.H.) No. 197982; 22.9 mm. **232b,** Paralectotype same Institution; 22.0 mm. **232c,** Lectotype of *T. apicina* Deshayes, British Museum (N.H.) No. 197981; 22.7 mm. **232d,** Lectotype of *T. lactea* Deshayes, British Museum (N.H.) No. 1979100; 19.6 mm. **233a-e,** *H. (I.) cinerea* (Born). **233a,** Holotype Natural History Museum, Vienna; 40.4 mm. **233b,** Paralectotype of *T. luctuosa* Hinds, British Museum (N.H.) No. 1968246; 33.0 mm. **233c,** Lectotype of *T. laurina* Hinds, British Museum (N.H.) No. 1968229; 67.6 mm. **233d,** Lectotype of *T. jamaicensis* C. B. Adams, Natural Museum Natural History, Washington No. USNM-6597; 30.6 mm. **233e,** Holotype of *T. exacuminata* Sacco (and *T. acuminata* Reeve), British Museum (N.H.) No. 1979117; 29.5 mm.

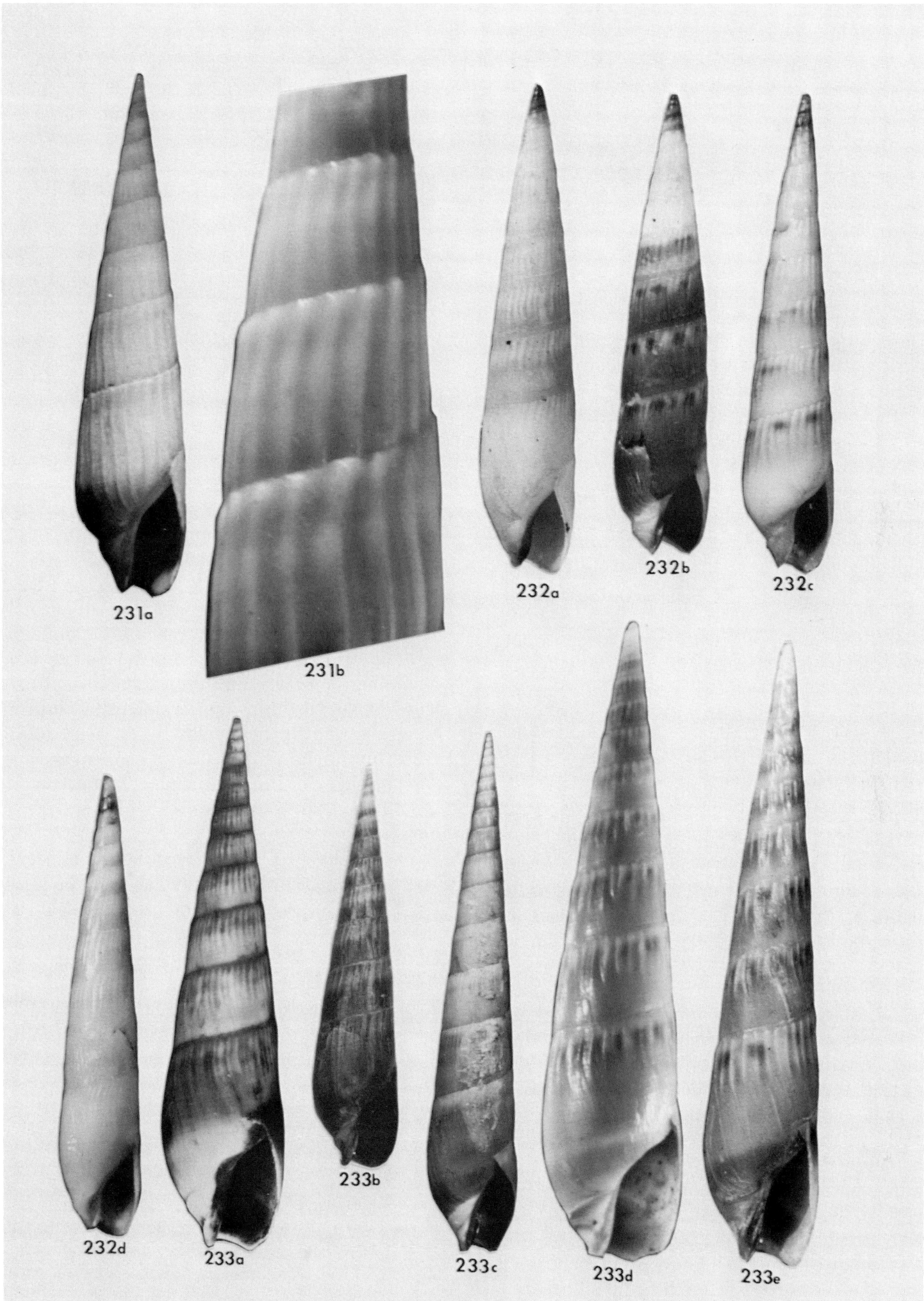

Plate 59. Indo-Pacific and Western Atlantic *Hastula (Impages)*.

235. *Hastula maryleeae* R. D. Burch, 1965

(Pl. 60, figs. 235a,b,c)

1965 *Hastula (Hastula) maryleeae* R. D. Burch, Veliger 7(4):242, pl. 31, fig. 4.

1968 *Hastula maryleeae* R. D. Burch, Morrison, Texas Conch. 14(9):67, fig. m, radula and microsculpture. 1971 Andrews, Sea Shells of the Texas Coast p. 124, text. fig.

1969 *Terebra tobagoensis* Usticke, A Supplementary Listing of New Shells: 26, pl. 5, fig. 1029.

1974 *Terebra (Hastula) maryleeae* (R. D. Burch), Abbott, American Seashells 2nd ed.: 260, fig. 2849.

Description: Shell to 25 mm; color shiny white to dark purple-brown with or without a row of brown spots and/or a dark band at the suture and at the periphery of the body whorl; outline of whorls slightly convex; protoconch of 2 whorls; subsutural groove absent; axial ribs usually rather coarse, on only the posterior third of the whorl, sometimes being replaced by a continuous swelling that forms a wide convex band, some individuals with fine, almost obsolete ribs persisting to the following suture; spiral sculpture absent; aperture with slight flair; columella short, faintly curved, with light parietal callus; the operculum is translucent yellow-orange.
Type locality: *Hastula maryleeae*, "Surfside Beach, Freeport, Texas (28°57′N, 95°38′W)"; *Terebra tobagoensis*, Tobago Is., Trinidad.
Distribution: From Galveston, Texas, to the Dominican Republic; usually in surf line.
Type: *Hastula maryleeae* holotype MCZ no. 251235; 23.7 × 5.0 mm; *Terebra tobagoensis* lectotype AMNH no. 195453.
Discussion: The aperture of this species is less flared at the bottom and hence looks less like those of the typical *Impages* group than the others that we have placed in this subgenus. Its hypodermic-like radular teeth are extremely small, about 0.153 mm long. This species was collected more than a century ago in both Texas and Vera Cruz but was not recognized as a distinct species until 1965. The animal is white to cream color (Sally Kaicher pers. comm.) The shells in the northern part of the range tend to be more slender and have heavy ribs on the upper portion of the whorl, causing a bulge in outline. Those farther south usually have weak, slender ribs, often from suture to suture. Both forms may be found in a single population. It was named after Burch's wife, Mary Lee.

Clench and Turner (1950) erroneously selected a specimen of this species as the neotype of *T. jamaicensis* C. B. Adams, whereas the emended lectotype selection by Morrison (1968) places *T. jamaicensis* C. B. Adams in the synonymy of *H. cinerea* (Born). Refer to *H. salleana* (**234**) for comparison.

236. *Hastula aciculina* (Lamarck, 1822)

(Pl. 61, figs. 236a-c)
(Color pl. D, fig. 3)

1822 *Terebra aciculina* Lamarck, Hist. Nat. Anim. S. Vert. 7:290; 1839 Kiener, Spéc. Gén. Icon. Coq. Viv. 9:18, pl. 7, fig. 13a, b; 1860 Reeve, Conch. Icon. 12: pl. 23, sp. 121a, c-f. [not fig. 121b].

1844 *Terebra micans* Hinds, Proc. Zool. Soc. Lond. pt. 11:157.

1859 *Terebra adansoni* Deshayes, Proc. Zool. Soc. Lond. pt. 27:291.

1983 *Hastula aciculina* (Lamarck), Bouchet, Boll. Malac. 18(9–12):204, figs. 14, 54–55; 1984 Aubry, Terebridae pl. 14.

Description: Shell to 39 mm; color white, gray, black, yellow or in combination, usually with a translucent look to the early whorls, often with a dark stripe showing through translucent enamel; outline of whorls flat; protoconch 2½ shiny embryonic whorls; axial plications fine, numerous, beginning at suture and continuing to flat callused band posterior to the following suture, band showing up more clearly on worn specimens; plications usually fading before reaching the periphery of the elongate body whorl; outer lip sturdy; columella recurved; siphonal fasciole quite smooth.
Type locality: *H. aciculina*, none. Senegal, West Africa, is here designated as the type locality; *T. micans*, none; *T. adansoni*, "Senegal, W. Africa."
Distribution: Mauritania to Angola, West Africa, intertidal and subtidal.
Type: *H. aciculina* lectotype MHNG no. 1102/

Plate 60
FIGS. **233f-i,** *Hastula (Impages) cinera* (Born). **233f,** La Playita, Boca del Yuma, Dominican Republic; 29.8 mm. 233g, Lower whorls. **233h,** El Dorado Bay, Manzanillo, Mexico; 27.6 mm (*luctuosa* form). **233i,** Middle whorls. **234a, b,** *H. (I.) salleana* (Deshayes). **234a,** Lectotype British Museum (N.H.) No. 197911; 25.3 mm. **234b,** Alvarado, Veracruz, E.Mexico; 27.3 mm. **235a-c,** *H. (I.) maryleeae* R. D. Burch. **235a,** Holotype Museum Comparative Zoology, Harvard No. 251235; 23.7 mm. **235b,** Dominican Republic; 12.4 mm. **235c,** Holotype of *Terebra tobagoensis* Usticke, AMNH No. 195453; 13.8 × 3.8 mm.

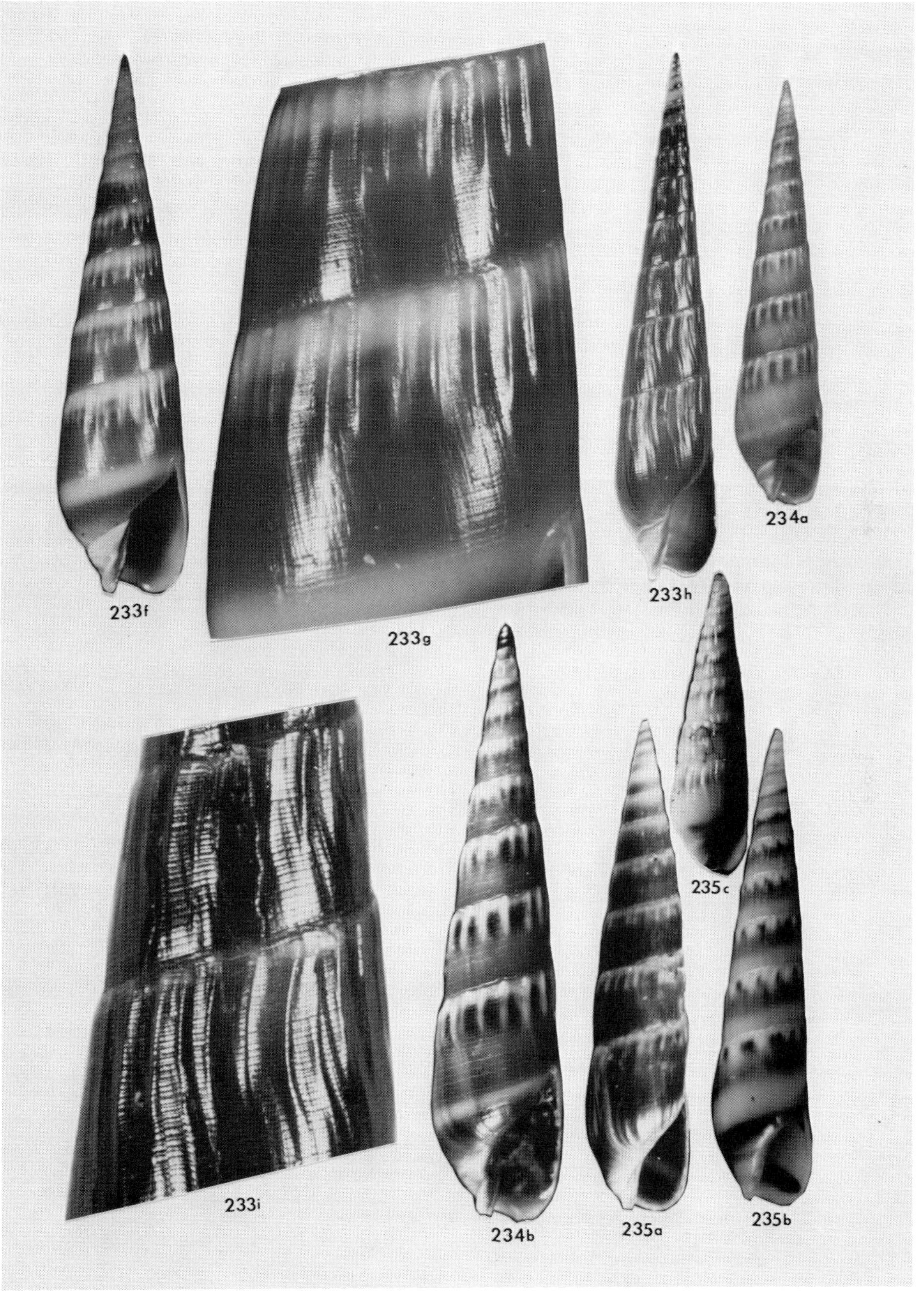

Plate 60. Western Atlantic *Hastula (Impages)*.

31/4, 33.8 mm; *T. micans* lectotype BM(NH) no. 1968248, 30.4 mm; *T. adansoni* lectotype MNHNP, 39.0 mm.

Discussion: The specimens of this species in the Lamarck collection illustrated by Kiener (1839) under "variety B" are white and somewhat worn. Had they been one of the other color forms with the stripe showing through, the species would not have remained undetermined for so long. When this species becomes worn, the unsculptured calluslike band posterior to the suture shows up plainly, although otherwise it is scarcely noticeable. *H. aciculina* has the smallest operculum of any species we have examined. It sometimes is confused with *H. bacillus* (**232**), a western Pacific species. They are similar in coloring and sculpture, but *H. bacillus* has a slightly more slender shell with a blue-black protoconch, a brown inner edge to the columella, and is found in a different faunal province.

Genus **Duplicaria** Dall, 1908

Type species, *Duplicaria duplicata* (Linnaeus, 1758)

The shell characteristics of the species we place in this genus are the following: shell axially ribbed; subsutural groove cutting through ribs; suture well marked; no spiral sculpture or very weak, macroscopic spiral lines.

We are using the generic division based entirely on shell characters for the convenience of identification. Some of the species listed under *Duplicaria* have no radula, others have a radula which differs from that of *Hastula*.

Without being able to correlate shell characters with anatomical characters, there may be species that could fall in more than one of our divisions. *D. raphanula* is one of those. It could as well be assigned to the genus *Hastula*, although recent authors have assigned it to *Duplicaria*.

Dall (1908, p. 125) and Bartsch (1923, p. 63) both erroneously attributed Lamarck as the author of *duplicata*.

237. *Duplicaria duplicata* (Linnaeus, 1758)

(Pl. 61, figs. 237a-d)
(Color pl. F, figs. 11–13)

1758 *Buccinum duplicatum* Linnaeus, Syst. Nat. ed. 10:742; 1956 Dodge, Bull. Amer. Mus. Nat. Hist. 111(3):230.

1839 *Terebra lamarckii* Kiener, Spéc. Gén. Icon. Coq. Viv. 9:30, pl. 9, fig. 19. [non Defrance, 1829].

1857 *Terebra reevei* Deshayes, Journ. Conchyl. 6:88, pl. 4, fig. 14.

1860 *Terebra duplicata* (Linnaeus), Reeve, Conch. Icon. 12: pl. 1, figs. 3a, b; 1917 Hirase, Terebridae Jap. Emp. p. 32, pl. 18, fig. 126; 1979 Hinton, Shells New Guinea & Cent. Indo-Pacific p. 28, pl. 24, fig. 10.

1923 *Myurella (Myurellisca) duplicatoides* Bartsch, Nautilus 37(2):64.

1908 *Duplicaria duplicata* Lamarck, Dall, Nautilus 21(11):125.

1961 *Noditerebra (Diplomeriza) duplicata* (Linnaeus), Oyama & Takemura, Moll. Shells, Resources Exploitation Inst. 5: *Noditerebra* pl. 2, figs. 1, 4.

1964 *Terebra (Diplomeriza) australis* (E. A. Smith) Shikama, Sel. Shells World Col. p. 124, pl. 101, fig. 11 [non E. A. Smith, 1873].

1964 *Terebra (Diplomeriza) duplicata* (Linnaeus), Shikama, Sel. Shells World Col. p. 124, pl. 101, fig. 12.

1967 *Diplomeriza duplicata* (Linnaeus), Habe & Kosuge, Shells World Col. 2:100, pl. 39, fig. 16.

1967 *Duplicaria duplicata* (Linnaeus), Cernohorsky, Marine Shells Pacific 1:208, pl. 53, fig. 395; 1982 Abbott & Dance, Compendium of Seashells p. 274; 1984 Aubry, Terebridae pls. 10, 15.

Description: Shell to 89 mm; color variable, white, cream, beige, gray, dark brown, beige with brown subsutural band, gray with brown blotches or gray mottled with cream and decorated with a row of small red-brown spots on each whorl; outline of whorls flat; protoconch of 1½ small whorls; subsutural band flat, slightly indented, with sharp ribs on early whorls, flat

Plate 61

FIGS. **236a-c,** *Hastula (Impages) aciculina* (Lamarck). **236a,** Lectotype Museum d'Histoire Naturelle, Geneva No. 1102/31/4; 33.8 mm. **236b,** Lectotype of *T. adansoni* Deshayes, Museum National d'Histoire Naturelle, Paris; 39.0 mm. **236c,** Senegal, West Africa; 26.9 mm. **237a-d,** *Duplicaria duplicata* (Linnaeus). **237a,** Probable type Linnean collection, London; 54.3 mm. **237b,** Lectotype of *T. reevei* Deshayes, British Museum (N.H.) No. 197951; 50.9 mm. **237c,** Holotype of *Myurella duplicatoides* Bartsch, National Museum Natural History, Washington No. USNM-348285; 55.1 mm. **237d,** Pathological specimen, Institut Royal Sciences Naturelles Belgique, Brussels; 47.7 mm. **238,** *D. concolor* (E. A. Smith). Holotype British Museum (N.H.) No. 1858.5.12.252.; 22.0 mm. **239a, b,** *D. bernardii* (Deshayes). **239a,** Lectotype British Museum (N.H.) No. 197955; 57.4 mm. **239b,** Middle whorls of paralectotype.

Plate 61. West African *Hastula (Impages)* and Indo-Pacific *Duplicaria*.

later, band defined by moderately narrow groove; axial ribs broad, sharp, becoming flat and broad with interspaces being merely thin axial grooves on mature whorls, ribs on band flattening first; aperture quadrate; columella almost straight; siphonal fasciole striate, with very broad keel.
Type locality: *D. duplicata*, "Indian Ocean." Ceram Is., Moluccas, Indonesia, is here designated as the type locality; *T. lamarckii*, none; *T. reevei*, "Moluccas," Indonesia; *M. duplicatoides*, "Ceylon."
Distribution: From the Red Sea to South Africa, Japan, and the Solomon Islands.
Type: *D. duplicata* probable type in Linnean Society, London coll., 54.3 mm; *T. lamarckii* holotype MNHG no. 152/74, 79.9 × 14.6 mm; *T. reevei* lectotype BM(NH) no. 197951, 50.9 × 10.9 mm; *M. duplicatoides* holotype USNM no. 348285.
Discussion: These animals are found on clean sand, particularly on the sides of channels, but not where the sand flats have been stabilized by sea grasses. This is the type species of the genus *Duplicaria* Dall, 1908, and also for the replacement name *Diplomeriza* Dall, 1919, a name given when he mistakenly thought that *Duplicaria* was preoccupied. It is also the type species for the subgenus of *Myurellisca* Bartsch, 1923 because *Terebra (Myurellisca) duplicatoides* Bartsch is *Duplicaria duplicata*. This species may be distinguished from all other *Duplicaria* by the mature whorls having broad, flat ribs separated by narrow grooves. The sharp angular ribs of the early whorls may not flatten out before 12 whorls are completed. For this reason young specimens frequently are mistaken for another species. The gray color form, sometimes mottled with cream and decorated with a row of small red-brown dots or a thin red-brown spiral line posterior to the suture, described by Kiener as *D. lamarckii*, appears to be found only in the Persian Gulf, Gulf of Oman and Red Sea area. We can find no consistent differences in shell morphology between it and other color forms of *D. duplicata*.

238. *Duplicaria concolor* (E. A. Smith, 1873)

(Pl. 61, fig. 238)

1873 *Terebra concolor* E. A. Smith, Ann. & Mag. Nat. Hist., ser. 4, 11:65; 1966 Cernohorsky & Jennings, Veliger 9(1):62, pl. 6, fig. 45 (illustrated holotype).

Description: Shell to 22 mm; color whitish; outline of whorls slightly convex; subsutural band moderately broad, axially ribbed, defined by narrow groove; axial ribs narrower than interspaces, about 18 on penultimate whorl; spiral sculpture absent; aperture quadrate; columella almost straight.
Type locality: None given.
Distribution: Unknown.
Type: Holotype BM(NH) no. 1858.5.12.252; 22.0 × 6 mm.
Discussion: No specimens have been found to match the holotype, which is in poor condition. This may prove to be a juvenile specimen of *D. duplicata* (**237**).

239. *Duplicaria bernardii* (Deshayes, 1857)

(Pl. 61, figs. 239a,b; pl. 62, fig. 239c)
(Color pl. F, fig. 8)

1857 *Terebra bernardii* Deshayes, Journ. Conchyl. 6:84, pl. 4, fig. 10; 1860 Reeve, Conch. Icon. 12: pl. 13, sp. 56; 1984 Aubry, Terebridae pl. 13.
1859 *Terebra addita* Deshayes, Proc. Zool. Soc. Lond. pt. 17:293.
1964 *Duplicaria (Duplicaria) bernardi* (Deshayes) R. D. Burch, Veliger 6(4):211.
1978 *Duplicaria bernardi* Deshayes, Hinton, Guide Austral. Shells pl. 56, fig. 3.

Description: Shell to 60 mm; color brownish gray and ivory blended in broad bands, often with ivory ribs and ivory stripe on body whorl, with protoconch and early whorls of teleoconch brown; outline of whorls convex on early whorls, becoming slightly convex on later ones; subsutural band narrow, noded, constricting, marked by shallow groove; axial ribs inflated on early whorls, straighter, about equal to interspaces, later; spiral sculpture absent; aperture brown with white stripe showing through, moderately quadrate; columella recurved.
Type locality: *D. bernardii*, "East Coast of Australia"; *T. addita*, "Tasmania," error.
Distribution: From N. W. Australia to southern Queensland, Australia.
Type: *D. bernardii* lectotype BM(NH) no. 197955, 57.4 × 14 mm; *T. addita* holotype BM(NH) no. 197978, 32.3 × 7.0 mm.
Discussion: The distinctive brownish gray and ivory color pattern is consistent in this species, although the areas of the two colors vary, with either predominating. It is easily distinguishable from *D. spectabilis* (**242**) which has a somewhat similar, though deeper and richer color, a much

deeper subsutural groove, and axial ribs which jut out and form a small node below the groove. *Voluta wisemani* Brazier has been observed feeding on this species. The species was named after A. Bernardi, an editor of the French *Journal de Conchyliologie.*

240. *Duplicaria jukesi* (Deshayes, 1857)

(Pl. 62, figs. 240a,b)

1857 *Terebra jukesi* Deshayes, Journ. Conchyl. 6:95, pl. 4, fig. 9; 1860 Reeve, Conc. Icon. 12, pl. 20, sp. 97.

Description: Shell to 34 mm; color cream to light gray with light ribs and a narrow light band at periphery of body whorl; outline of whorls extremely convex in early whorls becoming slightly convex later; protoconch of 1½ whorls; subsutural band convex, narrow on early whorls, decorated with small round nodes, becoming somewhat broader, with nodes lengthening into ribs, band defined by deep groove; axial ribs sharp, evenly spaced, narrower than interspaces; spiral sculpture absent; aperture quadrate; columella curved.
Type locality: "Port Essington, Northern Territory, Australia."
Distribution: Australia.
Type: Lectotype BM(NH) no. 197964; 32.8 × 8.0 mm.
Discussion: D. concolor **(238)** and *D. crakei* **(241)** may be separated from this species by having flat early whorls with straight, not arcuate ribs and flatter subsutural bands. *D. crakei* may be separated from *D. concolor* by its narrower subsutural band, sharper ribs, and pattern of two colors. *D. bernardi* Deshayes, also has inflated early whorls but lacks the deep subsutural groove and convex subsutural band. Deshayes named this species after the English shell collector, J. B. Jukes (1811–1869).

241. *Duplicaria crakei* R. D. Burch, 1965

(Pl. 62, fig. 241)

1965 *Duplicaria (Duplicaria) crakei* R. D. Burch, Veliger 7(4):245, pl. 31, fig. 1.
1971 *Duplicaria crakei* R. D. Burch, Wilson & Gillett, Austral. Shells p. 158, pl. 106, fig. 4.
1984 *Terebra crakei* (Burch, 1965), Aubry, Terebridae pl. 12.

Description: Shell to 25 mm; color shiny yellowish cream with lavender-gray base of body whorl and subsutural band, or dark gray with lighter gray; outline of whorls somewhat turreted, early whorls flat; protoconch of 2½ small amber whorls; subsutural band slightly convex, ribbed, defined by deep groove cutting through ribs; axial ribs narrow, straight, sharp, from suture to suture, about 14 to 17; spiral sculpture absent; aperture moderately quadrate, with base color of body whorl showing through outer lip as a brown stripe; columella recurved.
Type locality: "Cable Beach, Broome, W. Australia, 18°00′S, 122°15′E."
Distribution: Western Australia, intertidal.
Type: Holotype MCZ no. 251241; 23.0 × 5.5 mm.
Discussion: This is a striking, extremely glossy shell, usually with the subsutural band of a contrasting color, the grays usually having a lavender tone. While the color may vary, the sculpture is constant. It was named after Ted Crake who collected it in Australia.

242. *Duplicaria spectabilis* (Hinds, 1844)

(Pl. 62, figs. 242a-e;
pl. 63, figs. 242f-h)
(Color pl. E, fig. 24)

1844 *Terebra spectabilis* Hinds, Proc. Zool. Soc. Lond. pt. 11:150; 1844 Hinds in Sowerby, Thes. Conch. 1:157, pl. 44, figs. 88, 89; 1984 Aubry, Terebridae pl. 2.
1859 *Terebra geminata* Deshayes, Proc. Zool. Soc. Lond. pt. 27:296.
1860 *Terebra gracilis* Reeve, Conch. Icon. 12: pl. 24, fig. 131.
1910 *Terebra remanalva* Melvill, Ann. & Mag. Nat. Hist., 8th ser. 6(31):12, pl. 2, fig. 21.
1898 *Terebra edgarii* Melvill, Mem. Proc. Manchester Lit. Phil. Soc. 42(4):8, pl. 2, fig. 12.
1966 *Diplomeriza spectabilis* (Hinds), Habe & Kosuge, Shells World Col. 2:100, pl. 39, fig. 15.

Description: Shell to 50 mm; color brown with white subsutural band, and a white stripe posterior to the suture and at the periphery of the body whorl; outline of whorls convex; protoconch of 3 whorls; subsutural band convex with elongate nodes on early whorls, lengthening to ribs later, defined by narrow groove; axial ribs on early whorls inflated, wide spaced, becoming straight, evenly spaced, protruding at their posterior ends to form whitish nodes, 16 to 20 on penultimate whorl; spiral striae microscopic on most specimens; aperture elongate-quadrate; columella recurved.
Type locality: *D. spectabilis*, "Guinea," error, "And Sumatra." Sumatra, Indonesia is here designated as the type locality; *T. geminata*, "Cape

Natal," South Africa; *T. edgarii*, "Karachi," Arabian Sea; *T. gracilio*, "Africa"; *T. remanalva*, "Persian Gulf, Bundo, Abbas, and Bushire."

Distribution: Persian Gulf to South Africa and Indonesia; to 100 m.

Type: *D. spectabilis* lectotype BM(NH) no. 1968174, 49.7 × 9.2 mm; *T. geminata* lectotype BM(NH) no. 197995, 30.3 mm; *T. edgarii* lectotype BM(NH) no. 1898.7.5.32, 22.0 mm; *T. gracilis* holotype BM(NH) no. 1874.10.29.2, 22.0 mm; *T. remanalva* holotype BM(NH) no. 1911.6.21.11, 33.8 × 6.6 mm.

Discussion: This species may be compared with several other *Duplicaria*. *D. dussumierii* **(243)** has a larger shell with more numerous and more closely set axial ribs and a broader subsutural band. *D. jukesi* **(240)** lacks nodes at the posterior ends of the ribs and has almost flat whorls and subsutural band. *D. mozambiquensis* **(248)** has similar sculpture, but it lacks the color pattern, has a mamillate protoconch, and does not have direct development as this species has. *T. gracilis* Reeve, 1860 (non Lea, 1833) was erroneously included in the synonymy of *T. mozambiquensis* **(248)** in the original description. *T. edgarii*, a synonym, was named for Edgar A. Smith, once curator of Mollusca at the British Museum.

243. *Duplicaria dussumierii* (Kiener, 1839)

(Pl. 63, figs. 243a,b)
(Color pl. F, fig. 2)

1839 *Terebra dussumierii* Kiener, Spéc. Gén. Icon. Coq. Viv. 9:31, pl. 8, fig. 17; 1860 Reeve, Conch. Icon. 12: pl. 2, fig. 7; 1917 Hirase, Terebridae Jap. Emp. p. 17, pl. 8, fig. 127; 1974 Dance, Encycl. Shells, p. 217, center column, upper fig. left; 1982 Abbott & Dance, Compendium of Seashells p. 275; 1984 Aubry, Terebridae pl. 11.

1961 *Noditerbra (Diplomeriza) dussumieri* (Kiener), Oyama & Takemura, Moll. Shells, Resources Exploitation Inst. 5: *Noditerebra* (1) pl., figs. 9, 17.

1964 *Terebra (Diplomeriza) dussumieri* Kiener, Shikama, Sel. Shells World Col. 1:125, fig. 7.

1967 *Diplomeriza dussumieri* (Kiener), Habe & Kosuge, Shells World Col. 2:100, pl. 39, fig. 22.

1975 *Hastula nitida* Okutani, Gakken Illust. Nat. Encyclo. Shells Japan p. 155, fig. 11 [not *T. nitida* Hinds].

1978 *Duplicaria dussumieri* (Kiener), Hinton, Guide Austral. Shells pl. 56, fig. 1.

Description: Shell to 100 mm; color brown with grayish white or cream on anterior part of subsutural band and a narrow cream stripe posterior to suture, where it continues on body whorl at periphery into aperture, ribs usually lighter; outline of whorls almost flat, turreted, early whorls convex; subsutural band flat ribbed, defined by deep suture and deep groove; axial ribs numerous, crowded, closely set, straight, about 26 to 47 on the penultimate whorl; spiral sculpture absent; aperture elongate; columella straight; siphonal fasciole with several enlarged scalelike striations.

Type locality: "China."

Distribution: From Indonesia to China and Japan.

Type: Lectotype MHN Geneva no. 1152/75; 98.8 × 20.0 mm.

Discussion: This large handsome shell should be compared with those of two other species. *D. kirai* **(244)** has a similar size, shape and color but can be separated by its narrow subsutural band with round nodes, widely spaced ribs, and inflated, very convex early whorls, none of which occur on *D. dussumieri*. *D. hiradoensis* **(249)** has shorter whorls, fewer ribs, a monochromatic buff color and a subsutural space with just a hint of a narrow shallow groove (whereas *D. dussumierii* has a deep groove).

244. *Duplicaria kirai* (Oyama, 1962)

(Pl. 63, fig. 244)

1961 *Noditerebra (Diplomeriza) kirai* Oyama, Oyama & Takemura, Moll. Shells, Resources Exploitation Inst. 5: Noditerebra (1) pl., figs. 3, 4, 12 (nomen nudum); 1961 Oyama, Venus 21 (4): 446 (nomen nudum).

1962 *Noditerebra (Diplomeriza) kirai* Oyama, Venus, Jap. Journ. Malac. 22(1):57, text figs. 3, 4.

Description: Shell to 84 mm; color brown with beige axial ribs, beige between nodes on subsutural band, and a beige stripe posterior to suture and at periphery of body whorl; outline of whorls convex in early whorls, slightly convex later; protoconch of $1^1/_2$ whorls; subsutural band

---

Plate 62

FIGS. **239c,** *Duplicaria bernardii* (Deshayes). Holotype of *Terebra addita* Deshayes, British Museum (N.H.) No. 197978; 32.3 mm. **240a, b,** *D. jukesi* (Deshayes). **240a,** Paralectotype British Museum (N.H.) No. 197964; 34.2 mm. **240b,** Upper whorls of lectotype. **241,** *D. crakei* R. D. Burch. Holotype Museum Comparative Zoology, Harvard, No. 251241; 23.0 mm. **242a-e,** *D. spectabilis* (Hinds). **242a,** Lectotype British Museum (N.H.) No. 1968174; 49.7 mm. **242b,** Middle whorls. **242c,** Upper whorls of lectotype; **242d,** Holotype of *T. edgarii* Melvill, British Museum (N.H.) No. 1889.7.5.32; 19.0 mm. **242e,** Lectotype of *T. geminata* Deshayes, British Museum (N.H.) No. 197995; 30.3 mm.

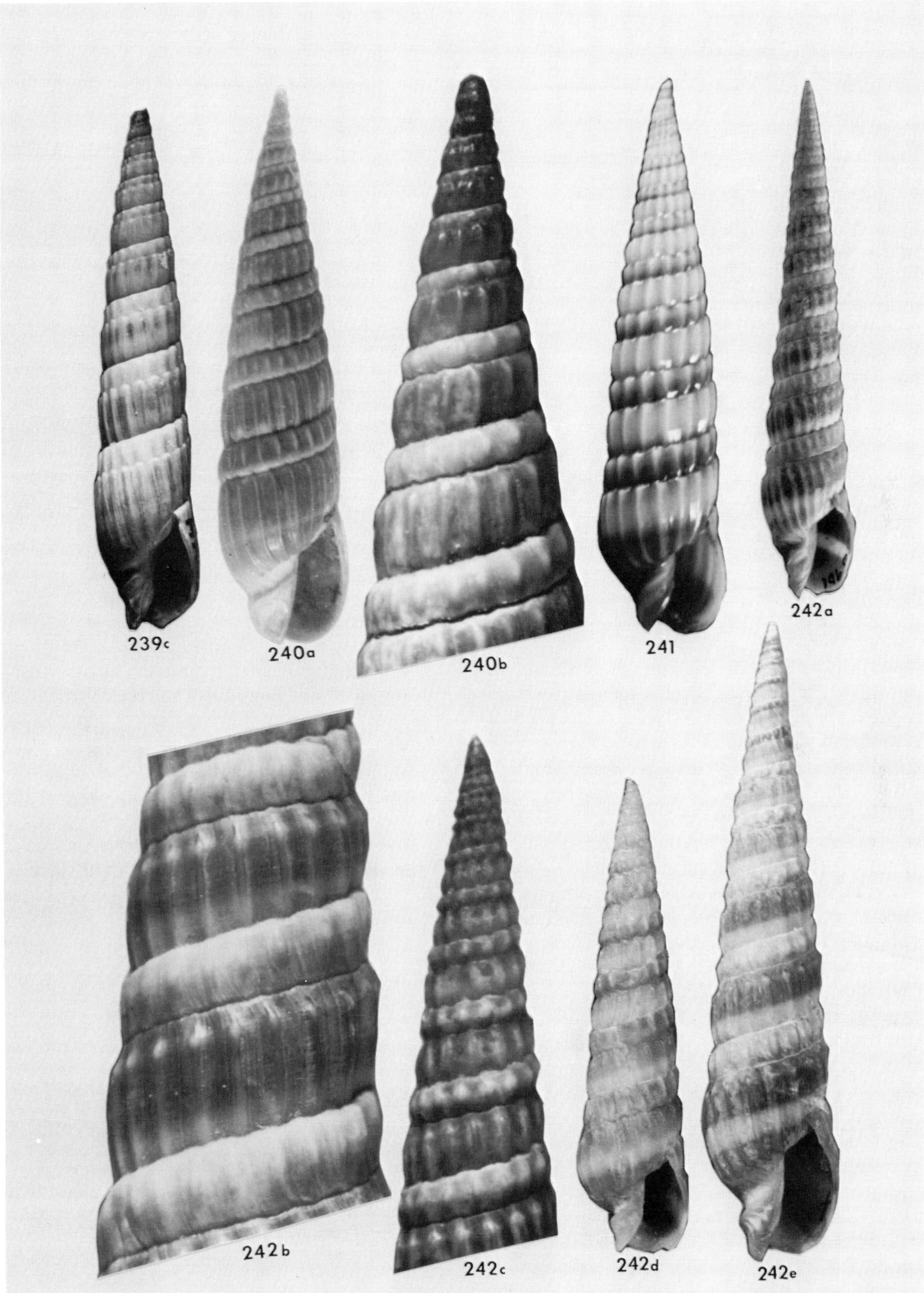

Plate 62. Indo-Pacific *Duplicaria*.

noded, defined by deep suture and subsutural groove in center of moderately broad depression; axial ribs unevenly spaced, narrower than interspaces; spiral threads in interspaces faint, lightly crossing ribs; aperture quadrate; columella curved.

Type locality: "Off Mie Prefecture, Japan, 100 m."

Distribution: Japan, subtidal.

Type: ? in collection of Oyama; 57.0 × 11.0 mm.

Discussion: Oyama mentioned this shell in the Japanese journal, *Venus*, a year before he described it in the same journal. Refer to *D. dussumieri* **(243)** for comparison.

245. *Duplicaria evoluta* (Deshayes, 1859)

(Pl. 63, figs. 245a,b)

1859 *Terebra evoluta* Deshayes, Proc. Zool. Soc. Lond. pt. 27:292; 1860 Reeve, Conch. Icon. 12:13, sp. 55; 1917 Hirase, Terebridae Jap. Emp. p. 16, pl. 4, fig. 57.

1961 *Noditerebra (Diplomeriza) evoluta* Deshayes, Oyama & Takemura, Moll. Shells, Resources Exploitation Inst. 5; *Noditerebra* pl. 1, figs. 5, 13.

1963 *Terebra (Diplomerize) evoluta* Deshayes, Shikama, Sel. Shells World Col. p. 124, pl. 101, fig. 13.

Description: Shell to 57 mm; color cream to brown or almost black; outline of early whorls straight, becoming convex later; protoconch of 1½ inflated whorls; subsutural band narrow with round nodes on early whorls, becoming progressively more elongate, band defined by broad shallow space between the subsutural nodes on band and the following ribs; axial ribs on remainder of whorl jutting out below subsutural space; ribs contiguous with nodes on band; interspaces smooth, wider than ribs; aperture moderately elongate; columella almost straight.

Type locality: "Japan."

Distribution: West Pacific.

Type: Lectotype BM(NH) no. 197992; 56.2 mm.

Discussion: The identity of this species and that of *D. australis* **(246)** have long been misunderstood and confused, as they are about the same size and shape and have the same color range. This species may be separated from *D. australis* by its fewer, more widely spaced ribs, its broad shallow space in place of a deep groove, and its more elongate aperture. Although smaller, *D. mozambiquensis* **(248)** is also somewhat similar, but it has a punctate subsutural groove, extremely inflated early whorls of the teleoconch, very rounded nodes on the subsutural band, and wider spaced and more arcuate ribs.

246. *Duplicaria australis* (E. A. Smith, 1873)

(Pl. 63, fig.246a;
pl. 64, fig. 246b)
(Color pl. E, fig. 4;
color pl. F, fig. 9)

1873 *Terebra australis* E. A. Smith, Ann. & Mag. Nat. Hist. ser. 4(11):264; 1938 Allan, Austral. Mus. Mag. 6(10):328, text fig.

1950 *Diplomeriza australis* E. A. Smith, Allan, Austral. Seashells p. 195, pl. 18, fig. 18.

1967 *Duplicaria evoluta* (Deshayes), Cernohorsky, Mar. Shells Pacific 1:208, pl. 53, fig. 396; 1971 Wilson & Gillett, Austral. Shells p. 156, pl. 105, figs. 10, 10a-c; 1978 Hinton, Guide Austral. Shells pl. 56, figs. 2, 2a; 1982 Abbott & Dance, Compendium of Seashells p. 274; 1984 Aubry, Terebridae pl. 11 [non *Terebra evoluta* Deshayes, 1859].

Description: Shell to 90 mm; color ivory or fawn to dark brown, usually with lighter ribs and nodes on subsutural band; early whorls weakly inflated; subsutural band narrow, with well-developed round nodes, defined and constricted by deep suture and deep, narrow subsutural groove; axial ribs jutting out from below subsutural groove and continuing to the following suture, 22 to 23 on penultimate whorl; interspaces smooth, about equal to ribs; curved.

Type locality: "Swan River and Paterson's Bay, Torres Strait, Northern Australia." The latter locality is selected as the type locality.

Distribution: Western and Northern Australia.

Type: Lectotype BM(NH) no. 1979159; 54.4 × 12.5 mm.

Plate 63

FIGS. **242f-h,** *Duplicaria spectabilis* (Hinds). **242f,** Holotype of *Terebra gracilis* Reeve, British Museum (N.H.) No. 1874.10.29.2.; 22.0 mm. **242g,** Holotype of *T. remanalva* Melvill, British Museum (N.H.) No. 1911.6.21.11.; 33.8 mm. **242h,** Paratype of *T. remanalva* Melvill, National Museum Wales, Cardiff; 20.3 mm. **243a, b,** *D. dussumierii* (Kiener). **243a,** Lectotype Museum d'Histoire Naturelle, Geneva No. 1152/75; 98.8 mm. **243b,** Middle whorls of lectotype. **244,** *D. kirai* (Oyama). Type-figure from Oyama & Takemura, 1961, (1) fig.4; 57.0 mm. **245a, b,** *D. evoluta* (Deshayes). **245a,** Lectotype British Museum (N.H.) No. 197992; 56.2 mm. **245b,** Middle whorls of lectotype. **246a,** *D. australis* (E. A. Smith). Lectotype British Museum (N.H.) No. 1979159; 54.4 mm.

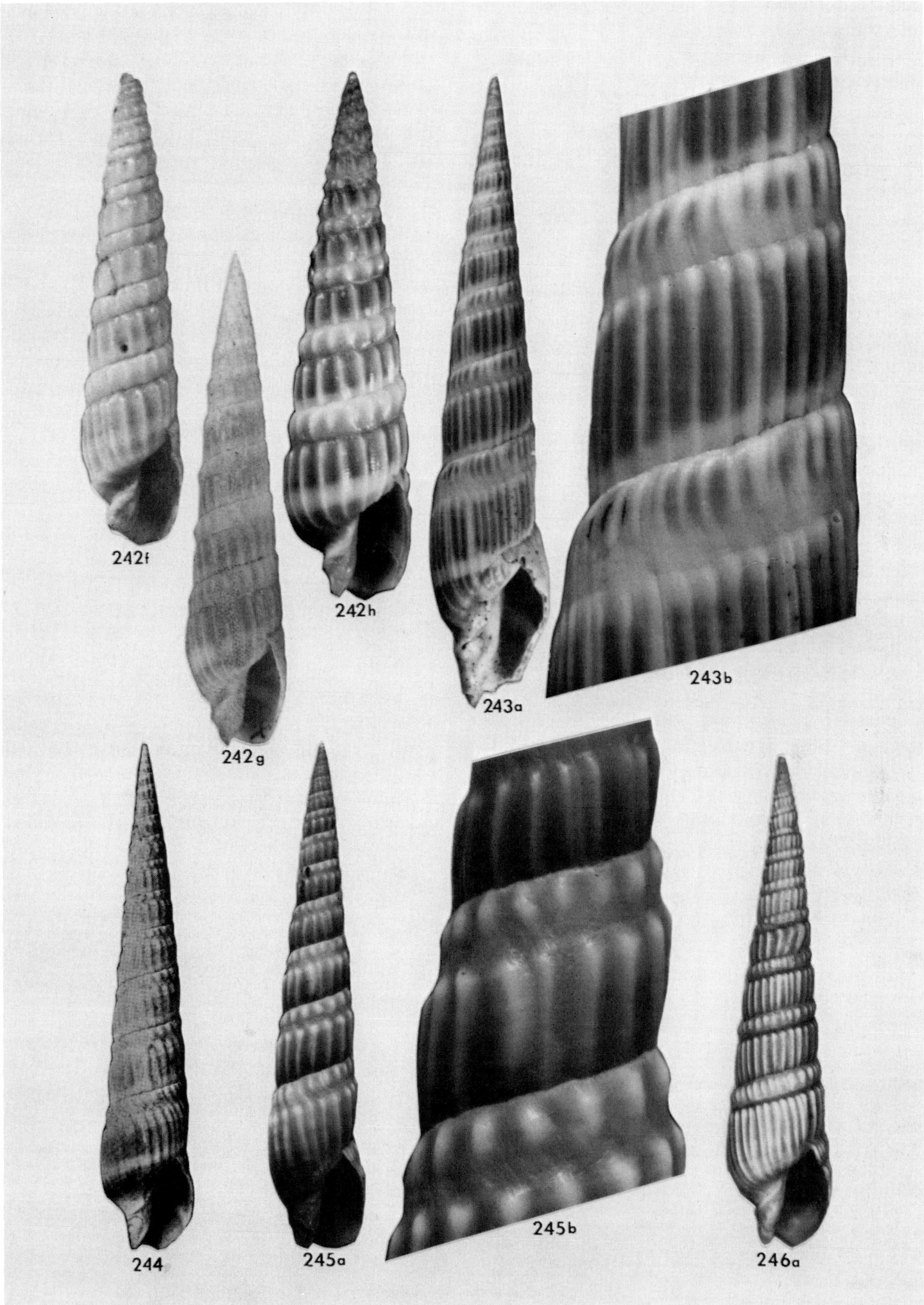

Plate 63. Indo-Pacific *Duplicaria*.

Discussion: This species has one of the deepest subsutural grooves and one of the narrowest subsutural bands of any of the *Duplicaria*. See *D. evoluta* (**245**) for comparison.

247. *Duplicaria similis* (E. A. Smith, 1873)

(Pl. 64, fig. 247)

1873 *Terebra similis* E. A. Smith, Ann. Mag. Nat. Hist. 4th ser., 11:265.

Description: Shell to 28 mm; color pinkish white with white stripe at periphery of body whorl; outline of whorls extremely convex on early whorls, moderately convex later, shouldered below subsutural band; protoconch of 1½ small, inflated whorls; subsutural band narrow on early whorls becoming moderately sized later, with elongate nodes, band defined by deep subsutural groove; axial ribs sharp, wide-spaced, arcuate on early whorls, straight later, off-set from the nodes on band, 17 on penultimate whorl; aperture elongate; columella recurved with two plications.
Type locality: Unknown.
Distribution: Unknown.
Type: Questionable holotype BM(NH) no. 1979135; 27.4 × 5.8 mm.
Discussion: E. A. Smith stated in his original description that the habitat was unknown and that the holotype measured 22 mm. The questionable type, measuring 27 mm, contains a label with Madras as the locality. The specimen resembles *D. australis* (**246**) except that the early whorls are more inflated, and the axial ribs are much fewer and more widely spaced. We have seen no other specimens of this species.

248. *Duplicaria badia* (Deshayes, 1859)

(Pl. 64, fig. 247)
(Color pl. E, fig. 23)

1844 *Terebra castanea* "Kiener," Hinds in Sowerby, Thes. Conch. 1:161, pl. 43, fig. 59 [non Kiener, 1839].
1859 *Terebra badia* Deshayes, Proc. Zool. Soc. Lond. pt. 27:300 [nom. nov. pro *T. castanea* Hinds, 1844]; 1885 Tryon, Man. Conch. 7:15, pl. 3, fig. 40.
1860 *Terebra ligneola* Reeve, Conch. Icon. 12: pl. 7, sp. 25 (Hab.: China; lectotype BM(NH); 63.0 mm).

Description: Shell to 63 mm; color light to very dark brown, sometimes with a narrow light stripe posterior to the suture; outline of whorls moderately convex; subsutural band defined by a broad space rather than a groove; axial ribs narrow, sharp; spiral sculpture absent; interspaces wider than ribs; body whorl with ribs ending at periphery, followed by 3 to 5 spiral rows of pustules; aperture brown, quadrate; columella recurved, brown; siphonal fasciole usually white.
Type locality: *D. badia* "Mauritius and Guinea," error. Reeve's indication of China is here designated as the type locality; *T. ligneola*, "China."
Distribution: China and Taiwan; 60 m.
Type: *D. badia* type lost. The specimen illustrated by Hinds in Sowerby (1844) on plate 43, fig. 59, is here designated as the illustrated lectotype of *T. castanea* Hinds and *T. badia* Deshayes; *T. ligneola* lectotype BM(NH) no. 197925, 63.0 mm.
Discussion: The spiral rows of pustules anterior to the periphery of the body whorl make this species easy to identify and separate from other species of *Duplicaria*.

249. *Duplicaria mozambiquensis* Bratcher & Cernohorsky, 1982

(Pl. 64, figs. 249a,b)

1982 *Duplicaria mozambiquensis* Bratcher & Cernohorsky, Nautilus 96(2):65, figs. 3, 8.

Description: Shell to 33 mm; color light brown, peach, or ivory, usually with ivory ribs and narrow peripheral stripe on body whorl; outline of whorls convex; protoconch with 1½ moderately inflated whorls; subsutural band of round

---

Plate 64
FIGS. **246b,** *Duplicaria australis* (E. A. Smith). Lower whorls of lectotype. **247,** *D. similis* (E. A. Smith). Questionable holotype British Museum (N.H.) No. 1979135; 27.4 mm. **248a, b,** *D. badia* (Deshayes). **248a,** Illustrated lectotype from Hinds in Sowerby, 1844, pl.43, fig.59. **248b,** Lectotype of *Terebra ligneola* Reeve, British Museum (N.H.); 63.0 mm. **249a, b,** *D. mozambiquensis* Bratcher & Cernohorsky. **249a,** Holotype Natal Museum No. H-7843/T-2541; 22.3 mm. **249b,** Upper whorls of holotype. **250a, b,** *D. hiradoensis* (Pilsbry). **250a,** Holotype Academy Natural Sciences, Philadelphia No. 81994; 55.4 mm. **250 b,** Paratype same Institution; 29.1 mm. **251,** *D. gouldi* (Deshayes). Lectotype British Museum (N.H.) No. 197962; 61.6 mm. **152 a, b,** *D. thaanumi* (Pilsbry). **252a,** Holotype Academy Natural Sciences, Philadelphia No. 117019; 59.9 mm. **252b,** Immature lectotype of *T. costellifera* Pease, Museum Comparative Zoology, Harvard No. 148800; 19.3 mm.

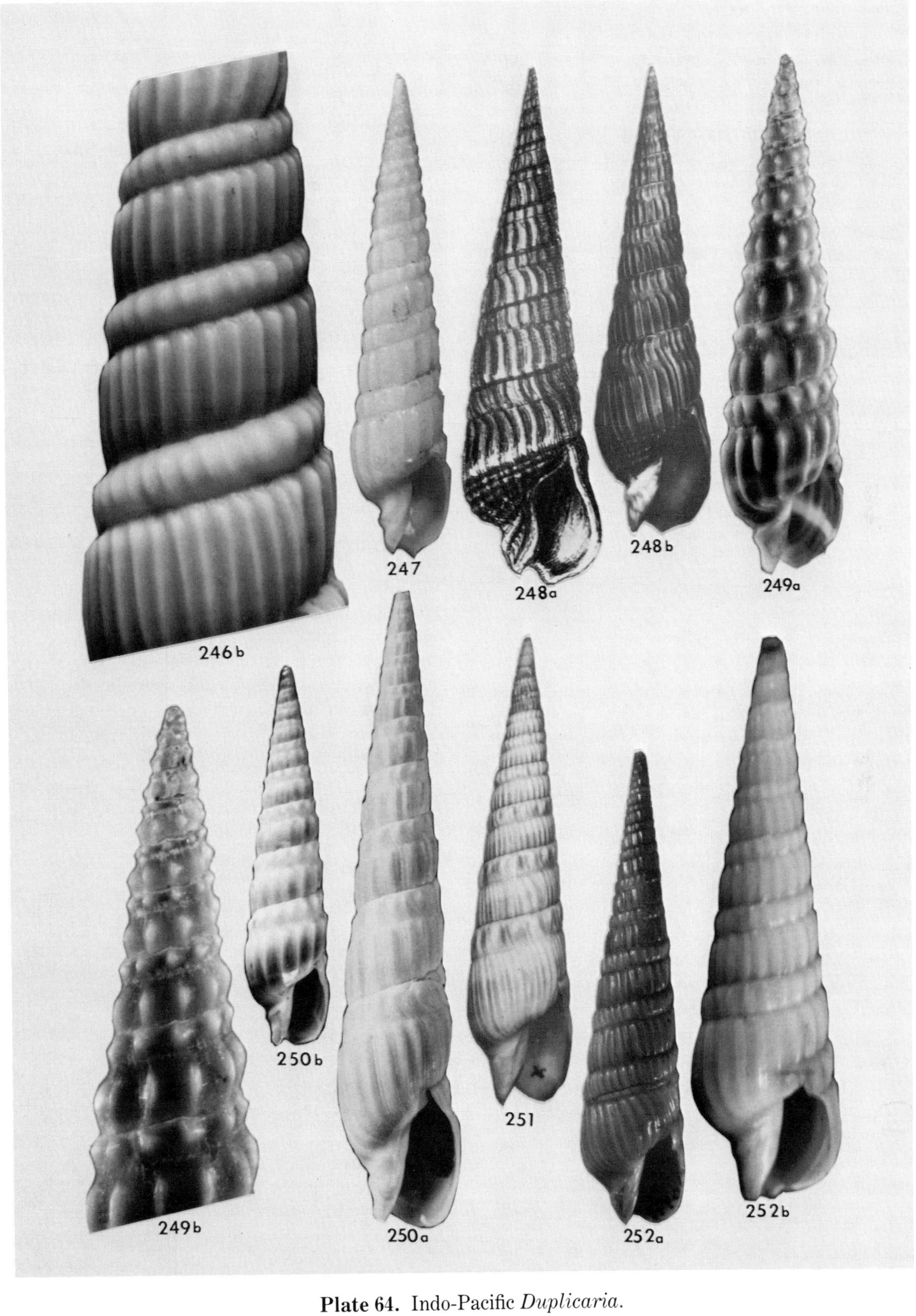

Plate 64. Indo-Pacific *Duplicaria*.

nodes, narrow on early whorls, becoming somewhat wider later, band defined by punctate groove; axial ribs widely spaced, arcuate, almost noded in center on early whorls, giving an angulate outline to whorls; ribs later become widely spaced, sharp, slightly noded at posterior end, but losing the angulate form, about 12 on penultimate whorl; aperture elongate; columella almost straight.
Type locality: "Off Chinde Is., Mozambique."
Distribution: From Durban, South Africa, to Mozambique.
Type: Holotype Natal Mus. no. H7843/T2541; 22.3 × 5.6 mm.
Discussion: Of the paratypes, four are light brown with ivory ribs, one is peach color, and two are ivory. For comparison see *D. spectabilis* **(242)** and *D. evoluta* **(245)**.

250. *Duplicaria hiradoensis* (Pilsbry, 1921)

(Pl. 64, figs. 250a,b)

1921 *Terebra dussumieri hiradoensis* Pilsbry, Proc. Acad. Nat. Sci. Philadelphia 72:309.

Description: Shell to 55 mm; color gray-beige; outline of whorls convex; protoconch of 1½ whorls; subsutural band defined by shallow inconspicuous groove in wider space; axial sculpture of wide spaced ribs, 15 on penultimate whorl, axial striae in interspaces; body whorl elongate; aperture moderately elongate; columella straight.
Type locality: "Hirado, Hizen, Japan."
Distribution: Japan.
Type: Holotype ANSP no. 81994; 55.4 × 11.0 mm.
Discussion: This species differs from *D. dussumieri* **(243)** Kiener by having longer whorls; fewer ribs; a subsutural space with just a hint of a narrow shallow groove; and no pale line at the periphery of the body whorl. *D. dussumieri* has a deep subsutural groove.

251. *Duplicaria gouldi* (Deshayes, 1857)

(Pl. 64, figs. 251)

1857 *Terebra gouldi* Deshayes, Journ. Conchyl. 6:89, pl. 5, fig. 2; 1860 Reeve, Conch. Icon. 12, pl. 7, sp. 24; 1960 Weaver, Hawaiian Mar. Moll. 1(1):2, pl. 1,, figs. top left; 1974 Quirk & Wolfe, Seashells of Hawaii p. 13, fig. 4; 1978 Salisbury, Hawaiian Shell News 26(8):8, text fig; 1979 Kay, Hawaiian Marine Shells p. 398, figs. 128, 129d; 1982 Abbott & Dance, Compendium of Seashells p. 276; 1984 Aubry, Terebridae pl. 11.

1978 *Duplicaria gouldi* (Deshayes), Cernohorsky, Trop. Pacific Mar. Shells p. 149, pl. 53, fig. 3.

Description: Shell to 70 mm; color shiny cream to dark tan with brown streaks and splotches; outline of whorls slightly convex; protoconch of 1½ dome-shaped whorls; subsutural band ribbed, defined by incised groove; axial ribs sharp, 28 to 41 on penultimate whorl; interspaces, about equal to ribs, smooth; spiral sculpture absent; aperture moderately elongate; columella slightly curved.
Type locality: "Hawaiian Islands."
Distribution: Endemic to the Hawaiian Islands, occurring from Midway to Maui, but not reported as yet from the Island of Hawaii; 2 to 150 m.
Type: Lectotype (here designated) specimen illustrated by Deshayes BM(NH) no. 197962; 61.6 mm.
Discussion: This species develops directly from the eggs without a planktonic stage. It feeds by engulfing its prey, having no radula. It could be confused with *D. thaanumi* **(251)** which has a shell of similar size, shape and sculpture. *T. thaanumi* has shorter whorls and a protoconch of 3½ whorls, and has planktonic development. This species was named for Augustus Addison Gould (1805–1866), a Harvard physician who made major contributions to malacology.

252. *Duplicaria thaanumi* (Pilsbry, 1921)

(Pl. 64, fig. 252a,b;
pl. 65, fig. 252c)

1869 *Terebra costellifera* Pease, American Journ. Conch. 5(2):66. *[nomen oblitum]*.

1921 *Terebra thaanumi* Pilsbry, Proc. Acad. Nat. Sci. Philadelphia 90:305, pl. 12, figs. 1, 2; 1974 Quirk & Wolfe, Seashells Hawaii p. 14, fig. 15; 1978 Salisbury, Hawaiian Shell News 26(8):9, text fig.; 1979 Kay, Hawaiian Marine Shells p. 402, fig. 129e; 1982 Abbott & Dance, Compendium of Seashells p. 275; Aubry, Terebridae pl. 13.

1978 *Duplicaria thaanumi* (Pilsbry), Cernohorsky, Tropical Pacific Marine Shells p. 149, pl. 53, fig. 3.

Description: Shell to 95 mm; color cream, pale orange, or light cinnamon brown, sometimes with darker subsutural band or with inconspicuous dark streaks; outline of whorls slightly turreted; protoconch of 3½ conical whorls; subsutural band of sharp ribs defined by deep groove;

axial ribs sharp, angulate, jutting out at a sharp angle anterior to subsutural groove; spiral sculpture absent; aperture quadrate; columella slightly curved.
Type locality: *D. thaanumi*, "Off Honolulu, Hawaii, 6 to 8 fms"; *T. costellifera*, "Oahu, Hawaii."
Distribution: The Hawaiian Is.; subtidal to 100 m.
Type: *D. thaanumi* holotype ANSP no. 117.019, 59.9 × 13.3 mm; *T. costellifera* lectotype MCZ no. 248800, 19.3 × 5.4 mm.
Discussion: This species may be separated from *D. gouldi* **(250)** by its shorter whorls, deeper subsutural groove, ribs which jut out at an angle anterior to the subsutural band, and a protoconch of 3½ whorls. *D. gouldi* has a protoconch of 1½ whorls and non-planktonic development in contrast to the planktonic development of this species. It was named for the well-known collector from Hawaii, Ditlev Thannum (1867–1963).

253. *Duplicaria trochlea* (Deshayes, 1857)

(Pl. 65, figs. 253a,b)
(Color pl. C, fig. 9)

1857 *Terebra trochlea* Deshayes, Journ. Conchyl. 6:89, pl. 5, fig. 6; 1860 Reeve, Conch. Icon. 12: pl. 13, sp. 54; 1969 Bratcher, Veliger 11(4):334, pl. 52, figs. 1–4, and replacement page in vol. 12, p. 144b; 1975 Salvat & Rives, Coq. Polynésie p. 362, fig. 3911; 1980, O. Schoenberg, Hawaiian Shell News 28(3):15, text. fig.; 1984 Salvat & Rives, Shells of Tahiti p. 134, pl. 34, fig. 9.
1888 *Terebra cochlea* Deshayes, Paetel, Cat. Conchyl. Samml. p. 255; 1969 Bratcher, Veliger 11(4):334

Description: Length to 79 mm; color fawn to pinkish brown with flammules of creamy white; outline of whorls slightly convex, turreted, and shouldered anterior to prominent convex subsutural band; band with axial ribs, defined by deeply impressed groove, microscopically punctate in early whorls; sutures deep and broad; axial ribs sharp, evenly spaced on early whorls, later becoming weaker, more unevenly spaced, and forming nodes anterior to the subsutural groove; spiral sculpture absent; aperture quadrate; outer lip thin with color pattern showing through; columella white, heavy, quite straight; siphonal fasciole with sinuous striations and an extremely sharp keel.
Type locality: "Zanzibar," error. The Marquesas Islands are here designated as the type locality.
Distribution: Endemic to the Marquesas Islands.
Type: Lectotype BM(NH) no. 197975; 65.3 × 13 mm.
Discussion: Although described in 1857, this beautiful large species was not rediscovered until 110 years later by Richard Sixberry. The name has sometimes been misquoted as *"cochlea."*

254. *Duplicaria baileyi* Bratcher & Cernohorsky, 1982

(Pl. 65, figs. 254a,b)

1966 *Duplicaria (Duplicaria)* sp. Cernohorsky & Jennings, Veliger 9(1):58, pl. 6, fig. 44.
1967 *Duplicaria australis* (E. A. Smith), Cernohorsky, Mar. Shells Pacific 1:208, pl. 52, fig. 392. [non *Terebra australis* E. A. Smith, 1873]; 1978 Salisbury, Hawaiian Shell News 26(8):8, text fig. [non *Terebra australis* E. A. Smith, 1873].
1978 *Terebra australis* E. A. Smith, Hinton, Guide Australian Shells, pl. 59, fig. 6. [non *Terebra australis* E. A. Smith, 1873].
1982 *Duplicaria baileyi* Bratcher & Cernohorsky, Nautilus 96(2):64, figs. 5, 11.

Description: Shell to 40 mm; color cream to beige, with a few reddish brown splotches and dots scattered at random, and with the same color on base of body whorl; outline of whorls slightly convex; protoconch 4½ pink, translucent conical whorls; subsutural band with axial ribs, band defined by a shallow groove with deeper punctations between ribs; axial ribs straight, contiguous with those on band, 20 to 29 on penultimate whorl; interspaces about equal to ribs, smooth; periphery of body whorl marked with light band between rib endings and light brown base; aperture elongate; columella recurved; siphonal fasciole striate, with moderate keel.
Type locality: "Guadalcanal, Solomon Islands, 9° 25′S, 159°56′E, 20 m."
Distribution: From Mozambique, East Africa, to the Fiji Islands; intertidal to 200 m.
Type: Holotype LACM no. 1970; 24.9 × 5.3 mm.
Discussion: This species has been in collections, usually without a name, or with an incorrect name, for a long time although it was not described until 1982. The brown splotches may vary in number and size, but they are inconspicuous. Refer to *D. teramachii* **(254)** for comparison. This species was named for Brian Bailey, a collector and professional scuba diver of the Solomon Islands.

255. ***Duplicaria teramachii*** R. D. Burch, 1965

(Pl. 65, figs. 255a-c)

1965 *Duplicaria (Duplicaria) teramachii* R. D. Burch, Veliger 7(4):247, pl. 31, fig. 10.

Description: Shell to 37 mm, slender; color shiny ivory to dark beige with elongate reddish brown blotches and streaks, sometimes forming an interrupted band; outline of whorls flat; protoconch of 4 pale lavender translucent whorls; subsutural band flat, defined by narrow groove cutting through ribs; axial ribs narrow, sharp, close-set, from suture to suture, 27 to 29 on penultimate whorl; spiral sculpture absent; body whorl elongate; aperture elongate; columella recurved, with well-developed parietal callus; siphonal fasciole shiny, with sharp keel.
Type locality: Off Tosa Bay, Japan; 70 fms."
Distribution: From Japan to New Caledonia and the Solomon Is.; 140 to 160 m.
Type: Holotype MCZ no. 251243; 37.0 × 5.5 mm.
Discussion: This is the most slender of all the *Duplicaria* species and is the most flat in outline. It differs from *D. baileyi* **(253)** by being flatter in outline, more slender, having brown streaks and splotches forming an interrupted band and having shorter whorls. This species was named for the well known Japanese shell dealer, Akibumi Teramachi.

256. ***Duplicaria raphanula*** (Lamarck, 1822)

(Pl. 66, figs. 256a-c)
(Color pl. E, fig. 5)

1822 *Terebra raphanula* Lamarck, Hist. Nat. Anim. S. Vert. 7:288; 1839 Kiener, Spéc. Gén. Icon. Coq. Viv. 9:21, pl. 10, fig. 20; 1844 Hinds in Sowerby, Thes. Conch. 1:159, pl. 44, fig. 94; 1860 Reeve, Conch. Icon. 12, pl. 17, sp. 77.

1961 *Subula (?) raphanula* (Lamarch), Oyama & Takemura, The Moll. Shells, Resources Exploitation Inst. 5, *Subula* pl., fig. 6.

1963 *Terebra (?Diplomeriza) raphanula* Lamarck, Burch, Veliger 5(4):158.

1964 *Terebra (Diplomeriza) raphanula* Lamarck, Shikama, Sel. Shells Col. 1:124, pl. 101, fig. 10.

1966 *Duplicaria (Duplicaria) raphanula* (Lamarck), Cernohorsky & Jennings, Veliger 9(1):58, pl. 6, fig. 46.

1967 *Duplicaria raphanula* (Lamarck), Cernohorsky, Mar. Shells Pacific 1:209, pl. 53, fig. 397; 1982 Abbott & Dance, Compendium of Seashells p. 276.

1984 *Duplicata raphanula* (Lamarck), Aubry, Terebridae pl. 8.

Description: Shell to 64 mm; color shiny ivory to beige with dark brown streaks and splotches on the subsutural band, and lighter brown rectangular blotches on remainder of the whorl; outline of whorls flat; protoconch of 3½ slender whorls; subsutural band axially ribbed, defined by moderately deep subsutural groove; axial ribs well developed or almost obsolete; spiral sculpture absent; body whorl elongate; aperture extremely elongate; columella straight.
Type locality: None. Rabaul, Papua New Guinea is here designated as the type locality.
Distribution: From South Africa to the Fiji Is.; subtidal to 36 m.
Type: The type of *T. raphanula* has not been found. In Delessert's collection are 4 specimens, none reaching the 52 mm indicated by Lamarck.
Discussion: This is the most graceful and beautiful of the genus, *Duplicaria*. It also fills our criteria for the genus, *Hastula*, being smooth and shiny, and having no spiral sculpture except for the subsutural groove. The identity of this species has never been in doubt and has been elucidated by subsequent authors such as Kiener (1839), Hinds in Sowerby (1844) and Reeve (1860).

257. ***Duplicaria vallesia*** Hedley, 1912

(Pl. 66, fig. 257)

1912 *Duplicaria vallesia* Hedley, Rec. Austral. Mus. 8(3):147, pl. 43, fig. 31.

Description: Shell to 27 mm, thin and delicate; color light orange with a cream band through center of whorls and dark orange on base of body whorl; outline of whorls convex in early whorls, flatter in later ones; protoconch of 2 whorls; subsutural band rather flat formed by groove cutting through ribs; axial ribs wide spaced,

---

Plate 65
FIG. **252c,** *Duplicaria thaanumi* (Pilsbry). Middle whorls. **253a, b,** *D. trochlea* (Deshayes). **253a,** Lectotype British Museum (N.H.) No. 197975; 65.3 mm. **253b,** Lower whorls. **254a, b,** *D. baileyi* Bratcher & Cernohorsky. **254a,** Holotype Los Angeles County Museum No. 1970; 24.9 mm. **254b,** Lower and middle whorls of holotype. **255a-c,** *D. teramachii* R. D. Burch. **255a,** Holotype Museum Comparative Zoology, Harvard No. 251243; 37.0 mm. **255b,** Off southern New Caledonia, 230 m; 35.1 mm. **255c,** Middle and upper whorls.

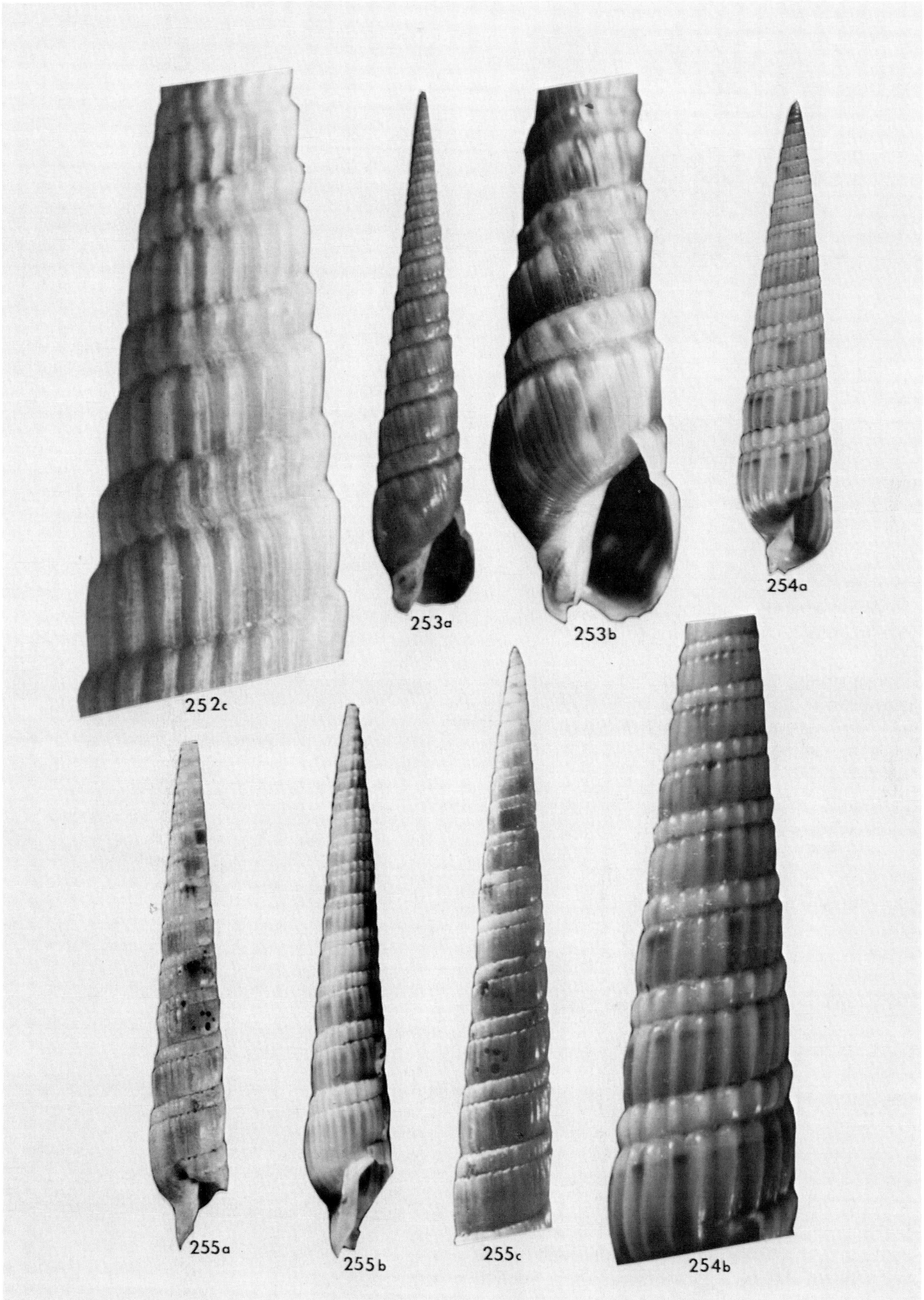

Plate 65. Indo-Pacific *Duplicaria*.

becoming weaker in later whorls and fading about center of body whorl; aperture quadrate; columella curved; siphonal fasciole with heavy striations and sharp keel.
Type locality: "Trial Bay, New South Wales, Australia."
Distribution: East Australia.
Type: Holotype AM Sydney no. C-28470; 27.2 × 6.4 mm.
Discussion: This thin and delicate shell is unlikely to be mistaken for another *Duplicaria*.

258. *Duplicaria ustulata* (Deshayes, 1857)
(Pl. 66, figs. 258a-c)

1857 *Terebra ustulata* Deshayes, Journ. Conchyl. 6:97, pl. 3, fig. 12; 1860 Reeve, Conch. Icon. 12: pl. 17, sp. 73.
1923 *Duplicaria ustulata* (Deshayes), May, Illust. Indx. Tasmanian Shells pl. 34, fig. 14.
1924 *Pervicacia ustulata* (Deshayes), Iredale 1850, Proc. Linnean Soc. N. S. W. 49(3):262, text fig. 44/45; 1952 Cotton, Trans. Royal Soc. S. Australia 75:39, pl. 3, fig. 2; 1959 Allan, Australian Shells p. 196, fig. 5; 1962 Macpherson & Gabriel, Mar. Moll. Victoria p. 240, fig. 278; 1976 Garrard, Journ. Malac. Soc. Australia 3(4):184, fig. 4; 1978 Hinton, Guide Austr. Shells, pl. 59, figs. 7, 7a.

Description: Shell to 37 mm; color beige to orange-brown, often with purple-brown anterior to periphery of body whorl, early whorls of teleoconch usually light brown; outline of whorls slightly convex, shouldered anterior to subsutural band; protoconch of 1½ whorls; subsutural band with axial ribs, defined by shallow groove; axial ribs on remainder of whorl contiguous with those on band; sutures deep; body whorl with ribs continuing to keel of siphonal fasciole; aperture quadrate; columella curved.
Type locality: "Van Diemen Land." Tasmania, Australia.
Distribution: Australia, from New South Wales to Tasmania; 20 to 80 m.
Type: Lectotype BM(NH) no. 197976; 35.0 × 10.7 mm.
Discussion: This is the type species of the genus *Pervicacia* Iredale, 1924, which Iredale separated from *Duplicaria* because it has only a basal twist of the columella to represent a fold. *D. fictilis* **(258)** may be separated from this species by being more gently rounded and less angulate below the subsutural band, the band being less well defined, and having a narrower apical angle.

259. *Duplicaria fictilis* (Hinds, 1844)
(Pl. 66, fig. 259a;
pl. 67, figs. 259b-d)

1844 *Terebra fictilis* Hinds in Sowerby, Thes. Conch. 1:183, pl. 45, figs. 109, 110; 1860 Reeve, Conch. Icon. 12, sp. 145.
1932 *Pervicacia fictilis* (Hinds), Cotton & Godfrey, S. Austral. Nat. 13(2):65; 1976 Garrard, Journ. Malac. Soc. Australia 3(3–4):182, fig. 1; 1978 Hinton, Guide Austral. Shells, pl. 59, figs. 10, 10a; 1984 Aubry, Terebridae pl. 9.
1952 *Pervicacia helenae* Cotton, Trans. Roy. Soc. S. Australia 75:40, pl. 3, fig. 1.

Description: Shell to 25 mm; color light gray to yellowish brown with base of body whorl darker; outline of early whorls straight, later ones convex; protoconch of 1½ whorls; subsutural band noded, defined by with constriction but no subsutural groove; axial ribs (about 14) contiguous with nodes on band; spiral sculpture of numerous fine grooves in interspaces; aperture quadrate; columella curved, brown.
Type locality: *D. fictilis*, "Australia, probably southern shores"; *P. helenae*, "Farme Beach, Coffin Bay, Eyre Peninsula," S. Australia.
Distribution: South Australia; subtidal to 40 m.
Type: *D. fictilis* lectotype BM(NH) no. 1968252, 14.8 mm; *P. helenae* holotype SAMA no. D-4436, 20.5 × 5.6 mm.
Discussion: The subsutural band may become quite narrow on some individuals of this species. Refer to *D. ustulata* **(257)** for comparison.

260. *Duplicaria kieneri* (Deshayes, 1859)
(Pl. 67, figs. 260a,b)

1839 *Terebra duplicata* var. junior Kiener, Spéc. Gén. Icon. Coq. Viv. 9: pl. 12, fig. 26a only [non *Buccinum duplicatum* Linnaeus, 1758].
1859 *Terebra kieneri* Deshayes, Proc. Zool. Soc. Lond. pt. 27:294; 1860 Reeve, Conch. Icon. 12, pl. 21, sp. 110.

Plate 66
FIGS. **256a-c,** *Duplicaria raphanula* (Lamarck). **256a,** Illustration from Kiener, 1839, pl. 10, fig. 20. **256b,** Rabaul, Papua New Guinea; 43.7 mm. **256c,** Middle whorls. **257,** *D. vallesia* Hedley. Holotype Australian Museum, Sydney No. C-28470; 27.2 mm. **258a-c,** *D. ustulata* (Deshayes). **258a,** Lectotype British Museum (N.H.) No. 197976; 35.0 mm. **258b, c,** Tasmania, Australia; 31.6 mm and 32.0 mm respectively. **259a,** *D. fictilis* (Hinds). Paralectotype British Museum (N.H.) No. 1968252; 12.5 mm.

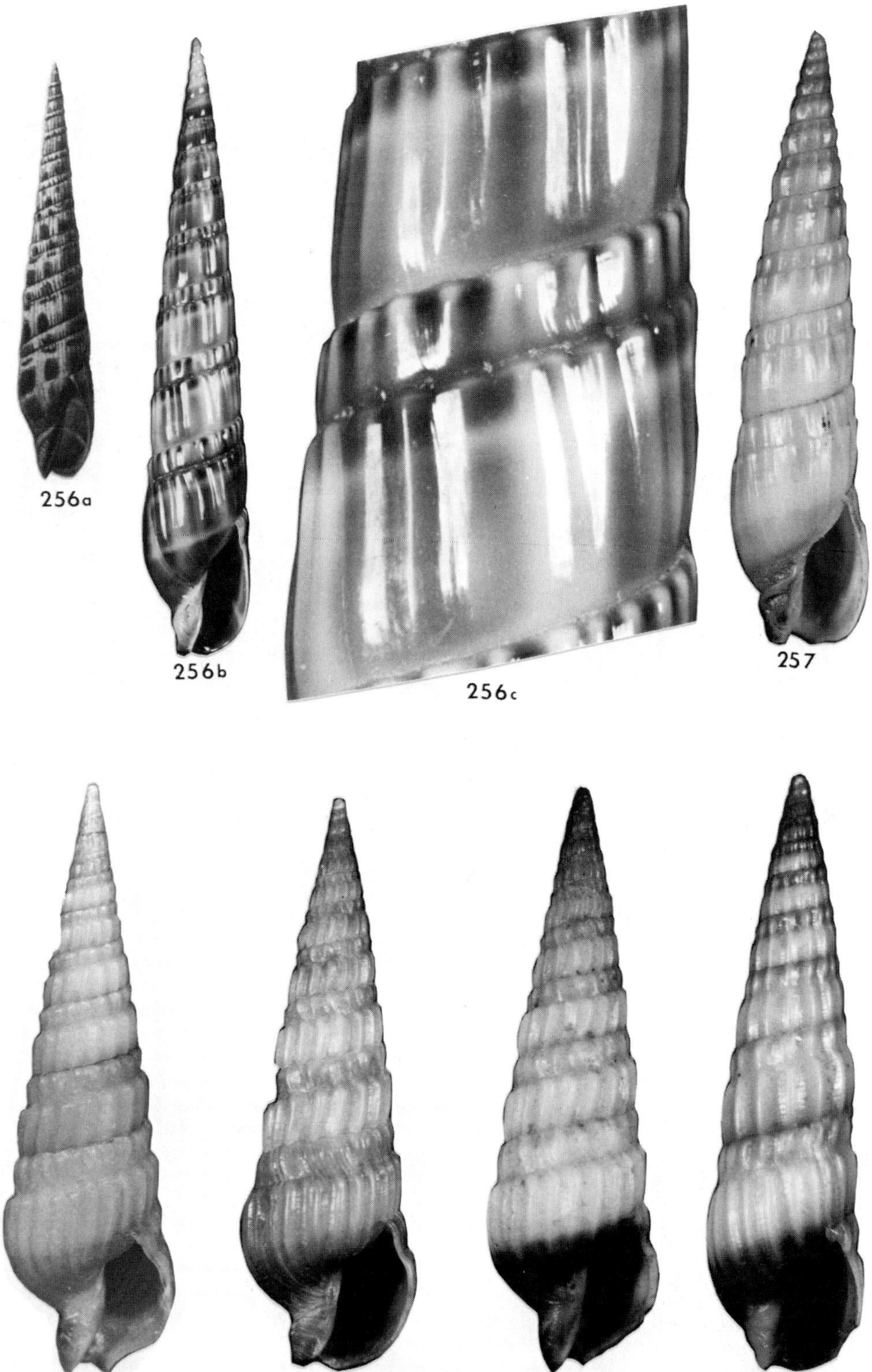

Plate 66. Indo-Pacific *Duplicaria*.

1962 *Pervicacia kieneri* (Deshayes), Macpherson & Gabriel, Mar. Moll. Victoria p. 240, fig. 279; 1876 Garrard, Journ. Malac. Soc. Australia 3(4):183, fig. 7; 1978 Hinton, Guide Austral. Shells pl. 59, figs. 9, 9a.

Description: Shell to 27 mm; color from white or cream to dark brown, sometimes with light nodes and ribs, white specimens usually having purple protoconch and early whorls of teleoconch; outline of whorls slightly convex; protoconch 1/2 to 2 whorls; subsutural band axially ribbed, marked by deep suture and deep subsutural groove; axial ribs aligned with those on band, some individuals having crowded fine ribs throughout, others with distant ribs in early whorls, crowded later, up to 40 ribs on penultimate whorl, 50 on body whorl; spiral sculpture absent; aperture quadrate; columella slightly curved, with light parietal callus.
Type locality: "Tasmania, Australia."
Distribution: Victoria to Tasmania and southern Australia.
Type: Lectotype BM(NH) no. 1979129; 21.2 × 6 mm.
Discussion: This species has numerous fine crowded ribs, making it easy to identify.

261. *Duplicaria albozonata* (E. A. Smith, 1875)

(Pl. 67, fig. 261)

1875 *Terebra albozonata* E. A. Smith, Ann. & Mag. Nat. Hist. ser. 4, 15:415; 1879 E. A. Smith, Proc. Zool. Soc. Lond. p. 185, pl. 19, fig. 5.

Description: Shell to 25 mm; color pale brown with a narrow white zone above the suture and at the periphery of the body whorl; outline of whorls almost flat except for inflated body whorl; subsutural band defined by narrow deep groove; axial ribs, narrower than interspaces, spiral sculpture absent; aperture broad, quadrate; columella recurved.
Type locality: "Matoza Harbour, Japan, 6 fms."
Distribution: Japan.
Type: Lectotype BM(NH) no. 1873.8.6.9; 24.7 × 8.0 mm.
Discussion: The two specimens in the type lot are worn and faded. The lectotype has a bad break on the body whorl which causes it to look more inflated than it might be if it were not damaged. The paralectotype is a very young specimen of 8 mm. We have not seen any recently collected specimens which would match the type of *T. albozonata*.

262. *Duplicaria morbida* (Reeve, 1860)

(Pl. 67, fig. 262)

1860 *Terebra morbida* Reeve, Conch. Icon. 12, pl. 24, sp. 133; 1885 Tryon, Man. Conch. 7:25, pl. 7, fig. 23.

Description: Shell to 14 mm, broad and sturdy; color pinkish beige with small orange-brown spots between ribs, mostly on dorsum, no spots on subsutural band; outline of whorls slightly convex with broad apical angle; protoconch of 1 1/2 whorls; subsutural band of elongate nodes, band occupying half of whorl; axial ribs broad, rounded, contiguous with nodes on band; interspaces narrow; spiral sculpture absent; aperture moderately elongate; columella straight.
Type locality: None given.
Distribution: Unknown.
Type: Holotype BM(NH) no. 1879.2.26.25; 14.1 mm.
Discussion: The holotype is an immature specimen with only seven whorls of the teleoconch and could be quite large at maturity because of the extremely wide apical angle. It has not been reported subsequent to the original description.

Plate 67
FIGS. **259b,d,** *Duplicaria fictilis* (Hinds). **259b,** Holotype of *Pervicacia helenae* Cotton, South Australian Museum, Adelaide No. D-4436; 20.5 mm. 259c, Paratype same Institution; 22.8 mm. **259d,** Farme Beach, Coffin Bay, Eyre Peninsula, Sth. Australia; 23.3 mm. **260a, b,** *D. kieneri* (Deshayes). **260a,** Lectotype British Museum (N.H.) No. 1979129; 21.2 mm. 260b, Aldinga, Sth. Australia; 25.4 mm. **261,** *D. albozonata* (E. A. Smith). Lectotype British Museum (N.H.) No. 1873.8.6.9.; 24.7 mm. **262,** *D. morbida* (Reeve). Holotype British Museum (N.H.) No. 1879.2.26.25.; 14.1 mm. 263a, b, *Terenolla pygmaea* (Hinds). **263a,** Illustrated lectotype from Hinds in Sowerby, 1844, pl. 45, fig. 112. **263b,** Port Moresby, Papua New Guinea; 8.7 mm.

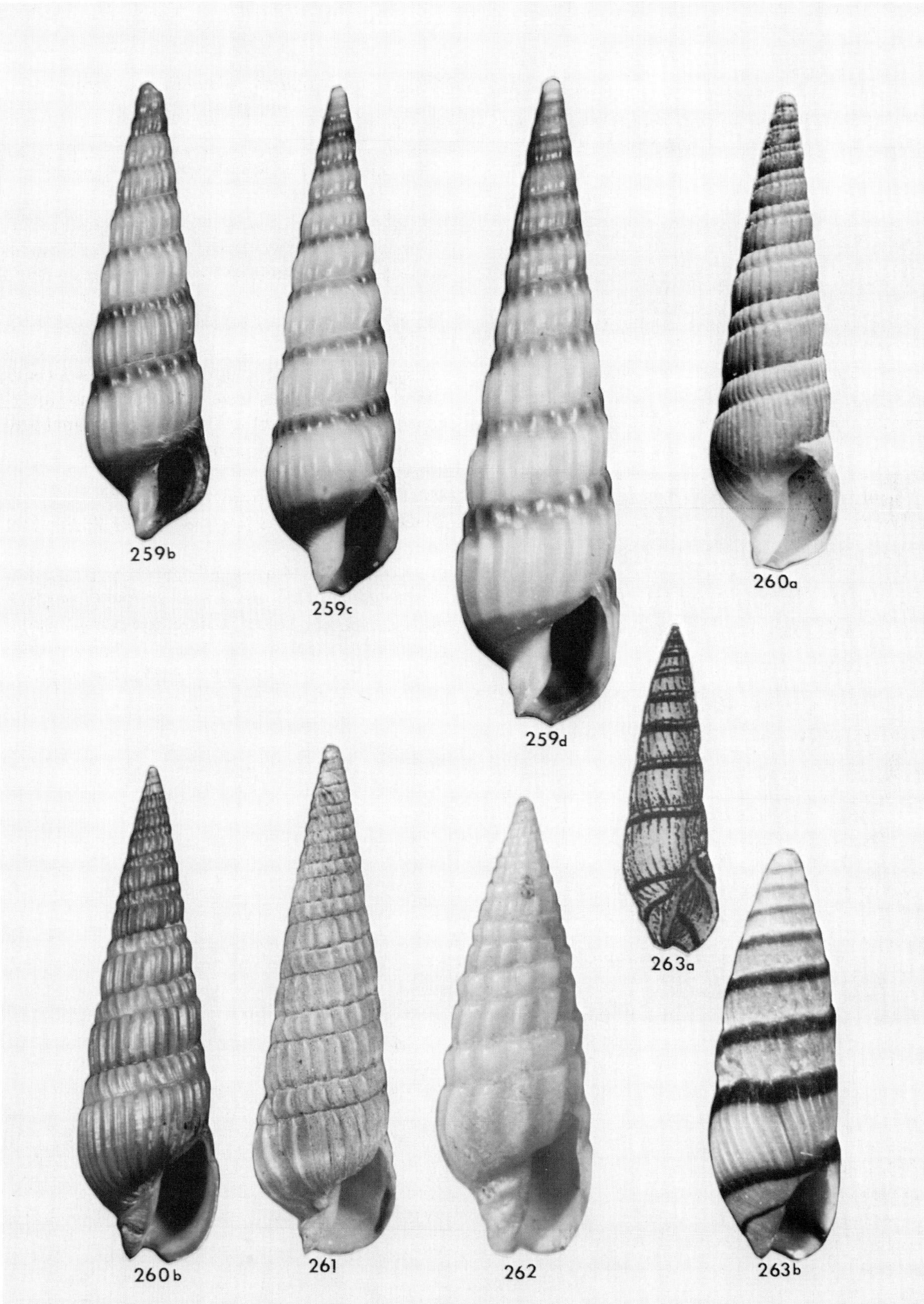

Plate 67. Indo-Pacific *Duplicaria* and *Terenolla*.

Genus *Terenolla* Iredale, 1929

Type species: *Terenolla pygmaea* (Hinds, 1844).

The shells of this genus are small and straight-sided with the outer lip and columella rimmed with brown. We know of no anatomical work done on this species, and it is questionable whether this genus belongs in the Terebridae, but we are tentatively retaining it there. It possibly is a columbellid.

263. *Terenolla pygmaea* (Hinds, 1844)
(Pl. 67, figs. 263a,b)

1844 *Terebra pygmaea* Hinds, Proc. Zool. Soc. Lond. pt. 11:158; 1844 Hinds in Sowerby, Thes. Conch. 1:184, pl. 45, fig. 112; 1860 Reeve, Conch. Icon. 12, pl. 27, sp. 149; 1984 Aubry, Terebridae pl. 8.

1905 *Turbonilla (?) princeps* Preston, Journ. Malac. 12(1):7, pl. 2, fig. 33 (Hab.: Ceylon; type?).

1929 *Terenolla pygmaea* (Hinds), Iredale, Mem. Queensland Mus. 9(3):282; 1961 Oyama & Takemura, Moll. Shells, Resources Exploitation Inst. 5: *Terenolla* pl., fig. 1; 1978 Cernohorsky, Trop. Pacific Mar. Shells p. 150, pl. 53, fig. 5; 1979 Kay, Hawaiian Mar. Shells, p. 403, fig. 130K.

Description: Shell to 11 mm; color pinkish beige to gray, with narrow brown stripe immediately anterior to suture, at periphery of body whorl, and posterior to siphonal fasciole; outline of whorls straight; protoconch of 2½ conical whorls; axial ribs numerous, irregular, closely set, quite evenly spaced; spiral grooves microscopic, not crossing ribs; aperture semi-quadrate; outer lip and columella rimmed with brown, columella curved.

Type locality: "Straits of Malacca, Indonesia, 17 fms."

Distribution: Throughout the tropical Indo-Pacific; intertidal to 31 m.

Type: Type lost. Hinds' figure 112 on plate 45 from Sowerby (1844) is here designated as the illustrated lectotype. Type unknown for *princeps*.

Discussion: This species is found in Queensland, Australia, in depths of 8 to 15 m among shell and bryozoa rubble under steep drop-offs. It appears not to be nocturnal, as it has been found crawling during daylight hours. This is a beautiful small shell with a good sheen when in fresh condition, but usually they are found eroded.

RECENTLY DESCRIBED SPECIES OF TEREBRA

264. *Terebra angelli* (Gibson-Smith & Gibson-Smith, 1984)
(Pl.68, fig. 264)

1981 *Strioterebrum quadrispiralis* (Weisbord) Petuch, Malacologia 20(2)2:337, figs. 103, 104 (non Weisbord, 1962).

1981 *Strioterebrum trispiralis* (Weisbord) Petuch, Malacologia 20(2):337, figs. 105, 106 (non Weisbord, 1962).

1984 *Strioterebrum angelli* Gibson-Smith & Gibson-Smith, Tulane Studies in Geology & Paleontology 18, (2):65, pl. 1, figs. 6, 7.

Description: Shell to 15 mm; color, cream with purplish brown anterior to periphery of body whorl and on subsutural band; outline of whorls convex; protoconch 1½ whorls; subsutural band indented, strongly noded; axial ribs sturdy, wider than interspaces, contiguous with nodes on band; spiral sculpture of 2 to 6 flat spiral cords; aperture quadrate; columella curved.

Type locality: Punta Mangle, Isla Margarita, Venezuela.

Distribution: This species appears to be limited to Venezuelan waters.

Type: Holotype USNM no. 784720; 15.4 × 4.2 mm.

Discussion: There are no Western Atlantic species with which this species might be confused.

Plate 68

FIG. **264:** Holotype of *Terebra angelli* Gibson-Smith & Gibson-Smith. USNM no. 784720; 15.4 × 4.27 mm. **265,** Holotype of *Terebra pseudopertusa* Bratcher & Cernohorsky, Museum National d'Histoire Naturelle Paris; 36.1 mm. **266,** Holotype of *T. reunionensis* Bratcher & Cernohorsky, Museum National d'Histoire Naturelle Paris; 13.2 mm. **267,** *T. connelli* Bratcher & Cernohorsky, Natal Museum no. B6307; 97.6 mm. **268,** Holotype of *T. riosi* Bratcher & Cernohorsky, Los Angeles County Museum of Natural History no. 1974; 9.8 mm.

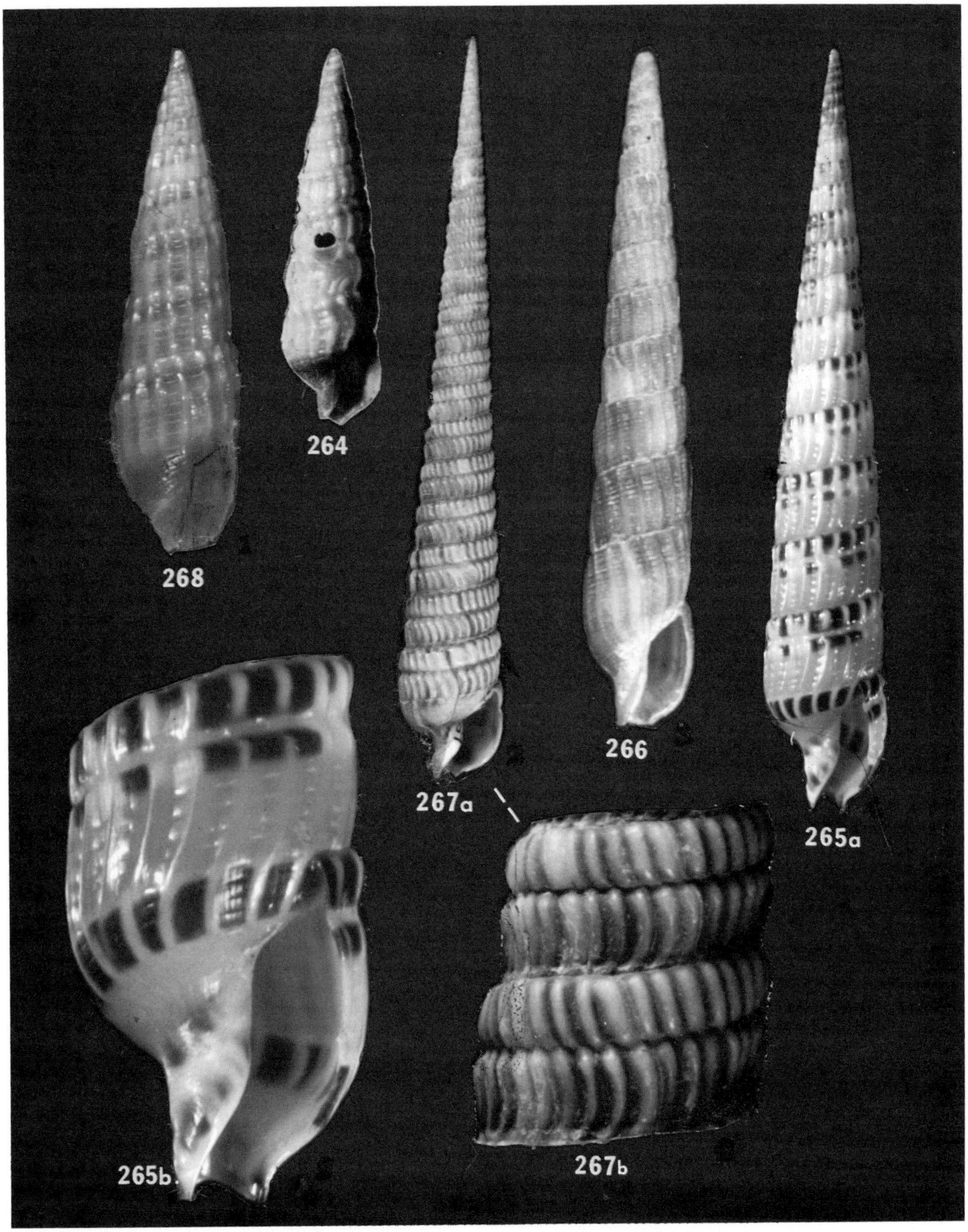

Plate 68. Recently Described *Terebra* (various oceans).

An Eastern Atlantic species, *T. fernandesi* Bouchet (**205**) bears a slight resemblance, but it is more slender and has less inflated whorls. Both have direct development, which indicates a short range of dispersal. It was named for the marine biologist Charles Angell, who collected numerous specimens, some from *Thalassia* mud flats.

265. *Terebra pseudopertusa*
Bratcher & Cernohorsky, 1985
(Pl. 68, Figs. 265, 265a)
(Color pl. E, fig. 12)

1985 *Terebra pseudopertusa* Bratcher & Cernohorsky, Nautilus 99(4):91, figs. 4, 5.

Description: Shell to 36 mm; color shiny yellowish with dark brown and white maculations between ribs on the substantial band, immediately posterior to the substantial band, at the periphery of the body whorl, and sometimes on the siphonal canal; protoconch of 1½ mamillate whorls; outline of whorls straight; subsutural band defined by punctations between ribs; axial ribs sharp, 15 on penultimate whorl, with wider interspaces; interspaces with 2 spiral grooves not crossing ribs; body whorl with ribs ending at periphery, smooth below; aperture quadrate; columella slightly recurved.
Type locality: Reunion Is., Indian Ocean (21° 06′S, 55°01″E) at 80-83 m depth.
Distribution: From South Africa to Papua New Guinea, in depths to 110 m.
Type: Holotype MNHNP; 36.1 × 6.2 mm.
Discussion: This species was originally thought to be a color form of *T. pertusa* (Born, 1778) (species, **77**), until a specimen was examined with its mamillate protoconch intact. All specimens with the extra row of maculations subsequently examined also posess paucispiral protoconchs. *T. pertusa* has a multispiral protoconch, always lacks the additional rows of maculations, and grows to a much larger size (97.5 mm) than any specimens of this species examined.

266. *Terebra reunionensis*
Bratcher & Cernohorsky, 1985
(Pl. 68, fig. 266)

1985 *Terebra reunionensis* Bratcher & Cernohorsky, Nautilus 99(4):93, fig. 3.

Description: Shell to 14 mm, slender; color light brownish beige; protoconch of 3 conical embryonic whorls; outline of whorls flat; no subsutural band; axial ribs faintly curved, about equal to interspaces, extending from suture to suture, about 18 on penultimate whorl; spiral cords in interspaces numerous, about 10 on penultimate whorl, lightly crossing ribs, fading below periphery of body whorl; aperture elongate; columella slightly curved.
Type locality: Reunion Island, Indian Ocean (21°05′S, 55°12′E), in depths of 170-225 m.
Distribution: Reunion Island, Indian Ocean, in depths of 110 to 225 m.
Type: Holotype, MNHNP; 13.2 × 2.2 mm.
Discussion: This is the most slender of the small Indo-Pacific terebrids. *Terebra contracta* (E. A. Smith, 1873) may be separated from this species by its brown aperture and columella, its turrited outline and more elongate body whorl.

There is little variation in the three specimens of this species except that the spiral cords on one specimen are missing from the anterior third of the last three whorls only.

267. *Terebra connelli*
Bratcher & Cernohorsky, 1985
(Pl. 68, figs. 267, 267a)
(Color pl. F, Fig. 1)

1985 *Terebra connelli* Bratcher & Cernohorsky, Nautilus 99(4):93, figs. 2, 6.

Description: Shell to 98 mm, robust; color brownish beige stained with brown and with a nebulous brown area anterior to periphery of body whorl; protoconch unknown; outline of whorls shouldered below convex subsutural band; band with round nodes on early whorls, becoming strong ribs later, occupying about ⅓ of the whorl, defined by a moderately deep groove; axial ribs on remainder of whorl rounded, strong, crowded, the same strength and width as those on band, straight on early whorls, curved later, about 31 on penultimate whorl; spiral sculpture absent except for a narrow cord below the axial ribs immediately posterior to the suture; body whorl short, with ribs ending at the periphery; aperture quadrate; columella recurved; siphonal fasciole striate, defined by keel.
Type locality: Off Umhlanga Rocks, Natal, South Africa, 116 m.
Distribution: Natal, South Africa.
Type: Holotype, NM no. C4715; 97.6 × 14.4 mm.
Discussion: The ribs on the band and remainder of the whorl have the appearance of polished antique ivory. The only species with which this

can be compared is *Terebra pretiosa* Reeve (43) which also has numerous heavy ribs, but that species has longer whorls, a flatter subsutural band defined by deep punctuations, and its color is yellowish beige marked with dark brown maculations. This species was named for Dr. Allan Connell, a South African collector who dredged the holotype.

268. ***Terebra riosi*** Bratcher & Cernohorsky, 1985

(Pl. 68, fig. 268)

1984 *Terebra doellojuradoi* Carcelles, Aubry, Terebridae pl. 9, upper left [non Carcelles, 1953].
1985 *Terebra riosi* Bratcher & Cernohorsky, Nautilus 99(4):93, fig. 1.

Description: Shell to 10 mm; color yellowish cream, almost translucent, with light purplish brown anterior to periphery of body whorl and nebulous purlish-brown subsutural line which appears to be under translucent enamel; protoconch of 3½ amber conical whorls; outline of whorls faintly turreted; subsutural band marked only by a nebulous purplish brown line on early whorls, 2 rows of weak nodes appear later with the nebulous line in the center, no subsutural groove; axial ribs sharp, narrower than interspaces, in early whorls unbroken from suture to suture, subsutural nodes developing later, about 15 on penultimate whorl; spiral grooves evenly spaced, about 5 on penultimate whorl, not crossing summit of ribs; body whorl with axial ribs ending at periphery, spiral grooves continuing to siphonal fasciole, aperture quadrate; columella dark brown, slightly curved.
Type locality: Buzios, Cabo Frio, Rio de Janeiro, Brazil, intertidal in sand.
Distribution: Brazil, intertidal to 30 m.
TypeL Holotype, LACM no. 1974; 9.8 × 2.5 mm.
Discussion: This species has been misidentified by dealers and Aubry (1984) as *Terebra doellojuradoi* Carcelles (species **194**) which has a protoconch of 1½ whorls, a teleconch with cancellate sculpture, a subsutural band marked by punctuations or square pits between the ribs, and no brownish subsutural line nor brownish area below the periphery of the body whorl. *T. riosi* was named for Prof. E. C. Rios of Brazil who first brought it to the authors' attention.

GLOSSARY

angulate: having angles or sharp corners.
anterior: situated at the front part or that nearest the aperture.
anterior canal: the channel on the dorsum or back of the shell.
aperture: the opening of the shell.
apex: the tip or pointed end of the shell formed earliest.
apical angle: the angle of the spire as the whorls increase.
auger: the common name for a *Terebra*.
axial: lengthwise of the shell; not spiral; parallel to the axis of coiling in a gastropod.
base: the bottom or lower part of the shell; opposite the apex.
body whorl: the last whorl, that containing the aperture.
buccal: pertaining to the mouth.
bulbous: inflated or swollen like a bulb.
callus: a thickening of surface material.
canal: a groove, tube, or duct, as for the siphon in gastropods.
cancellate sculpture: having striae, threads or cords crossing each other at right angles, dividing it like latticework.
carnivorous: flesh-eating; feeding on animals.
chitinous: horny or leatherlike (made of conchiolin).
coalesce: come together, fuse, or blend.
columella: the axial or central pillar around which the whorls are built, part of which can be seen at the inner edge of the aperture.
concave: hollowed out or excavated.
conspecific: members of the same species.
contiguous: touching or joining at the edge.
convex: curving or bulging out.
cord: rounded, string-like sculpture.
cotype: any one of a series of specimens to which equal weight has been given in describing a species (old term for syntype).
crenulated: wrinkled on the edges or delicately corrugated.
depressed: lowered in positions; sunk below the general surface.
detritus: a mass of disintegrated, fragmentary material, waste on the ocean floor or sediment on the under side of rock, coral etc.
diagnosis: a formal, usually abbreviated, statement of the characters distinguishing one taxonomic category from closely related categories.
divergent: extending in different directions from the same origin.
dorsum: the side opposite the aperture, the back.
elongate: lengthened.
embryonic whorls: those formed before hatching (at the apex of the shell)
endemic: having a habitat restricted to a certain geographical region.
eroded: worn away by physical means.
everted: turned backward or outward.
fathom: a nautical unit of measure, six feet or 1.83 meters.
fauna: the animals living within a given area or geological era.
flammules: spots of color shaped like small flames.
flaring: spreading from within outward.
flexuous: winding or bending gently to and fro, zigzag.
fold: a part folded over another part; a spiral ridge on the columella.
foot: the muscular undersurface of the body of the mollusk upon which the animal rests or moves; the fleshy organ of locomotion.
fossil: hardened remains or traces of animal or plant life of a previous geologic period.
fulvous: reddish yellow, reddish-brown, or rust color.
genus: a taxon including one to many species of presumed common phylogenetic origin, separated from related but similar taxa by a decided anatomical gap.
granulated: covered with minute grains or sandlike nodes.
habitat: the place where a race, species, or individual naturally or usually lives or is found.
heterostrophic protoconch: one which is turned in another direction for the first few turns.
homonym: identical name given to two or more different genera or species.
incised: sculptured with sharply cut grooves.
indigenous: native to or originating in a specified place or country (endemic).
inflated: swollen, distended.
intercostal: placed or occurring between ribs or ridges.
interspaces: spaces between the axial ribs.
interstices: the space between the axial ribs or ridges.
intertidal zone: the area between high and low tides.
juvenile shell: an immature, not fully grown shell.
keel: a carina or prominent ridge, usually marking an abrupt change in the slope of the shell outline.
lectotype: a specimen upon which a species is based selected from a syntypic series by a subsequent reviewer.
lip: the outer edge of the aperture, opposite the columella.
lirate: fine raised lines on the shell surface.
littoral: the shore area between high and low tide lines.
macroscopic: visible to the naked eye.
malacology: the branch of zoology which deals with mollusks.
mamillate: rounded, nipplelike in shape.
mantle: a membranous covering of a mollusk that secretes the shell with its glands.
mollusk: member of the phylum Mollusca, an invertebrate animal with a soft, unsegmented body, usually covered with a shell or having an internal shell.
microscopic: visible only with the aid of a microscope.
millimeter: one-thousandth of a meter; 0.039337 of an inch.
monograph: a treatise discussing a single subject in detail.
monotypic genus: a genus represented by a single species.
mottled: finely spotted or blotched with different colors.
multispiraled: with many whorls.

nebulous: clouded, misty, indistinct.
neotype: a type specimen of a species selected to replace the original type if lost or destroyed.
node: a knob, swelling, or tubercle.
nucleus: the tip or earliest part of the shell already formed when the egg is hatched.
obese: distended, enlarged, fat.
obsolete: obscure, not very distinct; atrophied.
operculum: a horny or shelly plate, usually serving to close the aperture when the animal is retracted.
paratype: a specimen other than the holotype which was before the author at the time of preparation of the original description and was so designated or indicated by the original author.
parietal callus: a thickening of the inner lip along the columella. When thin, it resembles a coat of varnish; when heavy, it extends like a plate beyond edge of the columella.
paucispiral: few-whorled.
penultimate whorl: the whorl before the body whorl; next to last whorl.
periostracum: a skin-like covering of the shell made of conchiolin.
periphery of body whorl: the point of greatest circumference of the last whorl.
plication: a small fold or corrugation.
planktonic: part of the group of minute pelagic animals.
polymorphic: having many forms or types of structure in a single species.
posterior: the part away from the aperture, situated away from the anterior end.
protoconch: the embryonic shell, usually different in texture from the subsequent whorls.
punctate: with small pinprick-like pits or punctures.
pustule: an elevation resembling a pimple or blister.
quadrate: somewhat rectangular or squarish in shape.
radula: toothlike processes, usually on a lingual ribbon, used in rasping or immobilizing prey.
recurved: curved and bent back.
retract: to draw back (within the shell).
rib: a long axial ridge or elevated strip of shell (usually from suture to suture).
s.s. (sensu stricto): in the strictest or limited sense.
sculpture: the raised or depressed markings on the surface of a shell.
serrated: marked with a saw-toothed edge.
shouldered: marked with a ridge.
sinistral: having the whorls of the spire turning to the left or counterclockwise.
sinuous: undulating, winding.
siphonal fasciole: curved growth lines, located at the lower left side of the aperture, marking the former position of the siphonal notch.
spiral sculpture: that revolving around the shell.
spire: all the whorls prior to the last or body whorl.
striae: extremely fine lines of sculpture.
subsutural band: the band immediately below the suture line.
subsutural groove: the groove marking the anterior or lower border of the subsutural band.
suture: the line of junction of seam between two whorls of the shell.
sympatric: coexisting in one place.
syntype: one of several specimens of equal rank upon which a species is based.
teleoconch: the entire shell, except the protoconch.
terminal: at the extreme end.
threads: fine thin elevations of sculpture, finer than cords.
translucent: semi-transparent.
tubercles: small nodes or bumps.
type: a specimen or specimens upon which a description of a new species is based.
type species: the species used by the author of a genus-group to characterize that particular genus or subgenus.
undulate: having a wavy appearance, sinuous.
veliger: a larval mollusk in the stage where it has a ciliated swimming membrane or membranes.
velum: a thin membranous covering.
whorl: one complete coil of a spirally coiled shell.

BIBLIOGRAPHY

We have tried to include in this bibliography everything that has been published with illustrations on the recent Terebridae. Without them one can never be sure of the actual species the author had in mind. In addition, we are including the literature containing descriptions of new species, whether illustrated or not and articles on terebrid animal morphology, habits, or radula. We have used the names under which authors publish, whether they gave full names or merely used initials, such as R. Tucker Abbott, A. Myra Keen or E. A. Smith.

Abbott, R. Tucker. 1952. A New *Terebra* from the Philippines. *Nautilus* **65** (3): 77-80, pl. 5.

1954. New Gulf of Mexico gastropods (*Terebra* and *Ocenebra*). *Nautilus* **68** (2): 37-44, pl. 2.

1968. *Seashells of North America.* Golden Press, New York, 296 p., text figs.

1974. *American Seashells.* The marine mollusca of the Atlantic and Pacific coasts of North America. (2nd ed.). Van Nostrand/Reinhold Co., New York, 633 p., 24 pls., text figs.

Abbott, R. Tucker and S. Peter Dance. 1982. *Compendium of Seashells.* E. P. Dutton, Inc., New York 411 p.

Adams, A., and L. A. Reeve. 1848-50. *The Zoology of the Voyage of H. M. S. Samarang;* Mollusca. Reeve, Benham and Reeve, London, x + 87 p., pls. 1-24.

Adams, Henry and Arthur Adams 1853-58. *The Genera of Recent Mollusca; arranged according to their organization.* 3 vol. London.

Adanson, M. 1757. *Histoire des coquillages.* 96 pp. + 19 pls. in *Histoire naturelle du Sénégal.* Paris.

Allan, Joyce. 1950. *Australian Shells.* Georgian House, Melbourne, p. XXI + 487, 44 pls., 112 text figs.

Andrews, J. 1975. *Seashells of the Texas coast.* Texas University Press, Austin & London. 298 p., text figs.

Angas, George French. 1867. Descriptions of thirty two new species of marine shells from the coast of New South Wales. *Proceedings of the Zoological Society of London.* pp. 110-117, pl. 13.

Aubry, Umberto. 1984. *Terebridae.* Umberto Aubry. Sorrento, Italy, p. 1-44, 15 pls.

Baird, W., in J. L. Brenchley. *Cruise of the H.M.S. Curacoa Among the South Sea Islands,* Section "Shells." Longmans, Green, and Co. London, p. 432-454, pls., 36-42.

Barnard, K. H. 1958. Contributions to the knowledge of South African marine Molusca. Part. 1. Gastropoda: Prosobranchiata: Toxoglossa. *Annals of the South African Museum.* **44** (4): 73-163, pl. 2, text figs.

Bartsch, Paul 1923. A key to the family Terebridae. *Nautilus* 37 (2): 60-64.

Bartsch, Paul and Harald Alfred Rehder. 1939. Mollusks collected on the presidential cruise of 1938. *Smithsonian Miscellaneous Collections,* **98** (10): 1-21, pl. 1.

Bean, W. 1844 in Thorpe's. *British Marine Conchology,* a supplement of new species. London: 263-267 pp.

Berry, S. Stillman 1958. Notices of New Eastern Pacific Mollusca.-II. *Leaflets in Malacology.* S. Stillman Berry, Redlands, Calif. **1** (15): 83-90.

1958a. West American Molluscan Miscellany.-II. *Leaflets in Malacology.* S. Stillman Berry, Redlands, Calif. **1** (16): 91-98.

1959. Notices of new Eastern Pacific Mollusca-III. *Leaflets in Malacology.* S. Stillman Berry. Redlands, Calif. **1** (18): 107-114.

Blainville D. in Thomas Wyatt. 1838. *Manual of Conchology,* according to the system laid down by Lamarck, with the latest improvements by de Blainville. Harper Brothers, New York. i-ix + 11-191, 36 pls.

Born, I. von 1778. *Index rerum naturalium Musei Caesarei Vindobonensis;* pars I, Testacea. pp.xlii + 458.

1780 *Testacea Musei Caesarei Vindobonensis.* Vienna, pp. xxxvi + 442; 18 pls.

Bosc, Loui Auqustin Guillaume. 1824. *Histoire Naturelle Des Coquilles.* Paris **5**: 10-22.

Bouchet, Philippe. 1981. Evolution of larval development in Eastern Atlantic Terebridae (Gastropoda), Neogene to Recent. *Malacologia* **21** (1/2): 363-369, text figs.

1983. Les Terebridae (Mollusca, Gastropoda) de l'Atlantique Oriental. *Bolletino Malacologico* **18** (9-12): 185-216. Milano.

Bouchet, P. & S. Golfas *"Terebra" cosentini* Philippi, 1836, an American columbellid species. *Nautilus* **97** (1); 26-28, text figs. 1, 2.

Bouvier, L., 1887. *Systéme nerveux, morphologie générale et classification des Gastérepods Prosobranches.* Paris.

Bratcher, Twila. 1969. Rediscovery of *Terebra cochlea* Deshayes. *Veliger* **11** (4): 334-335, pl. 52. [i.e., *trochlea*].

1973. Six *Terebra* look-alikes. *Hawaiian Shell News* **21** (4): 1, 3, text figs.

1977. Deshayes' terebrid types in Ecole des Mines, Paris. *Nautilus* **91** (2): 39-42, text figs.

1979a. Taxonomic changes in Eastern Pacific Terebridae, with the description of new species (Mollusca: Gastropoda). *Veliger* **22** (1): 61-64, text figs.

1979b. A new Indo-Pacific terebrid. *Veliger* **22** (1): 65-66, text fig.

1981. Four previously undescribed Indo-Pacific terebrids (Mollusca: Gastropoda). *Veliger* **23** (4): 329-332, text figs.

1983. *Terebra knudseni* sp. n. from North Borneo (Mollusca, Gastropoda, Terebridae). *Steenstrupia,* Zoological Museum, Univ. Copenhagen, 8 (15): 301-305.

Bratcher, Twila and R. D. Burch. 1970a. Five new species of

Terebra from the Eastern Pacific. *Veliger* **12** (3): 295-300, pl. 14.

1970b. Four new terebrid gastropods from Eastern Pacific Islands. *Contributions in Science*, Los Angeles County Museum of Natural History no. 188: 1-6, text figs.

1971. The Terebridae (Gastropoda) of Clarion, Socorro, Cocos and Galapagos Islands. *Proceedings of California Academy of Sciences* (ser. 4), **37** (2): 537-566, text figs.

Bratcher, Twila and R. D. Burch in Keen, 1971. *Seashells of Tropical West America*, ed. 2. Stanford University Press, Stanford, California. p. 670-686, 56 text figs.

Bratcher, T. and H. K. Mienis. 1978. On the identity and distibution of *Terebra castigata* Cooke, 1885. *Argamon*, Israel Journal of Malacology **6** (3-4): 21-25, text figs.

Bratcher, Twila and Walter O. Cernohorsky 1982. Six new species of Indo-Pacific Terebridae (Gastropoda). *Nautilus* **96** (2): 61-66, text figs.

Briggs, J. 1974. *Marine Zoogeography*. McGraw Hill. 475 pp.

Burch, F. D. 1965. New terebrid species from the Indo-Pacific Ocean and from the Gulf of Mexico, with new locality records and provisional list of species collected in Western Australia and at Sabah, Malaysia (Mollusca: Gastropoda). *Veliger* **7** (4): 241-253, pl. 31.

Campbell, G. Bruce. 1961. Four new Panamic gastropods. *Veliger* **4** (1): 25-28, pl. 5.

1963. Rediscovery of *Terebra formosa* Deshayes, 1857. *Veliger* **5** (3): 101-103, pls. 12, 13.

1964. New terebrid species from the Eastern Pacific (Mollusca: Gastropoda). *Veliger* **5** (3): 132-138, pl. 17.

Carcelles, Alberto R. 1953. Nuevas especies de gastropodos marinos de las Republicas Oriental del Uruguay y Argentina. *Communicationes Zoologicas del Museo de Historia* **70** (4): 1-16.

Carpenter, P. P. 1857. *Catalogue of the Collection of Mazatlan Shells in the British Museum: Collected by Frederick Reigen*. London (British Museum), pp. i-iv + ix-xvi + 552.

1864. Supplementary report on the present state of our knowledge with regard to the mollusca of the west coast of North America. *Report British Association for Advancement of Science* **33: 517-686.**

1865. Diagnosis of new forms of Mollusca, from the West coast of North America, first collected by Col. E. Jewett. *Annals and Magazine of Natural History*, 3rd Series, **15**: 394-399.

Cernohorsky, W. O. 1967. *Marine Shells of the Pacific*. Pacific Publications, Sydney. 248 pp., 60 pls.

1969. List of Type Specimens of Terebridae in the British Museum (Natural History). *Veliger* **11** (3) 210-222.

1978. *Tropical Pacific Marine Shells*. Pacific Publications, Sydney. 352 p., 68 pls.

1980. The taxonomy of some Indo-Pacific Mollusca. Part 8. *Records Auckland Inst. Mus.* **17**: 135-152, 35 text figs.

Cernohorsky, Walter O. and Twila Bratcher. 1976. Notes on the taxonomy of Indo-Pacific Terebridae (Mollusca: Gastropoda), with description of a new species. *Records Auckland Inst. and Mus.* **13**: 131-140, text figs.

1985. Three New Deep-water Indo-Pacific and One Intertidal Brazilian Species of *Terebra* (Gastropoda). *The Nautilus* **99:** 91-94.

Cernohorsky, Walter O. and Albert Jennings. 1966. The Terebridae of Fiji (Mollusca: Gastropoda). *Veliger* **9** (1): 37-67, pls. 4-7, text figs.

Chemnitz, J. H. 1780. *Neues Systematisches Conchylien-Cabinet* fortgesetzt durch Johann Hieronymus Chemnitz. G. N. Raspe, Nürnberg, 4: 1-344, pls. 122-159.

Chenu, J. C. 1842-53. *Illustrations conchyliologiques* ou Description et figures de toutes le coquilles conues, vivantes et fossiles. Paris 215 pp., 481 pls.

Clench, William J. 1938. A new species of *Oliva* from Santo Domingo, with notes on other marine forms, *Nautilus* **51** (4): 109-114, pl. 9.

Clench, William J., and C. G. Aguayo. 1939. Notes and descriptions of new deep water mollusca obtained by the Howard-Havana expedition off the coast of Cuba. *Memoirs of the Sociedad de Historia Natural*, Universidad de la Habana **13** (3): 189-197, pl. 28, 29.

Clench, William J., and Ruth D. Turner. 1950. The Western Atlantic marine mollusks described by C. B. Adams. *Occasional Papers on Mollusks*, Harvard University **1** (15): 233-403, pls. 29-49.

Coleman, Neville 1975. *What Shell is That?* Paul Hamlyn Ltd., Sydney, 308 p., 811 figs.

Conrad, Timothy Abbott 1845. Descriptions of new species of fossil and recent shells and corals. *Proceedings of the Academy of Natural Sciences of Philadelphia* **3** (1946): 19-27, pls. 1-2.

Cooke, A. H. 1885. Report of the Testaceous mollusca obtained during a dredging-excursion in the Gulf of Suez in the months of February and March 1869. By Robert McAndrew, republished, with additions and corrections, by Alfred Hands Cooke. *Annals and Magazine of Natural History* (5th ser.) **15**: 322-339.

Cossman, Alexandre Edouard Maurice 1896. *Essais de Paleoconchologie Comparee*, **2**: 46-56. Paris. 13 vols.

Cotton, Bernard C. 1952. Australian Recent and Tertiary Mollusca (Terebridae, Rissoinidae, Rissoidae, Fasciolariidae, Volutidae). *Trans. Royal Soc. South Australia* **75;** 38-54, pls. 3, 4.

Cotton, Bernard C. and Godfrey, F. C. 1932. South Australian Shells. *South Australian Naturalist* **13** (2): 35-86 pls. 1-3.

Dall, William Healey 1889a. Reports on the results of dredging in the Gulf of Mexico (1877-78) and the Caribbean Sea (1879-80), by the U. S. Coast Survey Steamer "Blake", *Bulletin Museum of Comparative Zoology*, Harvard. **18** (2): 1-492, pls. 1-40.

1889b. A preliminary catalogue of the shell-bearing marine mollusks and brachiopods of the southeastern coast of the United States, with illustrations of many of the species. *Bulletin of the U. S. National Museum* no. 37: 1-221, pls. 1-74.

1890. Scientific results of explorations by the U. S. Fish Commission steamer "Albatross." *Procedings of the U. S. National Museum*. **12** (773): 219-326, pls. 5-13.

1895. Diagnosis of new Tertiary fossils from Southern United States. *Proceedings of the U. S. National Museum* 18 (1035): 21-46.

1898. A new species of *Terebra* from Texas. *Nautilus* 12 (4): 44-45.

1901. The mollusca of Porto Rico. *Bulletin of the U. S. Fish Commission* no. 20: 351-516, pls. 53-58.

1902. Illustrations and descriptions of new unfigured or imperfectly known shells, chiefly American, in the U. S. National Museum. *Proceedings of the U. S. National Museum* **24** (1264): 499-566, pls. 27-39.

1908a. Subdivisions of the Terebridae. *Nautilus* **21** (11): 124-125.

1908b. Reports on the scientific results of the expedition to the eastern tropical Pacific in charge of Alexander Agassiz, by the U. S. Fish Commission steamer "Albatross" from October 1904, to March, 1905. *Bulletin of the Museum of Comparative Zoology,* Harvard **43** (6): 205-487; 22 pls.

1925 Illustrations of unfigured types of shells in the collection of the U. S. National Museum. *Proc. U. S. Nat. Mus.* **66** (2554): 1-41, pls. 1-36.

Dance, S. P. and F. E. Eames. 1966. New molluscs from the recent Hammar formation of south-east Iraq. *Proc. Malac. Soc. London* **37** (1): 35-43; pls. 2-5.

Dautzenberg, P. 1910. Contributions á la faune malacologique de l'Afrique que occidentale. *Actes Soc. Linn. Bordeaux* **64:** 1-174.

1913. Mission Gruvel sur la cote occidentale d'Afrique: Mollusques marins. *Ann. Inst. Oceanogr* **5** (3): 1-111.

1935. Résultats Scientifiques du Voyage aux Indes Orientales Néerlandaises de le Prince et la Princesse Léopold de Belgique. Gastéeropodes marins. Terebridae & Mitridae. *Memoirs de Muséum Royale d'Hist. Nat. Belgique* **2** (17): 1-43; pls. 1-4.

De Kay, James 1843. *Zoology of New York Fauna.* Albany: Carroll & Cook. Part 5, Mollusca, p. 1-271, plts. 1-40.

Deshayes, G. P. 1857. Description d'espéces nouvelles du genre *Terebra. Journal de Conchyliogie* **6:** 65-102, pls. 3-5.

1859. A general review of the genus *Terebra* and a description of new species. *Proc. Zool. Soc. London* (Pt. 27): 270-321.

Dillwyn, Lewis Weston 1817. *A Descriptive Catalogue of recent Shells.* John & Arthur Arch, Cornhill, London. **3:** 581-1092 + index.

Dodge, Henry 1956. A historical review of the mollusks of Linnaeus. Pt. 4. The genera *Buccinum* and *Strombus* of the class Gastropoda. *Bull. Amer. Mus. Nat. Hist.* 3, art. 4:157-312.

Dunker, Wilhelm 1853. Diagnoses molluscorum novorum. *Zeitschr. für Malakozool.* **10** (6): 95-96; 10 (7): 110-112.

1882. *Index Molluscorum Maris Japonici.* T. Fischer, Cassel, p. 1-301; pls. 1-16.

DuShane, Helen & Gale G. Sphon. 1968. A checklist of intertidal mollusks from Bahia Willard and the southwest portion of Bahia San Luis Gonzaga, State of Baja California, Mexico. *Veliger* **10** (3): 233-246, 1 pl.

Edmonson, Charles Howard 1933. *Reef and Shore Fauna of Hawaii,* ed. 2. Bernice P. Bishop Mus. special Pub. 22. III + 381 pp., 223 figs.

Fischer-Piette, E. 1942. *Les Mollusques d'Adanson. Journal de Conchyliologie* **85:** 103-374, 16 pls.

Fischer von Waldheim, 1807. *Museum Demidoff.* Moscow, 3: 178-181 (Terebridae only).

Garcia-Cubas, Antonio, 1981. Mollusks of a Tropical Lagoon System in the Southern Gulf of Mexico. *Inst. Cienc. del Mar Y Limnol. Univ. Nat. Auton, Mexico* **5**: 1-182.

Garrard, T. A. 1976a. Two species of *Terebra. Australian Shell News* no. 13: 31, text figs.

1976b. The genus *Pervicacia* (Terebridae: Gastropoda: Mollusca) in south-eastern Australia. *Journ. Malac. Soc. Australia* 3 (3-4): 79-185, text figs.

Gibson-Smith, J. and W. Gibson-Smith. 1984. A Revision of the Terebrid Gastropods of the Mare Formation, Cabo Blanco, Venezuela. *Tulane Studies in Geol. & Paleo.:* 18, (1&2): 61-66, pl. 1.

Gmelin, J. F. 1791. *Systema naturae* per regna tria naturae. Editio decima tertia, aucta, reformata Leipzig 1(6): 3021-3910.

Gould, Augustus A. 1861. Descriptions of New Shells Collected by the North Pacific Explor. Exped. *Proc. Boston Soc. Nat. Hist.* **7:** 323-340.

Grabau, Amadeus W. and Sohtsu King. 1928. *Shells of Peitaiho.* Peking Labor. Nat. Hist. ed. 2, pp. I-IV + 1-283, pls. 1-11.

Grant, Ulysses S., IV, & Hoyt Rodney Gale. 1931. Catalogue of the marine, Pliocene and Pleistocene mollusca of California and adjacent regions. *Mem. San Diego Nat Hist.* pp. 1-1036, 15 text figs; pls. 1-32.

Gray, J. E. 1834. Enumeration of the species of the genus *Terebra Proc. Zool. Soc. London* Pt. 2: 59-63.

1834a. in Edward Griffith and Edward Pidgeon. *The Animal Kingdom by Cuvier.* London: pp. viii + 601, 40 pls.

1853. On the division of ctenobranchous Gastropodous Mollusca into larger groups and families. Ann. Mag. Nat. Hist. ser. 2, 11 (62): 124-133, 10 text figs.

Griffiths, Edward and Edward Pidgeon. 1833(?)-34. The Mollusca and Radiata in E. Griffith, *The Animal Kingdom by Cuvier.* London: pp. viii + 601.

Guerin-Meneville, Félix E. 1854. Traveux inédits. *Revue et Mag. de Zool.*, ser. 2, vol. **6:** 218, pl. 4.

Habe, Tadashige 1964. *Shells of the western Pacific in color.* Hoikusha Publ. Co., Osaka, **2:** 1-233, pls. 1-66.

Habe, Tadashige and Sadao Kosuge. 1966. *Shells of the world in color.* Hoikusha Publ. Co., Osaka, **2:** 1-193, pls. 1-68.

1967. *The Standard Book of Japanese Shells in Color.* Japan **3:** 1-223, pls. 1-64.

Hanna, G. Dallas and Leo George Hertlein, 1961. Large *Terebras* (Mollusca) from the eastern Pacific. *Proc. California Acad. Sci.* (ser. 4) **30** (3): 67-80, pls. 6-7.

Hedley, C. 1912. Descriptions of some new noteworthy shells. *Rec. Australian. Mus.* 8 (3): 131-160; pls. 40-45.

1915. Studies on Australian Molusca, XII. *Proc. Linn. Soc. New South Wales* **39** (4): 695-755; pl.84.

Hedley, C. and W. F. Petterd 1906. Mollusca from three hundred fathoms, off Sydney. *Rec. Australian Mus.* **6** (3): 211-225, pls. 7-38.

Hertlein, Leo George & A. M. Strong. 1955. Marine mollusks collected during the "Askoy" expedition to Panama, Colombia, and Ecuador in 1941. *Bull. American Mus. Nat. Hist.* **107** (2): 159-318, pls. 1-3.

Hinds, R. B. 1844a. Descriptions of new shells, collected during the voyage of the Sulphur, and in Mr. Cuming's late visit to the Philippines. *Proc. Zool. Soc. London* (1843) Pt. 11: 149-168.

Hinds, R. B. in Sowerby. 1844b. Monograph of the genus *Terebra;* Bruguiere. *Thesaurus Conchyliorum.* G. B. Sowerby, London 1: 147-190, pls. 41-45.

Hinton, A. 1972. *Shells of New Guinea and the central Indo-Pacific.* R. Brown and Associates Pty. Ltd., Port Moresby, 94 p., 44 pls.

1978. *Guide to Australian shells.* R. Brown and Associates Pty. Ltd., Port Moresby, 77 pls., Index.

Hirase, Y. 1917. *Terebridae of the Japanese Empire.* Hirase Conch. Mus. Kyoto, 50 pp., 9 pls.

Homes, Peter B. 1974. A very good year in New Hebrides.

Hawaiian Shell News **22** (9): 11. Text figs.
Humfrey, M. 1975. *Sea shells of the West Indies.* Taplinger, New York. 351 pp., 32 pls., text figs.
Iredale, Tom 1924. Results from Roy Bell's molluscan collections. *Proc. Linn. Soc. New South Wales* **49** (3): 179-278, pls. 33-36.
1925. Mollusca from the continental shelf of eastern Australia. *Rec. Australian Mus.* **14** (4): 243-270; pls. 41-43; 1 map.
1929. Strange mollusks in Sydney Harbour. *Australian Zoologist* **5** (4): 337-352; 2 pls.
Kay. E. Alison 1979. *Hawaiian Marine Shells.* Reef and shore fauna of Hawaii. Section 4: Mollusca. Bernice P. Bishop Mus. Spec. Publ. 64 (4): 1-653, text figs.
Keen, A. Myra 1958. *Seashells of Tropical West America.* Marine mollusks from Baja California to Columbia. Stanford University Press, Stanford, Calif. i-xi + 624 pp., 10 colored pls, 1700 text figs.
1943. New mollusks from the round mountain silt (Temblor) Miocene of California. *Transactions of the San Diego Society of Natural History* **10** (2): 25-60, pls. 3, 4.
1966a. West American mollusk types in the British Museum (Natural History) II, Species described by R. B. Hinds. *Veliger* **9** (4): 265-275; pl. 47.
1966b. Moerch's West Central American Molluscan Types with proposal of a new name for a species of *Semele. Occ. Papers Calif. Acad. Sci.* no. 59: 1-33, 41 figs.
1968. West American mollusk types at the British Museum (Natural History) IV., Carpenter's Mazatlan collection. *Veliger* 10 (4): 389-439; plts. 55-59, 171 figs.
1971 *Seashells of Tropical West America,* marine mollusks from Baja California to Peru., ed. 2. Stanford University Press, Stanford, Calif. i-xiv + 1064 pp., 22 colored pls.; 3306 text figs.
Keep, Josiah and Joshua Bailey. 1935. *West Coast Shells.* Stanford University Press. xi + 350 pp., 334 text figs.
Kensley, Brian 1973. *Sea-shells of southern Africa* - Gastropods. Maskew Miller Ltd., Cape Town. 225 pp., text figs.
Kiener, Louis Charles 1839. *Spécies Géneral et Iconographie des Coquilles Vivantes.* Genre *Terebra.* Rousseau, Paris. **9:** 1-42, pls. 1-14.
Kira, T. 1962. *Shells of the Western Pacific in Color.* Hoikusha Publ. Co., Osaka, 1: 1-224, pls. 1-72, text figs.
Knudsen, Jorgen. 1952. *Marine Prosobranchs of Tropical West Africa collected by the "Atlantide" Expedition* 1945-46. Part 1. Vidensk. Medd. Dansk. naturh. Foren. 114: 129-185; pls. 1-3.
1956 *Marine Prosobranchs of Tropical West Africa.* Atlantide Report no. 4, Danish Science Press, Ltd., Copenhagen. pp. 7-110; pls. 1-4.
Kosuge, Sadao 1979. Report on the Mollusca on Guyots from the Central Pacific collected by 2nd and 3rd cruises of R/V Kaiyomaru in 1972 to 73 with descriptions of twelve new species. *Bull. Inst. Malac. Tokyo* **1** (2): 24-, pl. 5-6.
1981. Notes on newly recorded species of the Superfamily Conacea from Philippines with descriptions of new species of the Genera *Terebra, Conus,* and *Glyphostoma. Bull. Inst. Malac. Tokyo* **1** (6): 93-96, pls. 31-32.
Kuroda, Tokubei and Tadashige Habe. 1952. *Check List and Bibliography of the Recent Mollusca of Japan.* L. W. Stach, Tokyo, 210 pp.
Lamarck, B.P.A.deM. 1822. *Histoire Naturelle des Animaux sans vertèbres.* Mollusques. Paris. **7:** 1-711 (*Terebra* p. 283-291).
Lea, Henry Carey. 1843. Description of some new fossil shells from the Tertiary of Petersburg, Va. *Transactions of the American Philosophical Society.* Philadelphia 2nd. ser. (9): 229-274, pls. 34-37.
Lesson, R. P. 1832-35. *Illustrations de Zoologie,* ou recueil de figures d'animaux, Paris. 60 pls.
Lewy, Z. 1975. Molluscs distribution on the Atlantic continental shelf off southern Spanish Sahara, West Africa. *Meteor Forsch. Ergebn.* (C) **21:** 52-60.
Li, Chih Chang. 1930. Miocene and recent molluscs of Panama Bay. *Bull. Geo. Soc. China* **9;** 249-296; pls. 1-8.
Lightfoot, John. 1786. *A Catalogue of the Portland Museum,* lately the property of the Duchess Dowager of Portland, deceased, which will be sold at auction, by Mr. Skinner and Co., London. 8 + 194 pp.
Linnaeus, Caroli. 1758. *Systema Naturae* per regna tria naturae. Editio decima, reformata. Stockholm, pp. 1-824.
Lorois, M. 1857. Description d'espècies nouvelles. *Journ. de Conchyl.* **6:** 388-389 pl. 12.
1858. Description d'une spècies nouvelle. *Journ. de Conchyl.* **8:** 90, pl. 1.
Macpherson, J. Hope and C. J. Gabriel. 1962. *Marine Molluscs of Victoria.* Melbourne Univ. Press, xv + 475 pp.; 486 text figs.
Marcus, Evelyn and Ernst Marcus. 1960. On *Hastula cinerea. Bol. Fac. Fil. Zoologica Sao Paulo* no. 23: 25-54, 5 pl.
Martens, Edouard Carl von. 1880. *Beiträge zur Meeresfauna der Insel Mauritius und der Seychellen, Mollusken.* Berlin. pp. 181-352, pl. 19-22.
May, W. L. 1923. *An illustrated Index of Tasmanian Shells.* Tasmania. p. 1-100, pls. 1-47.
McLean, J.H. 1969 *Marine Shells of Southern California.* Sci. Ser. 24, Zool. no. 11, 104 p, text figs. Los Angeles.
Melvill, J. C. 1897. Descriptions of thirty-four species of marine mollusca from the Arabian Sea, Persian Gulf, and Gulf of Oman. *Mem. and Proc. Manchester Lit. & Phil. Soc.* **41** (7): 1-26, pls. 6-7.
1898. Further investigations into the molluscan fauna of the Arabian Sea, Persian Gulf, and Gulf of Oman, with descriptions of forty species. *Mem. & Proc. Manchester Lit. Phil. Soc.* **42** (4): 1-39, pls 1-2.
1910. Descriptions of twenty-nine species of marine mollusca from the Persian Gulf, Gulf of Oman, and North Arabian Sea, mostly collected by Mr. F. W. Townsend, of the Indo-European telegraph Service. *Ann. Mag. Nat. Hist.* (ser. 8) **6:** 1-17, pls. 1-2.
1912. Descriptions of thirty-three new species of Gastropoda from the Persian Gulf, Gulf of Oman, and North Arabian Sea. *Proc. Malac. Soc. London* **10** (3): 240-254, pls. 11-12.
1917 Description of a new species of *Terebra* from the Mekran coast, Arabian Sea. *Journ. Conch.* **15** (6): 188-189, text fig.
Melvill, J. C. and R. Standen. 1895-97. Notes on a collection of shells from Lifu and Uvea, Loyalty Islands, formed by the Rev. James and Mrs. Hadfield, with a list of species. *Journ. Conch.* **8:** 84-132; **8:** 273-315; 8:396-421; pls. 2-3; 9-11; map.
1901. The Mollusca of the Persian Gulf, Gulf of Oman, and

Arabian Sea, as evidenced mainly through collections of Mr. F. W. Townsend, 1893-1900; with descriptions of new species. *Proc. Zool. Soc. London* **2**: 327-460, pls. 21-24.

1903. Descriptions of sixty-eight new gastropods from the Persian Gulf, Gulf of Oman, and North Arabian Sea dredged by Mr. F. W. Townsend. *Ann. Mag. Nat. Hist.* (ser. 7) **12**: 289-324, pls. 20-23.

1917. A revision of the species of *Terebra* occuring in the Persian Gulf, Gulf of Oman, and Arabian Sea, as evidenced in the collection formed by Mr. F. W. Townsend, 1893-1914. *Journ. Conch.* **15** (7): 204-216.

Melvill, J. C. and E. R. Sykes. 1898. Notes on a second collection of marine shells from the Andaman Islands, with descriptions of new forms of *Terebra. Proc. Malc. Soc. London* 3 (1): 35-48, pl. 3.

Menke, C. T. 1843. *Molluscorum Novae Hollandiae.* A. Hahn, Hanover 46 pp.

Mienis, H. K. 1970. A checklist of Terebridae from the northern part of the Red Sea, with Notes on *Terebra areolata, T. consobrina* and *T. subulata. Argamon,* Journ. Israel Malac. Soc. **1** (2): 37-42, text figs.

1971. Additions to the checklist of Terebridae from the nothern part of the Red Sea. *Argamon,* Journ. Israel Malac. Soc. **3** (1-4): 27-32, text figs.

Miller, B. A. 1966. Studies on biology and ecology of some Indo-Pacific *Terebra* from Eniwetok Atoll. *Master Science Thesis,* Univ. Hawaii. Unpublished.

1970. Studies on the biology of Indo-Pacific *Terebra; Ph.D. dissertation,* Univ. New Hampshire, available on microfilm.

1971. Feeding Mechanisms in the family Terebridae. *Annual Report American Mal. Union for 1970:* 72-74.

1975. The biology of *Terebra gouldi* Deshayes, 1859, and a discussion of the life history similarities among other terebrids of similar proboscis type. *Pacific Science* **29**: 227-241.

1979. The biology of *Hastula inconstans* (Hinds, 1844) and a discussion of life history similarities among other *Hastula* of similar proboscis type. *Pacific Science* **33** (3): 289-306.

Mills P. 1979. Radular tooth structure of three species of Terebridae (Mollusca: Toxoglossa). *Veliger* 19 (3): 259-265, text figs.

Mollick, R. S. 1973. Some aspects of the biology of *Terebra dislocata* Say, 1822. (Gastropoda: Prosobranchia). *Veliger* **16** (1): 82-84.

Mörch, Otto Andreas Lowson 1852. *Catalogus Conchyliorum quae reliquit D'Alphonso D'Aguirra & Gadea Comes de Yoldi.* L. Klein, Hafniae **1**: 1-170.

Morris, Percy A. 1966. *A Field Guide to Shells of the Pacific Coast and Hawaii.* The Riverside Press, Cambridge. pp. xxxiii + 297; pls. 1-72.

Morrison, J. P. 1968. Four American *Hastula* species. *Texas Conchologist,* Houston Conchology Club **4** (9): 67-70, 1 pl.

Nicklés, Maurice. 1950. *Mollusques TestacésM marins de la Cote Occidentale de'Afrique.* Paris. 169 pp.; 464 text figs.

Nowell-Usticke, Gordon W. 1969. *A Supplementary Listing of New Shells (Illustrated) to be Added to the Check list of the Marine Shells of St. Croix.* 60 pp. G. W. Nowell-Usticke, U.S.A.

Oldroyd, Ida S. 1925-27. *The Marine Shells of the West Coast of North America.* 2 vols., (4 parts). Stanford Univ. Ser. Geol. Sci.

Olsson, Axel A. 1971. Mollusks from the Gulf of Panama collected by R/V John Elliot Pillsbury, 167. In *Studies in Tropical American Mollusks.* University of Miami Press, Florida, pp. 35-92, 103 text figs.

Orbigny, Alcide Dessalines d'. 1839. *Voyage dans l'Amérique Méridionale.* Paris. **5** (3): xliii + 758 pp.

Oyama, K. 1961. On some new facts of the taxonomy of Terebridae *Venus: Jap. Journ. Malac.* **21** (2): 176-189, text figs.

Oyama, K. and Y. Takemura. 1961. *The Molluscan Shells. Resource Exploitation Institute,* Tokyo **5**: 61 pls. (Terebridae pls.)

Pease, Wm. Harper. 1868. Descriptions of marine gasteropodae inhabiting Polynesia. *American Journ. Conch.* **4** (3): 103-132.

1869. Descriptions of new species of marine gasteropodae inhabiting Polynesia. *American Journ. Conch.* **5** (2): 64-79.

1871. Notes on the synonymy and distribution of marine gasteropoda. *American Journ. Conch.* **7** (1): 20-25.

Perry, G. 1811. *Conchology* or the natural history of shells, univalves. W. Miller, London, 61 pls. and explanations.

Perry, Louise M. 1940. Marine Shells of the Southwest Coast of Florida. *Bull. American Paleont.* **26** (95): 5-260; pls. 1-39.

Petuch, Edward J. 1974. A new *Terebra* from the coral reef areas off North Carolina. *Veliger* **17** (2): 205-206; 1 pl.

1981. A relict Neogene caenogastropod fauna from northern South America. *Malacologia* **10** (2): 307-347.

Philippi, R. A. 1846. Diagnosen neuer Conchylien-Arten. *Zeit. Malak.*, yr. 3: 49-55.

Pilsbry, H. A. 1891. New Japanese shells. *Nautilus* 5 (8): 93.

1892. New and hitherto unfigured Japanese mollusks. *Proc. Acad. Nat. Sci. Philadelphia* **43**: 471-473, pls. 17-19.

1895. *Catalogue of the Marine Mollusks of Japan with Descriptions of New Species and Notes on Others Collected by Frederick Stearns. Detroit, F. Stearns, iii-viii + 196 pp., pls. 1-11.*

1904. New Japanese Marine Mollusca. *Proc. Acad. Nat. Sci. Philadelphia* **56**: 3-6, pl. 1.

1921. Marine mollusks of Hawaii, VIII-XIII. *Proc. Acad. Nat. Sci. Philadelphia* **72**: 296-326, pl. 12, text figs.

1931. The Miocene and Recent Mollusca of Panama Bay. *Proc. Acad. Nat. Sci. Philadelphia* **83**: 427-440, pl. 41, text figs.

Pilsbry, H. A., and H. N. Lowe. 1932. West Mexican and Central American mollusks collected by H. N. Lowe, 1929-1931. *Proc. Acad. Nat. Sci. Philadelphia* **84**: 33-144, pls. 1-17.

Powell, A. W. B. 1971. New Zealand Molluscan systematics with descriptions of new species: Part 7. *Rec. Auckland Inst. & Mus.* 8: 209-288, text figs.

1979. *New Zealand Mollusca,* Marine, Land, and Freshwater Shells. William Collins, Ltd., Auckland, 500 pp., 82 pls., text figs.

Preston, H. B. 1908. Descriptions of new species of land, marine and fresh water shells from the Andaman Islands. *Rec. Indian Mus.*, Calcutta **2** (2): 187-210, pls. 14-17.

Pritchard, G. B. & J. H. Gatliff. 1902. On some new species of Victorian mollusca, no. 5. *Proc. Royal Soc. Victoria.* (New series) **14** (2): 180-184; pls. 9-10.

Quoy, Jean René Constant & Joseph Paul Gaimard. 1832-35. *Voyage de découvertes de l'Astrolabe* exécuté par ordre du Roi, pendant les années 1826-1829, sous le commandement de M. J. Dumont d'Urville. Zoologie. Paris **1**: i-L; 1-264; **3**: 1-954; Atlas, Moll. pls. 1-93.

Ray, H. C. 1968. Small marine shells obtained off Puri, India (with eight new species and one new variety). *Indian Mus. Bull.* (Calcutta) 3 (1/2): 14-32, text figs.

Reeve, L. 1842. *Conchologia Systematica* or complete system of Conchology. Longman, Brown, Green & Longmans, London 2: 1-337, pls. 130-300.

1860. *Conchologia Iconica;* monograph of the genus *Terebra.* L. Reeve, London, 27 pls., Index.

Rehder, Harald A. 1980. The marine mollusks of Easter Island. *Smithsonian Contrib. Zool.*, no. 289, U. S. Nat. Mus., Terebridae p. 92-94.

Rios, E. C. 1970. *Coastal Brazilian Seashells.* Fundacao Cidade do Rio Grande, Museu Oceanografico de Rio Grande, 225 pp., 60 pls.

1975 *Brazilian Marine Mollusks Iconography.* Fundacao Univ. Rio Grande, 331 pp., 91 pls.

Risbec, J. 1953. Observations sur l'anatomie du Terebridae Neo-Caledonnais. *Bull. du Mus. d'Hist. Nat. Paris* (ser. 2) **25**: 576-583, 1 pl.

Robert, Paul A. 1945. *Wonders of the Sea Shells.* Iris Books, Oxford University. pp. 1-11; pls. 1-15.

Röding, P. F. 1798. *Museum Boltenianum* sive catalogus cimeliorum e tribus regnis naturae quae olim collegerat Joa. Fried. Bolten. Pars secunda. J. C. Trappii, Hamburg, i-viii + 199 pp.

Rudman, W. B. 1969. Observations on *Pervicacia tristis* (Deshayes, 1859) and a comparison with other toxoglossan gastropods. *Veliger* **12** (1): 53-64, text figs.

Rutsch, R. 1934. Die Gastropoden der Punta Gavilan, Abhandlungen der Schweizerischen Palaeontologischen Gesellschaft. *Memoires de la Societe Paleontologique Suisse* (Basel) **55**: 106-109, plts. 8, 9.

Sacco, Fedrico. 1891. *I molluschi dei terreni terziarii del Piedmonte e della Liguria* 10: Cassididae (aggiunte), Terebridae e Pusionellidae. Clausen, Torino, pp. 1-66, pls. 1-2.

Salisbury, R. 1978. Hawaii's fifty-odd *Terebra. Hawaiian Shell News* **26** (8): 7-10 text figs.

Salvat, B. and C. Rives. 1975. *Coquillages de Polynèsie.* Editions du Pacifique, Tahiti, 391 p., 446 color text figs.

Say, Thomas. 1822. An account of some of the marine shells of the United States. *Journ. Acad. Nat. Sci. Philadelphia.* **2** (2): 221-248.

1827. Descriptions of marine shells recently discovered on the coast of the United States. *Journ. Acad. Nat. Sci. Philadelphia.* **1** (5): 207-208.

Schepman, M. 1913. *Siboga—Expeditie.* Toxoglossa. Monog. 49 I (5): 365-452, pls. 25-30.

Schoenberg, O. 1980. *Terebra trochlea* lost and found. *Hawaiian Shell News* **28** (3): 15, text fig.

Schwengel, Jeanne. 1940. New Floridan Marine Mollusks. *Nautilus* **56** (2): 109-110, 1 pl.

Shikama, T. 1963. *Selected Shells of the World Illustrated in colours*, Hokuryu-kan Publ. Co., Tokyo, 1: 1-154, pls. 1-102, text figs.

Smith, E. A. 1871. A list of species of shells from West Africa, with descriptions of those hitherto undescribed. *Proc. Zool. Soc. London* pp. 727-739. pl. 75.

1873. Remarks on a few species belonging to the family Terebridae, and descriptions of several new forms in the collection of the British Museum. *Ann. & Mag. Nat. Hist.* (ser. 4) **11**: 262-271.

1875. A list of the gastropoda collected in Japanese Seas by Commander H. C. St. John, R. N. *Ann. & Mag. Nat. Hist.* (ser. 4) **15**: 414-427.

1877. Conidae and Terebridae. *Ann. & Mag. Nat. Hist.* (ser. 4) **19**: 223-231.

1879. On a collection of Mollusca from Japan. *Proc. Zool. Soc. London* p. 181-218, pls. 19, 20.

1880. Descriptions of twelve new species of shells. *Proc. Zool. Soc. London* p. 480-481, pl. 48.

1899a. Notes on some marine shells from north-west Australia, with descriptions of new species. *Proc. Malac. Soc. London*, **3** (6): 311-13, text fig.

1899b. Descriptions of new species of South African marine shells. *Journ. Conch.* **9** (8): 247-252, pl. 5.

1901. On South African marine shells, with descriptions of new species. *Journ. Conch.* **10** (4): 104-116, pl. 1.

1903. Descriptions of marine shells from South Africa. *Proc. Malac. Soc. London* **5** (6): 360-361, pl. 5.

1904. On a collection of Marine Shells from Port Alfred, Cape Colony. *Journ. Malac.* **11** (2): 30, pl. 2.

Smith, M. 1912. On a collection of marine gastropods from Aden with descriptions of new forms. *Nautilus* **26** (6): 74-69; 1 pl.

Sowerby, G. B. 1825. *Catalogue of the Shells Contained in the Collection of the Late Earl of Tankerville.* London, p. i-vii + 1-92, appendix, i-xxxiv, 9 col. pls.

Springsteen, F. J. and F. M. Leobrera. 1986. *Shells of the Philippines.* Carfel Seashell Museum, Manila, 377 pp., 100 pls.

Suter, H. 1913. *Manual of the New Zealand Mollusca.* Wellington. xxiii + 1120, Atlas (1915), plts. 1-72.

Sutty, L. 1986. *Seashell Treasures of the Caribbean.* E. P. Dutton, N.Y. pp. 128, 139 pls.

Talavera, F. 1975. Moluscos de sedimentos de la plataforma continental de Mauretania. *Bol. Inst. Espanol. Oceanogr.* **192**: 1-18.

Tenison-Woods. 1878. On some new marine shells. *Proc. Linn. Soc. New South Wales* for 1977/78, **2**: 262-266. (not illustrated)

1880 On some new marine shells. *Proc. Linn. Soc. New South Wales* for 1879, **4**: 2-24, pl. 4.

Thiele, J. 1925. *Wissenschaftliche Ergebnisse der deutchen Tiefsee-Expedition auf dem Dampfer "Valdivia"* 1898-1899. Gastropoda II. Teil. G. Fischer, Jena, **7** (2): 38-382, pls. 1-34.

Tinker, S. W. 1958. *Pacific Sea Shells*, ed. 2. Tokyo. pp. 1-240, text figs.

Tomlin, J. R. le B. 1944. Deshayes' Review of *Terebra. Journ. Conch.* **22** (5): 104-108.

Troschel, F. H. 1865. *Das Gebiss der Schnecken zur Begrüendung einer natürlichen Classification.* Nikolaische Verlagsbuchhandlung, Berlin, 2 (1): 1-48, pls. 1-4.

Tryon, G. W. 1885. *Manual of Conchology;* structural and systematic. With illustrations of the species. G. W. Tryon, Philadelphia 7: 1-64, pls. 1-12 (Terebridae only).

Turton, W. H. 1932. *The Marine Shells of Port Alfred, S.*

Africa. Oxford University Press, London, 331 pp., 70 pls.

Vanatta, E. G. 1924. Four new species of shells. *Proc. Acad. Nat. Sci. Philadelphia* **76**: 423-425, 5 text figs.

Vermeij, G. J., E. Zipser, and E. C. Dudley. 1980. Predation in time and space: peeling and drilling in terebrid gastropods. *Paleobiology* **6** (3): 352-364. text fig.

Warmke, Germaine L. & R. Tucker Abbott. 1961. *Caribbean Seashells* pp. 1-348. Livingston Press, Narbeth, Pa.

Watson, R. B. 1886. *Report on the Scientific Results of the Voyage of H.M.S. Challenger During the Years 1873-76*, Zoology XV. H. M. Govt. Publ. London, **15** (42): i-v + 1-756, pls. 1-50.

Weaver, C. S. 1960. Hawaiian Marine Mollusks, recent and new species. *Suppl. Hawaiian Shell News*, 36: pp. 9 pls.

1965. Comparing the large Hawaiian auger (*T. achates* Weaver) with related species. *Hawaiian Shell News* **13** (6): 6, 7 text figs.

1966. Small terebrid species found in Hawaii. *Hawaiian Shell News*. **14** (6): 2, text fig.

Webb, W. F. 1959. *Handbook for Shell Collectors*. Lee Pub., Wellesley Hills, Mass. 264 pp.; 112 pls.

Wenz, W. 1941. *Handbuch der Paläozoologie*. Gebrüder Borntraeger, Berlin. Prosobranchia. **6** (1): 1478-1488, text figs. (Terebridae only).

Whitney, R. A. 1976. A new species of *Terebra*. *Veliger* **18** (4): 381-382; text figs.

Wilson, B. R., and K. Gillett. 1971. *Australian Shells*. A. H. & A. W. Reed, Sydney. 168 pp., 106 pls.

Wood, W. 1825. *Index Testaceologicus;* or a catalogue of shells, British and foreign, arranged according to the Linnean system. W. Wood, London, 188 pp. 39 pls.

1828. *Supplement to the Index Testaceologicus;* or a catalogue of shells, British and Foreign. W. Wood, London, 59 pp., 8 pls.

INDEX
GUIDE TO NAMES

In lieu of a standard index, this alphabetical listing of all names in the Terebridae will take the reader directly to either the species number in the text or to an illustration of that species. The disposition of other names, whether synonyms or dubious species, is indicated.

Species No.

Species No.

TEMPERATURE
degrees Celsius are used

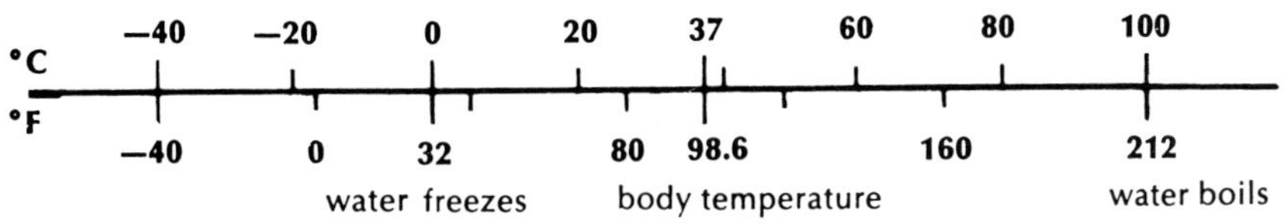

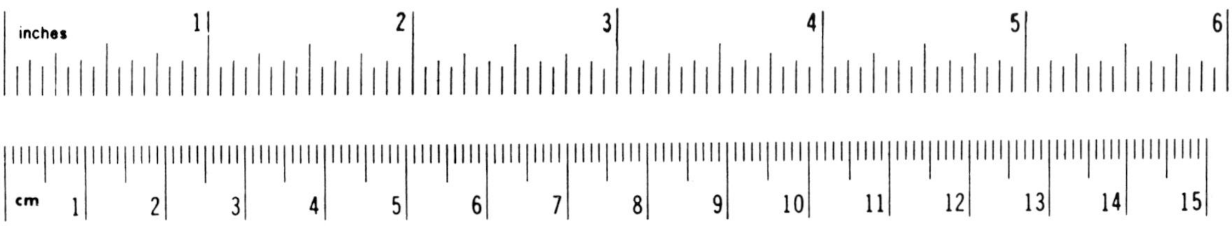

Approximate Conversions to Metric Measures

Symbol	When You Know	Multiply by	To Find	Symbol
		LENGTH		
in	inches	2.5	centimeters	cm
ft	feet	30	centimeters	cm
yd	yards	0.9	meters	m
mi	miles	1.6	kilometers	km
		AREA		
in^2	square inches	6.5	square centimeters	cm^2
ft^2	square feet	0.09	square meters	m^2
yd^2	square yards	0.8	square meters	m^2
mi^2	square miles	2.6	square kilometers	km^2
	acres	0.4	hectares	ha
		MASS (weight)		
oz	ounces	28	grams	g
lb	pounds	0.45	kilograms	kg
	short tons (2000 lb)	0.9	metric ton	t
		VOLUME		
tsp	teaspoons	5	milliliters	mL
Tbsp	tablespoons	15	milliliters	mL
in^3	cubic inches	16	milliliters	mL
fl oz	fluid ounces	30	milliliters	mL
c	cups	0.24	liters	L
pt	pints	0.47	liters	L
qt	quarts	0.95	liters	L
gal	gallons	3.8	liters	L
ft^3	cubic feet	0.03	cubic meters	m^3
yd^3	cubic yards	0.76	cubic meters	m^3
		TEMPERATURE (exact)		
°F	degrees Fahrenheit	5/9 (after subtracting 32)	degrees Celsius	°C

Approximate Conversions from Metric Measures

Symbol	When You Know	Multiply by	To Find	Symbol
		LENGTH		
mm	millimeters	0.04	inches	in
cm	centimeters	0.4	inches	in
m	meters	3.3	feet	ft
m	meters	1.1	yards	yd
km	kilometers	0.6	miles	mi
		AREA		
cm^2	square centimeters	0.16	square inches	in^2
m^2	square meters	1.2	square yards	yd^2
km^2	square kilometers	0.4	square miles	mi^2
ha	hectares (10,000 m^2)	2.5	acres	
		MASS (weight)		
g	grams	0.035	ounces	oz
kg	kilograms	2.2	pounds	lb
t	tonnes (1000 kg)	1.1	short tons	
		VOLUME		
ml	milliliters	0.03	fluid ounces	fl oz
l	liters	2.1	pints	pt
l	liters	1.06	quarts	qt
l	liters	0.26	gallons	gal
m^3	cubic meters	35	cubic feet	ft^3
m^3	cubic meters	1.3	cubic yards	yd^3
		TEMPERATURE (exact)		
°C	Celsius temperature	9/5 (then add 32)	Fahrenheit temperature	°F

radula: a tooth found on the lingual ribbon. Pleural: radulae. Found in all classes of mollusks, except bivalves.

range: geographic area in which a species lives.

Recent: the present geologic epoch. The word is usually capitalized.

sculpture: the surface markings on a shell, such as ribs, scales, cords, etc.

selenizone: a narrow, spiral band of growth scars indicating where a natural slit occurred in the body whorl (as in *Pleurotomaria*).

sinistral: coiling counterclockwise, or "left-handed." When a sinistral gastropod is held with the apex up and the aperture towards the viewer, the aperture is on the left side.

species: a distinct, genetically isolated series of actually or potentially interbreeding populations.

subspecies: a geographical, somewhat isolated, race with minor differences.

syntype: one of several specimens used in an original description, formerly called a cotype.

taxon: any taxonomic unit such as a species, a subgenus, genus or family. Pleural: *taxa.*

taxonomy: the study of classification and identification.

Tertiary: geological era extending from the Paleocene to the Pleistocene (about 60 millions years duration).

topotype: specimens collected at the type locality, usually at a later time.

type: specimen(s) used in describing a new species. (see holotype and paratype).

type species (TS): the species selected to characterize a genus.

variety: a minor form, such as a special color or shape, but not a subspecies.

veliger: molluscan larval stage, usually free-swimming.

vernacular name: the name of a taxon in any language other than the language of zoological nomenclature. Popular or common name.

whelk: any large snail shell, usually a *Buccinum, Colus* or *Busycon.*

NOTES

[Field notes and new literature information.]

NOTES

[Field notes and new literature information.]

NOTES

[Field notes and new literature information.]

NOTES

[Field notes and new literature information.]